ENVIRONMENTAL ARCHAEOLOGY

APPROACHES, TECHNIQUES & APPLICATIONS

KEITH WILKINSON & CHRIS STEVENS

ENVIRONMENTAL ARCHAEOLOGY

APPROACHES, TECHNIQUES & APPLICATIONS

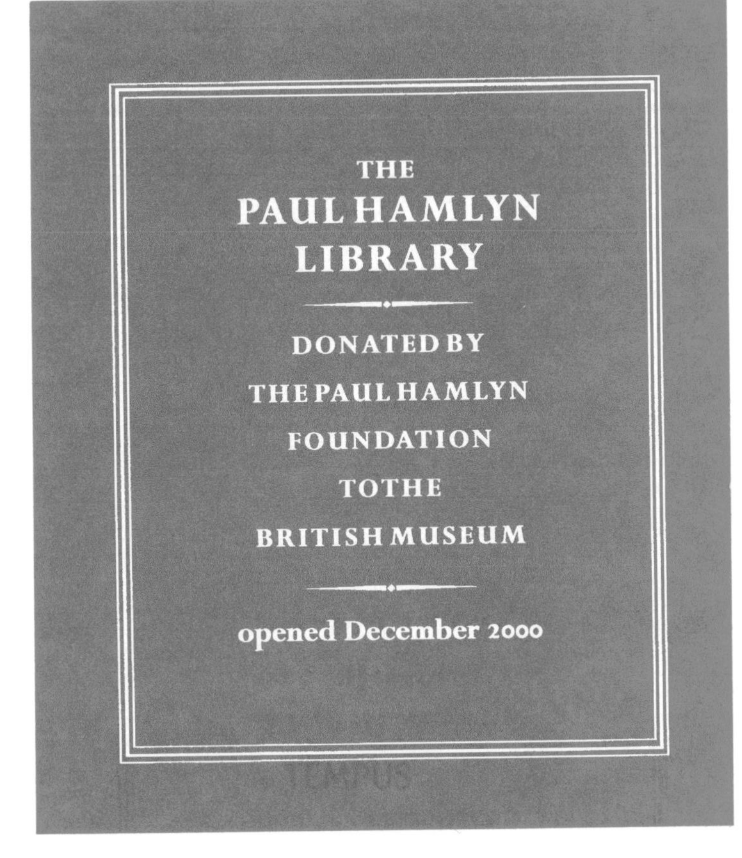

TEMPUS

This book is dedicated to our parents, Jo and Brian Wilkinson, and Pauline and Frank Stevens, without whose help neither of us would have been able to study archaeology at university. They are, therefore, ultimately responsible for the book!

First published 2003

PUBLISHED IN THE UNITED KINGDOM BY:
Tempus Publishing Ltd
The Mill, Brimscombe Port
Stroud, Gloucestershire GL5 2QG
www.tempus-publishing.com

British Library Cataloguing in Publication Data.
A catalogue record for this book is available from the British Library.

ISBN 0 7524 1931 5

Typesetting and origination by Tempus Publishing.
Printed in Great Britain by Midway Colour Print, Wiltshire.

CONTENTS

ACKNOWLEDGEMENTS

Certainly the largest debt we owe in producing this book is to those who stimulated our interest in environmental archaeology as undergraduate students at London's Institute of Archaeology in the late 1980s. As 18 year olds we separately went to study at that institution, then recently merged with University College London, with only a passing interest in the natural sciences. However, three years later we emerged, having spent most of our time studying ancient soils, sediments, seeds, insects, mollusc shells and bones, inspired by the information that examination of such material could provide on past environments and cultures. For this we thank Tony Barham, Don Brothwell, David Harris, Gordon Hillman, Simon Hillson and Ken Thomas, all then members of staff of the Institute's Department of Human Environment. Since 1989 we have gone our separate ways and have been lucky in that we have always been able to make our living by either teaching or researching environmental archaeology. In this time we have met many individuals with similar interests, several of whom have either knowingly or unknowingly contributed data and ideas that we have used in writing this book. We would therefore like to thank some of them; Stephen Rippon for providing material and commenting on the text on the North Somerset Levels, Martin Bell for the same on Goldcliff and Barry Kemp for providing data on Amarna. Martin Jones has been a valuable source of archaeobotanical information over the years, shedding light on how environmental data can be integrated with anthropological and sociological studies. Sue Colledge is thanked for discussing Neolithic Italian Lake Burials with us, while we must also thank Dorian Q Fuller and Alan Clapham for valuable discussions relating to various matters archaeobotanical. We wish to thank Mike Allen for outlining his ideas regarding the phenomenology of the Dorset *cursus* and Mim Bower for pointing out to us to some of the finer details of biomolecular studies. Chris Stevens would like to express his gratitude to Kate and Harry Fuller for their very generous hospitality when writing part of this book.

Without funding from a number of bodies it would not have been possible for us to carry out the analysis that enabled us to write some of the case studies. We therefore wish to record our thanks to the Arts and Humanities Research Board; the British School at Athens; the Egyptian Exploration Society; English Heritage; King Alfred's College, Winchester; the McDonald Institute for

Archaeological Research, University of Cambridge; Oxford Archaeology and the Science and Engineering Research Council. We also wish to thank our employers, King Alfred's College, Winchester and Wessex Archaeology, for their support during the time we spent researching and writing.

Several individuals and organisations have helped us to obtain illustrations. In particular we would like to acknowledge the Archeologisch en Bouwhistorisch Centrum, Gemeente Utrecht; Martin Bell; Rachel Ballantyne; Robin Bendry; Pete Clarke; Michelle Collings; Cotswold Archaeology; Huib de Groot; King Alfred's College, Winchester; Marco Mandella; Stephen Rippon; Vanessa Straker and Heather Tinsley while the Egypt Exploration Society have allowed us to reproduce photographs from Amarna, and Canterbury Archaeological Trust, the photograph of the Dover Boat. Otherwise, except where we state to the contrary in the text, all the illustrations were produced by the authors. We would like to thank Myra Wilkinson van Hoek for proofreading the text, suggesting more lucid ways of explaining some of the more complex ideas and most of all for compiling the index. Finally we would like to thank Mick Aston for the original idea for the book and Peter Kemmis Betty, Tim Clarke, Emma Parkin and Michelle Burns of Tempus for ensuring that we completed it.

PREFACE

This book was first conceived in 1998, when one of us (KW) was present at a conversation between Mick Aston and the present publishers, which lamented certain 'obvious' gaps in the archaeological literature. One of these was held to be 'an uncomplicated book on environmental archaeology that could be recommended to all students of archaeology'. Being the only scientist in the room all looks turned to me, and I had to answer that this was the situation and yes it would be a good idea if someone wrote just such a book. I was duly 'volunteered' and in turn pressed an old friend Chris Stevens into helping out.

During the prolonged period of gestation through which the book then went, two text books appeared that specifically dealt with environmental archaeology; *Environmental Archaeology: Principles and Methods*, by John Evans and Terry O'Connor in 1999 and *Environmental Archaeology: Principles and Practice*, by Dena Dincauze in 2000. At first sight such a proliferation in the literature appeared to remove the need for such a book as this. We prepared to abandon the project and return to our 'day jobs', both books becoming compulsory reading on the courses that we taught. However, through the use of these books as texts we soon realised that they were both written primarily for students doing specialist courses in environmental archaeology, rather than the archaeology student who does not specialise in the natural sciences, or indeed the interested 'lay' reader and in particular the amateur archaeologist. It was also the case that neither dealt with the relationship between archaeological theory and environmental archaeology. So, we resumed work and five years after its first conception we completed the text. This book has been written entirely to the 'brief' first outlined by Mick Aston in 1998 and deals with environmental archaeology from first principles, using non-technical language wherever possible. It is a pragmatic guide to the subject, taking the reader step-by-step through approaches, methods, theory and in particular, interpretation. This account is complementary to Evans & O'Connor's and Dincauze's recent books, and more akin to John Lowe and Michael Walker's (1997) *Reconstructing Quaternary Environments*, but written for archaeologists rather than physical geographers. The book can be used as the basis of exploring any of the techniques of environmental archaeology, as a companion volume when reading archaeological reports written by natural scientists, and even as a field or laboratory manual.

As befits a text introducing environmental archaeology, no prior knowledge of either biological or geological sciences by the reader has been assumed in this book and we have tried to explain all ideas, techniques and interpretary frameworks from scratch. We do, however, assume that the reader will have some familiarity with archaeological principles, practice and chronology, as the book is aimed at a diverse, but essentially archaeological audience.

Throughout the book we have taken a case study approach to explanation, believing that this is the clearest means of communicating ideas and most likely to catch in the reader's memory. We have tried to make the case studies as wide-ranging as possible, in terms of both their chronological and spatial coverage. Nevertheless, examples will inevitably reflect the interests of the authors, and we must confess now that this is why so many of our data relate to European (and particularly British) prehistory and the Near East. We have also divided the task of writing the first five sections of the book according to our particular areas of expertise. Sections 1 and 2 were written by Keith Wilkinson and sections 3, 4 and 5 by Chris Stevens. We co-wrote section 6, which brings the previous discussions together. Finally we should add that we do not discuss the study of human skeletal material (osteoarchaeology) at all. Although often considered to be part of environmental archaeology, osteo-achaeology has a well-developed, and separate, literature of its own. Accounts such as Don Brothwell's (1981) *Digging up bones* and Simon Mays' (1998) *The archaeology of human bones* provide excellent introductions to the subject.

In the pages that follow we try our utmost to communicate to the reader, the importance of archaeological investigation using the natural sciences. We hope to get over the thrill we first experienced when realising how much could be said about the environment and activities of ancient people from such apparently insignificant remains as snail shells, pollen grains and the properties of the geological layers they are found in.

Keith Wilkinson
Winchester

Chris Stevens
Salisbury

July 2003

SECTION 1

APPROACHES TO ENVIRONMENTAL ARCHAEOLOGY

Introduction

Environmental archaeology as a (sub-)discipline and environmental archaeologists in particular have been accused by Julian Thomas in his paper 'Silent running: the ills of environmental archaeology' of '. . . not always being conversant with current developments in the mainstream of archaeology', whilst being 'equipped only with the practical skills and epistemology [the theory of method] of the natural sciences,' therefore, 'the environmental archaeologist tends to come to conclusions about human activity which are naïve in the extreme', and consequently suggests that, 'environmental archaeology should be abolished: there should simply be archaeologists who specialise in the analysis of snails, pollen and seeds.'

Although these are opinions expressed by just one archaeologist, many in mainstream archaeology would agree with the sentiments. In the next 320 pages we aim to demonstrate the value and importance of the natural sciences in archaeological investigations of the twenty-first century and indeed suggest how environmental archaeology can use archaeological theory of the type advocated by Julian Thomas. Our purpose in writing this book is, however, a great deal broader than simply addressing and providing the background to the argument of a single individual. Our prime objective is to explain the concepts that underpin environmental archaeology and outline how those who call themselves environmental archaeologists work. We will explain the principles that lie behind the techniques used by environmental archaeologists, the way that data collected by environmental archaeologists are quantified and presented, and perhaps most importantly how the resultant information is interpreted to present a picture of the human past. In other words, our objective is to show how biological and geological material observed and recovered from an excavation (**1**) can be used to reconstruct the landscapes and activities seen in figure **2**. In order to do this we have split the book into six sections. In this the first, we set the scene by examining what environmental archaeology is and how it has developed out of 'mainstream' archaeology. We also discuss the key principles that underlie the interpretation of biological remains and geological strata in archaeological contexts. In other words, how biological and geological material is transported to, and

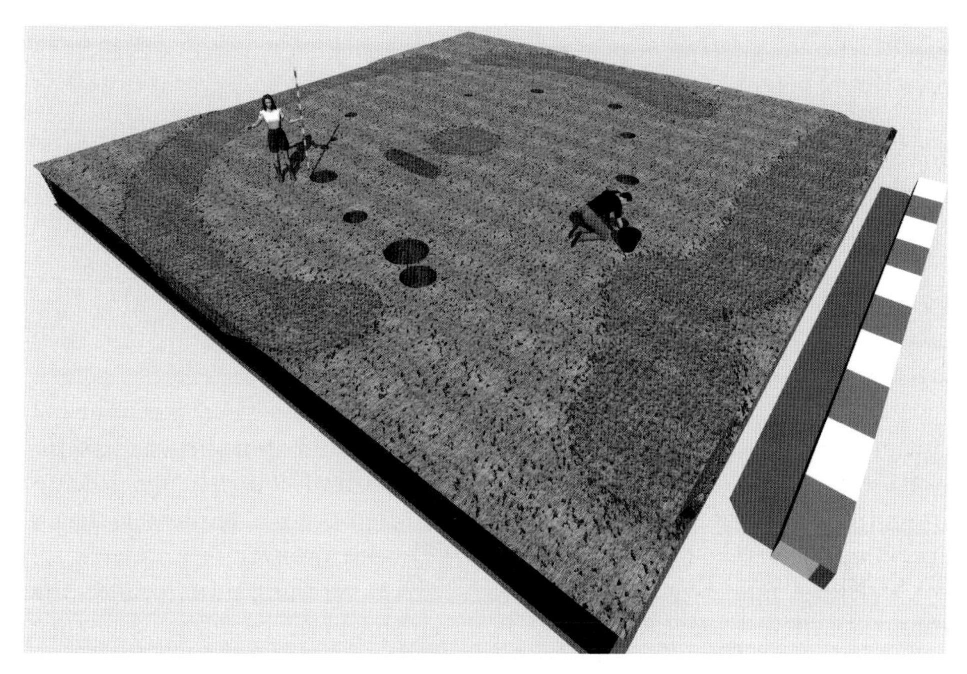

1 *An Iron Age round house revealed by excavation*

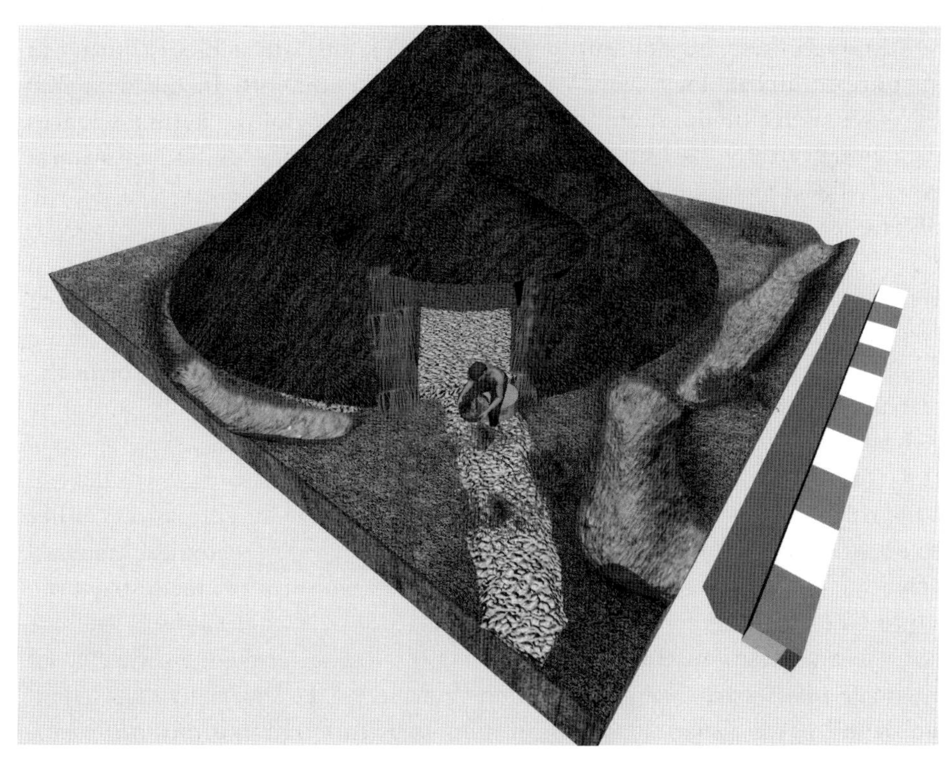

2 *A reconstruction of an Iron Age round house of the type shown under excavation in* **1**, *based on environmental archaeological data from Mingies Ditch, Oxfordshire*

deposited on, a site and is then preserved, how it relates to the environment in which it was originally produced, and how it can be interpreted to reconstruct that ancient environment. In sections 2 and 3 we look at the approaches and techniques that have been used by archaeologists to reconstruct ancient landscapes and economies respectively. In section 4 we look at alternative explanations for biological data that are conventionally interpreted in a functionalist economic manner, namely the vexed subject of ancient ritual (or as we term it, 'ideology'). Section 5 looks in more fundamental detail at how archaeological scientists interpret data, the problems of using scientific approaches to interpret culture and society, and alternative ideas that environmental archaeologists can adopt in examining the archaeological past. The last section (6) brings all that is discussed in the first four together. It seeks to explain how data produced by environmental archaeologists are used to formulate reconstructions of the human past and then how these ideas are communicated to the public at large. We do this by using a series of case studies drawn from Britain and the Near East, and covering the Neolithic to Roman periods. Finally we include in the single appendix contact details for a number of organisations concerned with the study of biological and geological remains from archaeological contexts. We have explained key terms in the text (they are in **bold** at the first mention), and there is thus no glossary. There are also no in-text citations, or indeed a single bibliography at the end of the book, as we have striven for a continuous, uninterrupted and therefore, hopefully, an absorbing narrative. Consequently, in place of Harvard-style references and a bibliography, we have included a list of the main sources that we used in writing the book, together with some other accounts which cover each of the subjects with which we deal.

What is 'environmental archaeology'?

Archaeology is a broad church; it encompasses many areas of study, from deciphering Bronze Age Linear B scripts from Crete to establishing the 'meaning' of near-contemporary rock art of Australia's native peoples, from theorising on the cognitive ability necessary to manufacture a Lower Palaeolithic Acheulian handaxe to determining the source of clay used in Roman pottery. Indeed so eclectic is archaeology as it is practised today, that it is probably the only academic discipline that spans the hard sciences (both physical and natural), social sciences, humanities and arts. Within this wider 'ministry', the sub-discipline of **environmental archaeology** is just a small, but nevertheless highly significant part. It is an area of archaeology that lies very firmly within the realms of the natural sciences, more specifically **biological** and **earth sciences**. Hence, if not trained as 'archaeologists', those who call themselves 'environmental archaeologists', might just as easily have originally been botanists, geologists, pedologists (soil scientists) or zoologists. However, the term 'environmental archaeology' is seen by many as a convenience to collectively describe a group of archaeological studies carried out using biological and geological materials, rather than a coherent sub-discipline with its own singular theory, method and approach. It is certainly from this perspective that we approach environmental archaeology in this book.

Defining environmental archaeology

Given the diversity that is modern archaeology, many attempts have been made at defining what archaeology is. The most common definition is: 'the study of people and past society from their material remains'. Accepting both this statement and environmental archaeology as an umbrella term, then the latter can be considered to be the study of the landscapes that were inhabited by past human populations and the economies they constructed, on the basis of preserved biological remains and geological phenomena. In other words, environmental archaeology concerns the study of vegetation (**flora**) and animals (**fauna**), which lived in association with the people of the past, and the way in which humans interacted with these other living organisms. Perhaps more fundamentally, environmental archaeology is also about reconstructing the physicality of the landscapes in which people lived, hunted and farmed (**geomorphology**). In other words, assessing where rivers and streams ran, the shape and slope of hills, the fertility of soils, the depth of lakes, the distance to the sea, and the countless other variables of the physical environment. The other major aspect of environmental archaeology according to our definition is more directly associated with past human society. This is a subject that we shall return to in much greater detail in sections 4 and 5, but suffice it to say for the present that environmental archaeology should also be interested in the activities carried out by past populations, of which by far the most important one is **subsistence**. In simpler language this means what food people ate, how they obtained it and how they processed it once it had been collected. Subsistence of course is only one aspect of society and, on occasion, environmental archaeologists have moved beyond reconstruction of subsistence to look at trade – after all, food stuffs and other biological materials were imported from outside a particular locality in the past, just as in the present – construction materials and aspects of **ritual**.

The sources of information on which all of the previously stated areas of investigation are based are the preserved remains of plants and animals, and the characteristics of the layers of sediment or soil that enclose them. Whether such information is actually obtained from an archaeological site or not is immaterial, as long as the biological and geological data can be used to reconstruct some aspect of the human past. Accepting these definitions, we can term the study of past floras, faunas and geomorphology associated with past people **palaeoenvironmental studies**, and that of diet, trade, building materials and the like **palaeoeconomic studies**. These form the basis of sections 2 and 3 respectively. We have separated our account in this way because of the very different approaches to data interpretation taken by those working in the two areas. Those who undertake palaeoenvironmental studies are usually providing an environmental background in which past populations act, albeit that those people often modify that background thereby providing feedback. Nevertheless natural (i.e. non-human) processes are pre-eminent when studying palaeoenvironment. This is because the biological and geological materials studied by a palaeoenvironmentalist are, even in the most humanly modified situation, unrecognised secondary outcomes of people's actions. For example a newly arriving species of land snail characteristic of open conditions is not an intended human 'product'; rather its appearance is an indirect result of forest clearance. After all from the snail's point of view, it matters little whether that forest opening was created by people, or by over grazing by deer, it is simply an ecological opportunity. Therefore, the approach taken in palaeoenvironmental studies is little removed from that of the parent disciplines of geology, botany and zoology.

In contrast to palaeoenvironmental approaches, palaeoeconomic studies tackle biological material collected from archaeological contexts (which are commonly termed **ecofacts**). Such ecofacts are produced as a result of human action, either deliberately, as for example cereal grains in a storage pit, or as a recognised by-product, for example animal bones in a rubbish pit. In the latter case it is the meat on the bones that was desired by people, but there is of course no meat without bones, and those bones are a known by-product. As palaeoeconomic studies are carried out on archaeological material that happens to be biological, the approach is a combination of archaeology and the biological sciences. It is for this reason that developments in archaeological theory have much greater relevance to palaeoeconomic studies and this is why in section 5 we explore in much greater detail the tensions that have resulted from interpreting scientific interpretation of palaeoeconomic evidence.

Subdivisions of environmental archaeology

The definitions we outline above are based upon the purpose to which environmental archaeological studies are put and the approach taken by those who carry them out. However, few practitioners of environmental archaeology would think of themselves as palaeoenvironmentalists or palaeoeconomists, but would instead group themselves by the type of materials they look at. To better understand how environmental archaeological study works within a larger archaeological project, it is useful to divide the sub-discipline along technical grounds; in other words, based on the methodologies employed by particular groups of specialists (**3**). The most obvious of these subdivisions is on the basis of the branch of natural science in which the study is based: biological or earth science. The study of once-living organisms in archaeology is termed **bioarchaeology** and that of geological materials **geoarchaeology**. The last useful tier of division is on the basis of the type of biological organism or geological phenomenon that is studied. We thus have **zooarchaeology** and **archaeobotany** for the study of ancient animal and plant remains respectively. The 'geoarchaeology' branch is, however, more difficult to subdivide. While there is a group of researchers who work predominantly on archaeological soils and who might be conceived as working in **archaeopedology**, no terminology has been applied to the study of archaeological sedimentology, geomorphology or stratigraphy. Indeed it is very common that geoarchaeologists will work across all three of these areas. Therefore for convenience, if not for technical correctness, the term **archaeosedimentology** can be used to define studies of this nature. The impression we have given perhaps overplays the differences between areas of study and in practice it is common for an individual to work across the 'boundaries' that have been outlined here, for example working on both pollen (archaeobotany) and stratigraphy (archaeosedimentology).

Problems with the previous definitions

Before we continue, we should concede that the subdivisions presented and their terminology are not universally accepted. In part this is a cultural phenomenon relating to the history of study in a particular region of the world, but there are also differences of a technical nature. For example, archaeobotany is the European term for the archaeological study of plant remains, while in North America, what is essentially the same area of study (albeit with a greater emphasis on using data of recent or extant 'primitive' peoples) is termed **paleoethnobotany**. Indeed, in a recent essay Ken Thomas has questioned whether the term 'environmental archaeology' is in itself useful. He sees bioarchaeology and geoarchaeology as being mature disciplines in their own right, while the term 'environmental archaeology' has run its course and is now both ambiguous

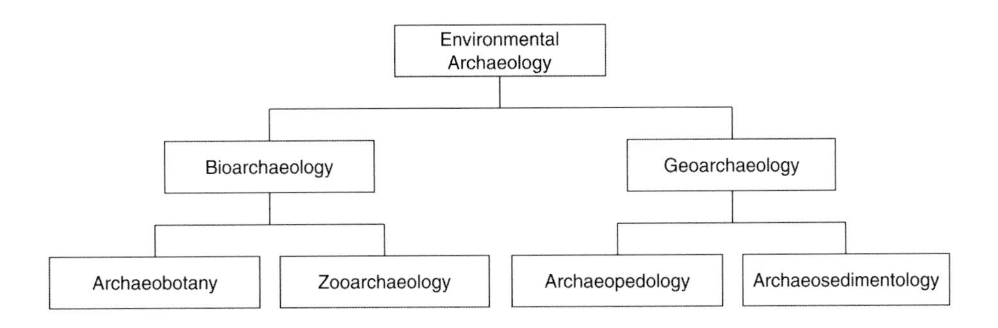

3 *Subdivisions of environmental archaeology*

and redundant. From what has been written above, it is obvious that we would agree with this analysis. On the other hand, some North American geoarchaeologists, such as George Rapp, see many of the bioarchaeological areas of specialism to be components of geoarchaeology, and hence see that sub-discipline as being the same as environmental archaeology as we define it in this book. However, the relatively long history of 'environmental archaeology' as meaning the application of natural sciences to archaeology indicates that it is still a term comprehended by the majority of archaeologists. Therefore, while accepting the intellectual baggage attached to 'environmental archaeology' we have decided to propagate its continued use in this book.

Time definitions

One final definition needs to be introduced at this stage to aid readers in their understanding of later sections; that of time divisions. Many of the examples we use relate to the time of the Palaeolithic, an archaeological period that dates before our present interglacial period (i.e. the warm geological stage that we are presently in). There are many different systems by which these earlier geological stages are named, but the system we use in the text is the one used for the northern European mainland. This scheme sees the present Holocene interglacial preceded by the Weichselian cold stage, in turn predated by the Eemian warm stage and so on, as is illustrated in figure **4**. In order to make comparison easier with studies in Britain and North America, figure **4** also correlates terms used in these areas with those of northern Europe. In writing this text we have used the timescale most familiar to archaeologists of years AD/BC for the Mesolithic and later periods, and years BP ('before present', which is taken as being 1950 for radiocarbon dates, and the literal present for other dating techniques) for the Palaeolithic. We have calibrated all radiocarbon dates post-dating *c.*22,000 radiocarbon years BP according to the most recent (INTCAL 98) curve. Therefore, although we do not use the convention, 'cal. years BP/BC/AD', all quoted date ranges which post-date this are in calendar years.

The development of environmental archaeology

The history of the subject we now know as archaeology spans the last 300 or so years, while the academic discipline with the same name has been taught in British universities since at least the beginning of the twentieth century. Environmental archaeology has a much shorter history, the first use of the term for this sub-discipline only occurring in the 1950s. The same decade also saw the first environmental archaeological post at a university, when Friedrich Zeuner became Professor of Environmental Archaeology at the Institute of Archaeology, London in 1952. Nevertheless, the history of natural sciences in

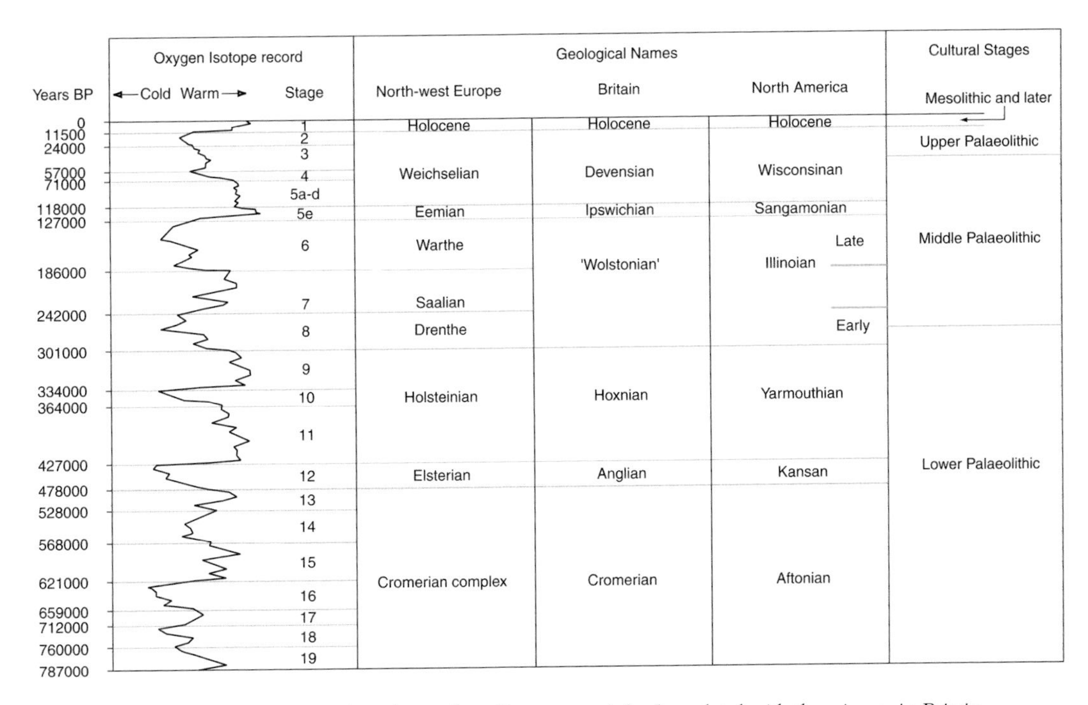

4 *Geological period names used on the northern European mainland correlated with those in use in Britain and North America. The oxygen isotope curve (from Pacific Ocean deep sea core V28-238) is a proxy record of ice cap size and hence climate*

archaeology extends a great deal further back – natural scientists were routinely working with archaeologists by the middle of the nineteenth century.

In their recent book *Geoarchaeology: the Earth-Science Approach to Archaeological Interpretation*, George Rapp and Christopher Hill divide that sub-discipline into three phases of development, and it is helpful to consider the history of environmental archaeology according to the same headings. In the first, or **foundation phase**, which covers the nineteenth century and before, there was no real boundary between archaeology and the natural sciences, and indeed individuals worked across what we would now see as a range of disciplines. Many of these early researchers were, in northern Europe at least, members of the professional or upper classes, who had time on their hands to pursue their interests. Early examples include John Frere, who in 1797 noted Palaeolithic tools at a brick pit in Hoxne, Suffolk (**5**) associated with river gravel, which he correctly interpreted as having formed at a period of 'great antiquity'. Later examples included Jacques Boucher de Crèvecoeur de Perthes and Sir Charles Lyell, who in the 1850s recognised that at sites in southern England and the Somme valley in north-eastern France Palaeolithic tools were often found in association with extinct mammals such as mammoths. Not only were they able to 'read' the stratigraphy and suggest how the gravels in which the tools that were found had formed, they were also able to argue that the artefacts had been deposited in the distant past, as animals such as mammoths had not been found together with artefacts of any culture then known.

The mid-nineteenth century is also important for developments in the biological sciences, such as the studies of evolution by Charles Darwin and Alfred Russell Wallace. They famously demonstrated that species gradually evolved in form over time, but also suggested that the geographical distribution of particular species also altered in response to outside effects such as climate, predation and colonisation by other species. This was a highly significant discovery, as it suggested that past environments were not necessarily the same as those found in the present. Previously, it had been assumed that environments had been unchanging over archaeological timescales. By the last decades of the nineteenth century, those working in archaeology, geology and biology were going their separate ways, with the latter two rarely using their knowledge for an archaeological end. However, occasionally natural scientists were asked to apply their specialism to a particular site; for example Ferdinand Keller in his investigations of Swiss lake villages from 1853 frequently involved

5 *The stratigraphic sequence of Hoxne, Suffolk at the time of a Quaternary Research Association visit in April 2000. The tools seen by Frere in 1797 came from a brick pit containing a similar stratigraphy some 100m to the west*

geologists, botanists and zoologists to work on those specialist aspects of his sites. It is rather a pity that he did not pay so much attention to the cultural aspects of the sites, interpreting structures dating from the Neolithic to later Iron Age as 'Celtic'.

The second phase of development recognised by Rapp and Hill is termed **collaboration**, and is characterised by natural scientists working together with archaeologists to research problems of interest to both. As we have seen, at the end of the nineteenth century both archaeologists and natural scientists became ever more specialised, a single individual rarely being an expert in more than one discipline. This collaboration phase lasted for the twentieth century prior to the Second World War. In northern Europe it is typified by A.S. Kennard's work on land snails from excavations carried out by E.C. Curwen of Neolithic causewayed enclosures in Sussex, or by Hans Helbaek's analysis of grain impressions from daub in British Iron Age sites. In North America, a different tradition emerged as geoarchaeology began to develop. Two geologists, Ernst Antevs and Kirk Bryan, worked extensively on the stratigraphic context of Palaeo-Indian sites, focusing particularly on correlating between sites from distinct geological strata and using archaeological artefacts as a means of relatively dating sites. In addition to collaborating with archaeologists, natural scientists were making discoveries and developing techniques that would later prove of great importance for archaeological study. For example, the Swedish geologist Lennart von Post had developed pollen analysis by 1916, as a means of reconstructing past vegetation. Similarly in 1908 two Danish geologists, Blytt and Sernander, had published an account of their use of macroscopic plant remains to reconstruct the past environments of bogs in the south of Denmark.

The final, or **integration**, phase of environmental archaeology runs from the end of the Second World War to the present. The development of ideas during this period was profound and ultimately led to the development of the archaeological science we know today. It is for this reason that we devote section 5 of this book to an in-depth discussion of the subject. Nevertheless, ideas were not the only developments that occurred during this time. Indeed, advances in archaeological theory in the post-war years were arguably only possible because New Archaeological techniques had been formulated. Possibly the most important was **radiocarbon dating**, developed as a by-product of the research on atomic weapons by Willard Libby. It not only revolutionised prehistoric chronology, but for the first time allowed different world cultures to be compared. More prosaically it removed at a stroke the need for geologists or those working with pollen to date sites by these means. Previously pollen had been used to date Holocene sites in northern Europe by reference to pollen zones developed on the premise that vegetation constantly changed due to the gradual influx of deciduous tree species following de-glaciation. Specialists with such expertise were therefore able to focus their efforts on other problems, such as the impact of people on the environment or the provenance of materials used to make artefacts. In North America in particular, there was a growing awareness that people interacted with their environment, and were to an extent dependent upon it – ideas that were encompassed within the term **cultural ecology**, first coined by Julian Steward in the 1950s. In Britain, broadly similar developments took place as archaeologists tried to move away from a dependence on artefacts for interpretation. In this respect Grahame Clark's excavations in the early 1950s at Star Carr in Yorkshire were groundbreaking. There were few structures at Star Carr, but there were extremely well-preserved biological remains, which were studied in meticulous detail. Through pollen analysis of the site carried out by Harry Godwin, Clark was able to look specifically at the interaction of people and environment in the early Mesolithic, demonstrating Mesolithic forest 'management' for the first time. Later, Clark was to work with Eric Higgs in developing models that enabled the reconstruction of natural resources around sites based on present-day geographical variables. These models formed part of an approach termed **economic prehistory** that saw prehistoric populations as being entirely dependent upon the landscape for all elements of subsistence and, consequently, society.

In the 1960s and 1970s, archaeological methodologies developed rapidly, in part as a result of the widespread rescue excavations that were being carried out in much of Europe. Perhaps the best known technique to appear at this time was that of **flotation**, which is now ubiquitously used to recover charred plant remains (see section 3). This was also a time of major development in archaeological theory in both Britain and North America as we shall see in section 5. Although primarily aimed at changing the way archaeological data were interpreted, other important aspects of the so-called **New** or **processual archaeology** were to improve scientific rigour of investigation and to introduce an element of anthropology into archaeology. Thus from the 1960s there was a greater focus on quantifying archaeological data, while a further emphasis was placed on making use of the world's 'primitive' peoples as a living laboratory. Processual archaeologists also realised that the archaeological record was a biased one, as not all remains from a past society survive to the present day. Thus a further result of the New Archaeology was the development of studies into **site formation processes** and **taphonomy**, which have become particular foci of geoarchaeology.

In some ways techniques and approaches developed in the 1980s and 1990s have merely refined those of earlier decades. The major change in archaeological thinking, so-called **post-processual** theory, has hardly impacted the way that environmental archaeology is carried out, except to change the way data are interpreted. Perhaps the single largest change of the last two decades has taken place in the minds of those who work as environmental archaeologists. Today environmental archaeologists are both archaeologists and natural scientists and have (or should have) been trained in both areas. In this respect the situation has reverted to that of the nineteenth century. The most important implication of this change is that those practising environmental archaeology are in theory fully conversant with the archaeological process and its associated problems, and can therefore take them into consideration in their sampling and analysis.

Material studied by environmental archaeologists

The situation for all who study the past is the same; it is not possible to directly observe the period of interest. This is just as true of an environmental archaeologist as a historian. Instead, the materials that have survived from that past must be utilised to reconstruct ancient environments and economies. These indirect records of a past situation are termed **proxies** and are the raw materials with which environmental archaeologists deal.

Bioarchaeology

In the case of bioarchaeology proxies are plant and animal remains preserved on, or in association with, archaeological sites, or in certain cases impressions of those bio-remains. The latter may include for example impressions of cereal grains and processing waste in pottery or burnt daub where the original biological fragment has been combusted during firing. As the data produced by bioarchaeologists are not direct observations that can simply be described, but rather proxy records, the latter have to be interpreted in order to study human palaeoenvironments or palaeoeconomies. We will discuss the basis for bioarchaeological interpretation in a later part of this section.

Biological material preserved in ancient soils and sediments does not survive in exactly the same condition as when the particular organism was alive. No matter what the preserving medium is or what the organism itself is made of, some decay will have taken place between the moment the organism died, or was deposited, and the point at which it became incorporated in a bioarchaeologist's sample. Part of this decay cycle is often chemical replacement within the bio-remain, for example changes in the composition of calcium carbonate in a mollusc shell. It may also manifest itself in the selective removal of one or more chemical compounds to leave others, as is the case with vertebrate bones (in most environments) where collagen is progressively lost. However, archaeological timescales are rarely sufficient for total chemical replacement to take place, which would lead to the formation of the **fossils** that we see forming in geological timescales. Instead, partial replacement or removal will convert biological organisms to **sub-fossils**, and it is these that are studied by bioarchaeologists.

At this point it is worthwhile examining the range of biological proxies that are used by environmental archaeologists, although again we will return to look at many of these in greater detail in sections 2 and 3. While the obvious division of proxy data is along the plant-animal line, another important consideration is size, or rather whether a microscope is needed to both find and identify a given type of remain. There is no hard and fast rule of how small a particular organism or its sub-fossil remains have to be to require a microscope for *all* sorting and identification tasks, but experience would suggest that a microscope is vital for sizes finer than 0.2mm (200μm). Hence sub-fossils that have an average size finer than 200μm can be termed **microfossils,** and those larger, **macrofossils**. Although there are several exceptions, low-powered binocular microscopes are used for identifying macrofossils and high power, epi-illuminating types for microfossils. **Table 1** outlines the range of biological remains commonly employed in bioarchaeological studies.

Taxonomy

Before leaving the material studied by bioarchaeologists we should consider how biological organisms are named, as it is the scientific (often termed 'Latin') names that are commonly used by specialists. There are two reasons for using

Class of remain	Description	Used to reconstruct:	
		Palaeoenvironment	Palaeoeconomy
Plant macrofossils			
Terrestrial and aquatic plants	Seeds and vegetative parts	✓	✓
Plant microfossils			
Flowering plants and ferns	Pollen and spores	✓	
Grasses	Phytoliths (silica skeleton)	✓	✓
Diatoms	unicellular algae preserved as silcica frustules	✓*	
Animal macrofossils			
Land and freshwater molluscs	Carbonate shells	✓	
Marine molluscs	Carbonate shells	✓*	✓
Insects	Invertebrates with chitinous exoskeletons	✓	✓
Mammals and fish	Bones and teeth	✓	✓
Animal microfossils			
Ostracods	bivalve crustaceans preserved as calcite/silica carapace	✓*	
Foraminifera	unicellular marine invertebrates preserved as carbonate test	✓*	

* Primarily used to reconstruct sea-levels and palaeosalinity

Table 1 *The main classes of biological remain studied by bioarchaeologists and the uses to which resultant data are commonly put* scientific rather than common names. Firstly, scientific names are the same everywhere in the world, so a botanist working in Korea for instance will know what *Quercus* is, although he or she may not know the English word 'oak'. Secondly, many organisms – especially very small ones – simply do not have common names in any language, and there is no option but to use the scientific term. Therefore, understanding biological conventions regarding naming and organisation (**taxonomy**) is extremely important for all those who work with biological material.

Taxonomic structure is hierarchical: all plants and animals are divided into **species**, groups of related species form **genera** (singular '**genus**'), several related genera comprise a **family** and so on (**Table 2**). There are also conventions in the way a name is written down in print. Genera and species names are always italicised or underlined, while only the first word (i.e. the genus name) has an initial capital letter, for example *Rangifer tarandus* (reindeer) or *Pinus* sp. (a species of the pine genus). Names at higher taxonomic levels are never italicised/underlined and always begin with a capital letter, for example Ranunculaceae (the buttercup family).

Geoarchaeology

The material preserved from the past and studied by geoarchaeologists comprises layers of sediments and soils which collectively form stratigraphy. Often these soils or sediments form the matrix in which archaeological artefacts or biological material are preserved, although this need not always be the case (**6**).

The terms '**soil**' and '**sediment**' are not interchangeable – soils and sediments form by significantly different processes. Sediments are the end result of the weathering ➤ erosion ➤ transport ➤ deposition cycle, and therefore always derive from a different location to where they are found. Different types of transportation and deposition processes lead to the formation of varied sedimentary bodies. For example, the sea may weather a

Taxonomic hierarchy	Wolf	Common garden snail
Kingdom	Animalia	Animalia
Phylum	Chordata	Mollusca
Subphylum	Vertebrata	
Class	Mammalia	Gastropoda
Order	Carnivora	Stylommatophora
Family	Canidae	Helicidae
Genus	*Canis*	*Helix*
Species	*lupus*	*aspersa*

Table 2 *An example of the taxonomic classification of the wolf and the common garden snail. Note that the species name is never used without the genus name*

coastal exposure of rock through the direct impact of waves. These rock particles are then transported by offshore currents and the tide to another coastal location and then deposited. As marine transportation and depositional processes tend to sort individual particles by size, our original rocks are likely to end up with others of broadly similar characteristics. The resultant beach is thus comprised of sediments, in this case of cobble-sized rocks (**clasts**). From the definition we have given of sediments we can see that, in a sense, they represent instability, the removal of material from one place in the landscape and its emplacement in another.

Soils on the other hand form in stable environments as a result of the weathering *in situ* of existing sediment deposits (which may be the bedrock), termed the **parent material**. The resultant weathering products are then manipulated by soil-dwelling plants and animals to produce something that has very different properties to the parent material. In other words: a soil forms at the surface of a sediment and then develops downwards, from the stable surface, modifying the original deposit to produce distinct **horizons.** These typically include an organic-rich A-horizon, a B-horizon rich in clay and carbonate, and a C-horizon which is the unmodified parent material. There is also a difference in the terminology used by **sedimentologists** (those who study sediments) and **pedologists** (those who study soils), for whereas superimposed layers of sediment are termed **sequences**, a series of soil horizons comprises a **profile**.

Environments rarely remain constant over archaeological timescales; stable conditions may change to unstable ones. If this were to occur, sediments may be deposited on top of a soil, and if the former are of a sufficient thickness they prevent the continuation of soil formation. A soil isolated from soil-forming processes and then preserved is termed a **palaeosol**, an example of which can be seen in figure **7**. Palaeosols are of great significance for archaeology as a whole, as stable land surfaces (soils only form in terrestrial situations) are precisely where humans are likely to have concentrated their activities.

As well as soils and sediments, the other main source of data for geoarchaeologists is the geomorphology of a landscape, as this will often provide significant information on past environments. For example, an up-standing **terrace** (i.e. the sedimentary remains of a former river bed) formed by an ancient river will indicate that in the past stream flow was at a higher topographic level than at present. There are two types of geomorphological features: depositional features, which are comprised of sediments and occasionally palaeosols – for example the previously mentioned river terraces – and erosional features, where sediment and soil stratigraphy have been removed during the formation of channels and other similar features. An example of the latter is the gully that can be seen in figure **6**, which formed as a result of the concentration of rainwater along a single weak seam in the underlying Jurassic limestone rock.

6 *A Bronze Age site in the Sierra de Moncayo, near Zaragoza, Spain. Erosion has removed all stratigraphy down to the bedrock, leaving the archaeological remains (ceramic and flint artefacts) as a lag. A gully has been cut into the bedrock by water action*

7 *A section through the fills of a dry valley near Brighton, southern England. The picture shows a recent soil profile, on top of a sequence of hillwash sediments of Roman and medieval date, sealing a Bronze Age palaeosol, which developed in a sequence of marls from the Weichselian Late Glacial period. The scale is 2m long*

Having discussed the types of remains that are studied by environmental archaeologists, we can move on to look at how a suite of such remains can provide us with palaeoenvironmental and palaeoeconomic information.

Interpreting proxy data

We suggested at the beginning of this section that there are differences in how those who study palaeoenvironment and those studying palaeoeconomy interpret their data. Although this is certainly the case, there are obvious similarities in the material studied, which in turn mean that there has to be a certain commonality of approach. In the remaining parts of this section we will examine these common elements, while highlighting the differences in sections 2 and 3.

The phrase 'the present is the key to the past' is often used to introduce a key concept that is applied in all aspects of natural science relating to the past, but particularly to the **Quaternary** (approximately the last two million years). This concept is termed **uniformitarianism** and is certainly the most important idea underpinning all studies of ancient geologies and organisms. Usually associated with the geologist Sir Charles Lyell, the ideas behind uniformitarianism were actually developed by another geologist, James Hutton, in the later eighteenth century. Although simple, the idea that the natural processes that operate in the landscapes of today also operated in landscapes of the geological past and produced the same effect was perhaps the most important discovery ever made in the natural sciences. As the phrase quoted above suggests, knowledge of the products – in terms of distinct geological deposits – of processes that take place in the present day can allow ancient sedimentary deposits to be explained. For example, present-day meandering rivers (**8**) are characterised by multiple layers of sand that dip in the direction of flow (**cross bedding**). If deposits of a similar character were found in an archaeological or geological section, it could be inferred, by analogy with the present-day environment, that a meandering river had existed in the location of that section at some time in the past, and the direction of its flow could be determined (**9**). Although initially only used in geology, by the middle of the nineteenth century, uniformitarianism was being applied to the study of fossil plant and animal remains, by Darwin amongst others. There the principle works in the exactly the same way: proxy records can be compared to the products of modern animal or plant communities and the past environment can be reconstructed by a similar process of analogy.

Despite the fact that uniformitarianism is fundamental to modern geology, **palaeobotany** (the study of fossil plant remains) and **palaeontology** (the study of fossil animals), as well as environmental archaeology, there are problems surrounding the concept. The most obvious is the assumption that

8 *The Evrotas, a modern meandering river in the Peloponnese, Greece. The labels indicate the main structural features discussed in section 2* (**27**, p.72)

9 *Deposits of Roman and medieval date from Utrecht in the Netherlands. The cross-bedded sands at the based of the photograph are characteristic of meandering rivers. Scale bar divisions are in 0.5m intervals.* Fotodienst Gemeente Utrecht all ancient environments have modern analogues. Without a present day analogue, a past plant or animal community or geological context can only be reconstructed in the most basic sense. The assumption that analogues exist for all past landscapes is becoming increasingly controversial, especially for the older archaeological periods. For example at the end of the Weichselian, during the period 15-13,000 BP, tundra environments similar to those found in northern Russia existed across the southern parts of Britain. The land snail assemblages that characterised these Late Glacial environments comprise around 14 species, some of which are today found in areas of permafrost in northern Scandinavia, but others of which have a distinctly central or even southern European distribution (**Table 3**). This particular combination, characteristic of Late Glacial Britain, does not exist in any modern tundra environment. If it were not for the presence of other proxy records (e.g. pollen) that indicate the presence of such an environment, tundra conditions could not be inferred from the mollusc data.

A further example of problems with modern analogues is that of pre-modern arable environments. Modern mechanised agriculture, combined with the use of pesticides, has simply killed off many species deemed to be weeds (which as we shall see in section 3 are often a key to how and where crop plants were cultivated), but which were common during prehistory. Obvious ways around this problem are to attempt to recreate prehistoric practice, as has been attempted at Butser Iron Age experimental farm in Hampshire, or by using arable activities as practised by peoples who do not use mechanised or agro-chemically enhanced agriculture as an analogy. However, there are problems with both these approaches. In the case of experimental farms such as Butser, it is by no means certain how long weed floras of characteristic prehistoric type will take to 'return' to land which was previously used for modern farming. There is also a certain circularity of argument: farms of this type are based on reconstructions obtained partially from environmental archaeological data, which themselves are interpreted using uniformitarian concepts. Using analogies from so-called traditional societies is also problematic. For the analogy to be truly appropriate, the archaeological site to which it is applied would have to be in exactly the same climate zone as the modern analogue. However, most populations who work without the aid of pesticides and mechanised forms of cultivation inhabit areas which we group under the title 'the Third World'. Therefore such ethnographic data do not provide reliable analogues for those of us who work inside Western Europe or North America.

As a result of these problems, few scientists apply uniformitarian principles rigidly; instead an approach termed by Stephen J. Gould, **methodological uniformitarianism**, is used; an approach which allows for such difficulties. As the name suggests, methodological uniformitarianism treats uniformitarianism as a method that can be applied when deemed appropriate rather than a piece of underlying theory which is universally applicable. It also recognises

Species	Present distribution			
	Britain	Northern Scandinavia	Central Europe	Southern Europe
Catinella arenaria		✓		
Oxyloma pfeifferi	✓		✓	✓
Cochlicopa lubricella	✓		✓	✓
Vertigo pygmaea	✓		✓	✓
Vertigo genessi		✓		
Vertigo geyeri		✓	✓	
Pupilla muscorum	✓		✓	✓
Vallonia costata	✓		✓	✓
Vallonia pulchella	✓		✓	✓
Punctum pygmaeum	✓	✓	✓	✓
Vitrina pellucida	✓	✓	✓	✓
Nesovitrea hammonis	✓	✓	✓	✓
Euconulus fulvus	✓	✓	✓	✓
Arianta arbustorum	✓	✓	✓	

Table 3 *An assemblage of land snail species dating between* c.*15,000 BP and 13,000 BP from Richard Preece's investigations at Holywell Coombe, Folkestone, Kent and their present distribution*

that the scale of processes that operated in the past, both spatially and temporally, may have been very different from those observed at the present. We will return to look at uniformitarianism as it applies to past economies in more detail in section 3.

Ecology and palaeoecology

When undertaking palaeoenvironmental studies, the ultimate goal is to determine all aspects of a particular past landscape, including its vegetation, animal population and soil type, as well as its topography. Nevertheless, adverse preservation conditions (see 'Survival of ancient biological remains' below) often mean that there are few sources of proxy data for the environmental archaeologist to exploit. This type of holistic environmental reconstruction would therefore seem to be impossible. However, nature serves those studying past environments well in this respect, as processes of energy exchange inextricably link all elements of a particular landscape – both biological and **abiotic** (i.e. non-living, e.g. the sun, the geology) in time and space. For example, carnivorous animals eat, and thereby gain energy from, herbivorous animals, which eat plants, which in turn obtain their energy from the sun and nutrients from the soil (**10**). It is possible to extend these energy cycles still further, as the soil is dependent on parent material (geology), and obtains its nutrients from by-products (urine, faeces) of animals, as well as decaying plant and animal matter. As all elements of a particular landscape system are intimately linked in this way,

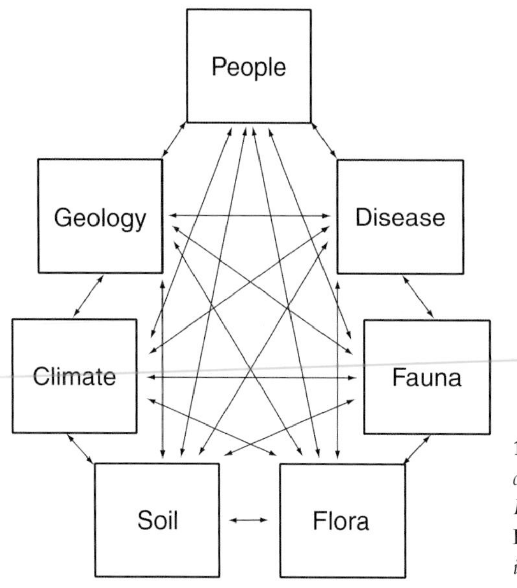

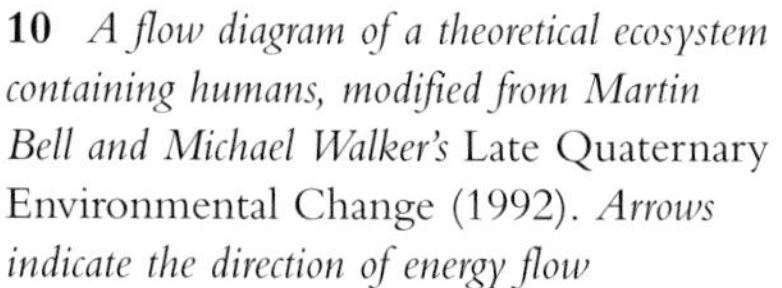
10 *A flow diagram of a theoretical ecosystem containing humans, modified from Martin Bell and Michael Walker's* Late Quaternary Environmental Change (1992). *Arrows indicate the direction of energy flow*

in theory, it is possible to infer the presence of one element in a past landscape from other elements, even if material evidence of the first is missing. For example, in the 1970s little was known of the Holocene vegetation of south-east England both during and before the Neolithic, as the calcareous geologies that characterise the area were thought to preclude pollen preservation. However, buried soils were known from beneath several Neolithic monuments such as South Street longbarrow in Wiltshire. The soils are of a distinct type called Brown Earths which today are associated with forests, while land snail assemblages from the palaeosols are characteristic of woodland. This combined evidence was used to suggest that the vegetation of south eastern England during the early and middle Holocene was wooded. During the 1980s and later, pollen sites were found, and the data from these studies confirmed that deciduous forests had indeed extended across much of southern England in the Neolithic.

Ecosystems

Systems of interacting plants, animals and abiotic components occupying defined spatial areas are called **ecosystems** (**10**) and their study is termed **ecology**. The British ecologist A.G. Tansley first proposed the term 'ecosystem' in 1935, but the idea that there are mutually dependent communities of plants and animals inhabiting distinct physical environments extends back much further. In 1877, Karl Mobius wrote about a community of organisms in an oyster reef as a 'biocoenosis', while ten years later S.A. Forbes termed the interacting **biota** (i.e. plants and animals) of a lake a 'microcosm'.

As explained above, ecosystems function by energy transfer, which occurs through the food chain, but also as a result of nutrient cycling. A hierarchical system, termed a **trophic** structure, exists within each ecosystem. In its most basic form, this trophic 'pyramid' consists of plants, above which are herbivores, while above herbivores are carnivores. As energy transfer from one trophic level to the next is inefficient (80-90% of available energy being lost as heat), the biomass (literally 'living weight') at each level is proportionally reduced. This translates into an ever-greater decline in the number of individual organisms at increasingly higher trophic levels. For example, within an ecosystem 10,000 plants may equate with 100 herbivores and one carnivore.

Part of the ecosystem definition relates to space. There are no constraints on how large or small any individual system has to be to qualify as an ecosystem. An ecosystem can be the size of a single pond just 20m across, or as large as a whole desert consisting of thousands of km^2. One of the most important spaces within an ecosystem is the edge. In fact, few ecosystems are divided by sharp boundaries; rather one ecosystem will gradually merge into another. The area where two ecosystems join thus is called an **ecotone**. It is an extremely important space from an archaeological point of view. Because ecotones contain many of the plants and animals of two different ecosystems, species diversity is higher than in either parent ecosystem, especially as some organisms live only in ecotones. This has obvious resource implications as far as humans are concerned and it is, therefore, hardly surprising that many past cultures have sited their settlements in ecotonal environments. Mesolithic sites in Scotland are clustered in coastal locations to take advantage of both terrestrial and marine ecosystems, as well as the ecotonal area (the intertidal zone) in between.

Palaeoecology

In studying palaeoenvironments, environmental archaeologists are essentially trying to piece together past ecosystems. The study of past ecosystems is sometimes referred to as **palaeoecology**, although in practice this term tends to be associated with reconstruction of past biota, particularly flora. Palaeoecology introduces the fourth dimension into ecosystem studies: time. Ecosystems are rarely static through time, as both intrinsic and extrinsic factors (i.e. those operating from within and without respectively) can cause change. The former become increasingly important at smaller spatial scales and might include increased predation, or, within a human ecosystem, changes in land-use, while the latter could consist of climate change. Dynamism within ecosystems can be considered in relation to two concepts: **resilience** and **persistence**. Resilience can be defined as how fast the biotic community returns to the 'normal' situation (or rather: **equilibrium**) following a change, such as might be caused by disease. Ecosystems will vary in their resilience; an ecosystem may be more resilient to some perturbations than to others. For example a Mediterranean oak forest ecosystem is adapted to fire and will readily regenerate following events of this nature. It is, however, not resilient to human forest clearance and subsequent agriculture, the landscape changing to *garriga* or *macchia* (shrub) ecosystems once abandoned by humans. Persistence is the length of time a particular type of ecosystem endures before it is replaced by another, which in the previous example is of the order of 7,000 years for surviving oak forest in Greece.

Death and the passage of time transform a 'living' ecosystem into sub-fossil biological material together with an associated sedimentary and/or soil matrix. In untangling the original ecosystem from these remains environmental archaeologists rely on a number of assumptions in addition to those previously considered under uniformitarianism. First amongst these is that the ecological requirements of the plants and animals that are found as sub-fossil **death assemblages** are known. This depends in turn upon the quality of modern ecological work on different groups of organisms, which is variable depending upon the type (and often size) of an organism, and its geographic location. Thus the ecology of most plant and many animal species in the northern hemisphere is relatively well understood, but the situation is more patchy to the south of the equator.

A second assumption is that all species within an ecosystem are in equilibrium with their environment, in other words that species are not living outside their ecological tolerances. There are several lines of evidence to suggest that on occasion this assumption cannot always be made. For example the Bristlecone Pine, a species from north-west America, famed for its usefulness in helping to calibrate radiocarbon dates, is known to have lived for 200 years or more without reproducing as a result of a short-lived and adverse climatic change. A second example, somewhat in contrast to the resilience of the Bristlecone Pine, is the consequences of climate change at the end of the Weichselian period. At the Pleistocene-Holocene boundary, *c.*11,500 BP, reconstructions of temperature based on sub-fossil insects (insects, particularly beetles, are extremely sensitive to even the most minor changes in climate and have been used extensively to model Pleistocene conditions) suggest that average annual temperatures were increasing at around 1°C per decade. Such was the pace of change it is thought that no ecosystem close to the former ice masses could operate in equilibrium, and indeed the botanical proxies indicate that vegetation appeared to have lagged behind climate and animal colonisation. Despite these examples to the contrary however, the bulk of present evidence suggests that ecosystems do normally exist in equilibrium. If this argument is accepted past systems can be reconstructed from partial evidence and using uniformitarian concepts.

Survival of ancient biological remains

We have already touched on the survival of biological remains in the archaeological record in the preceding text. If all biological material from the archaeological past survived in perfect condition and in exactly the locations where it had been originally placed or where it fell, interpretation using the previously discussed concepts would be relatively straightforward. Unfortunately, situations such as these are extremely uncommon – even at sites where preservation is exceptionally good, such as Goldcliff and Amama (discussed further in sections 2 and 6 respectively), some categories of biological materials have totally disappeared through decay. Survival of biological material in primary context is rather more common, but in order to interpret its palaeoenvironmental or palaeoeconomic meaning, it is very important that we are certain it has not been moved. Therefore, until it is understood how each deposit from which biological remains are to be extracted formed, and how it was modified since itsoriginal placement, none of the uniformitarian or palaeoecological models that we have discussed can be used. Before these ideas can be applied, we have to be able to assess the effects of processes operating during deposition and following burial, on the preservation and **integrity** (i.e. the likelihood of all material at a particular level having been deposited at the same time) of biological assemblages. Knowledge of such processes is also important at an earlier stage of study, namely that of sampling for biological remains (sampling strategies are discussed in greater detail in sections 2 and 3). This is because the processes that operate to form stratigraphy will determine the **resolution** (i.e. the thickness of deposit forming within a unit of time) and the integrity. There is little point, for example, in taking closely spaced samples from a deposit which formed extremely rapidly and which has been heavily modified by animal burrowing following burial (**11**). Therefore, from all points of view, it is important that for each site they study environmental archaeologists understand how the stratigraphic record has developed with time, and what processes have shaped this development.

Site formation

The mechanisms by which archaeological site stratigraphies form are termed **site formation processes**, while those that specifically relate to the modification of the site following abandonment are called **post-depositional processes**. The latter is also known as **diagenesis** in the geological literature, while both archaeologists and geologists use the term **taphonomy** to explain the processes that lead to partial fossilisation of biological remains. Before the 1960s these processes were rarely considered in interpreting archaeological remains, except in superficial detail, and certainly not by the biologists who worked on materials that came from the sites. As we have seen, it was with the advent of the New Archaeology that site formation processes were first studied, and it is largely through the pioneering work of Karl Butzer and Michael Schiffer that a framework for their understanding is in place. Although of special importance to bioarchaeological interpretation Butzer sees the study of site formation processes to be the prime task of geoarchaeology. However, despite this, it is obviously vital that bioar-

11 *A thick sequence of alluvial sediments of Bronze Age date exposed in a spring cutting in the Sierra de Moncayo, Zaragoza, Spain. There is little point in sampling the top of the sequence to recover biological remains, as it has little integrity due to the intensity of animal burrowing. The gravel layers are thought to have formed during single flood events, therefore closely spaced samples through them would be a waste of effort. On the other hand, the fine-grained alluvial layers towards the base of the section are thought to have formed gradually, and as there are no animal burrows they are being sampled for high-resolution analysis*

chaeologists understand the implications of the action of particular site formation processes on the class of remains they are studying.

Mechanisms of site formation are the subject for a whole book rather than a single sub-section. Indeed such an account has been published in Michael Schiffer's *Formational Processes of the Archaeological Record*. There are two aspects to consider when discussing how an archaeological site develops, namely structural and depositional evidence. Structures are what people construct; most obviously buildings, walls and fences, wells and similar features, but they can also include so-called negative features, such as ditches and pits. Deposits result either from deliberate human action, as for example in disposing of rubbish, or they may be the result of indirect processes that are not intended by humans, such as silting of ditches, accumulation of animal dung within a pen, etc. Deposits will also contain the vast majority of biological remains that are studied by environmental archaeologists. According to a framework developed by Kar Butzer, these remains can be considered as belonging to one of three categories depending on how they reached the site. **Primary material** was deliberately introduced onto a site by people. The most obvious examples is artefacts, used on the site. Also included, and of particular importance to bioarchaeologists, are raw materials for artefact manufacture, fuel and foodstuffs. **Secondary materials** are alteration products deriving from on-site processing or biochemical decomposition of primary materials. Included within this category would be the animal bones which result from butchery, fibres following clothes manufacture, charcoal from burnt firewood, as well as faeces, flint debris and pot sherds. **Tertiary materials** consist of secondary materials that have been removed, transported and deposited through human or other means. The deposits of rubbish of which so much of the archaeological resource consists would fall into this category. Material in each category will have a different interpretative value; the lower the level, the more difficult it is to associate the material with human activity (**12**).

Processes of decay will begin to operate while a site is still occupied. In fact, as soon as structures are completed they are exposed to the elements. This means that site formation processes can, and most frequently do, operate at the same time as post-depositional processes and the results of both may be found interdigitating in the stratigraphic record (**13**). So, for example, material relating to the weathering of a roof may become incorporated between two successive floor levels. The main agents of post-depositional deterioration are wind and water, which will affect structures in varying ways depending upon the material that structure is made of. Hence roofs, which are frequently made of organic materials and are also the area of a building most exposed to the elements, will deteriorate the most rapidly. Observation of buildings abandoned in recent times has demonstrated that as soon as the roof has been removed, walls quickly begin to deteriorate rapidly too, as they no longer have any protection.

12 *An Anglo-Saxon sunken building reconstructed at West Stowe, Suffolk. Primary material consists of building materials and tools. Secondary material would include charcoal from fires and by-products from construction, while tertiary materials consist of re-worked primary and secondary material, for example, blown thatch*

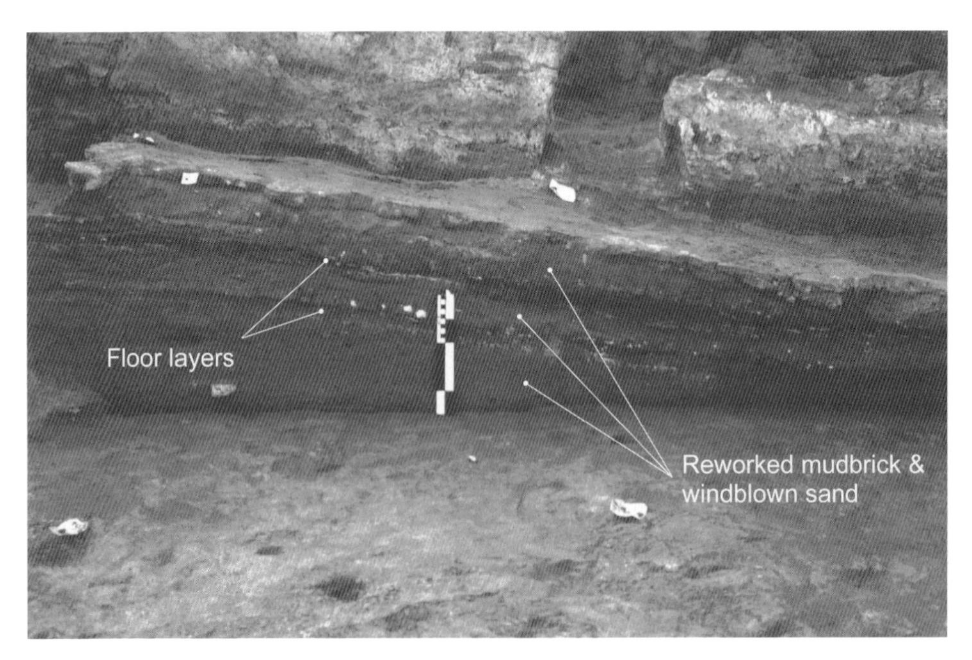

13 *Successive floors with layers of collapsed and eroded mud brick in between, from the early Neolithic site of Jeitun, in the Karakum desert of Turkmenistan. Scale divisions are in multiples of 1, 5 and 10cm*

Negative features commonly become infilled following their abandonment. No matter what the purpose for which a ditch is constructed, it is invariably kept reasonably open during its period of use. A great deal of experimental work has been carried out on the mechanisms by which negative features are infilled, not least at the experimental earthworks at Wareham and Overton Down in southern England. Sediments which accumulate within negative features left open at abandonment appear to form in three distinct stages (**14**). The **primary fill** forms from the weathering of the sides and the contemporary surface soil, very soon after the feature is dug (and is thus a deposit that could usefully be sampled in order to reconstruct the environment at the time of construction). The **secondary fill** forms after abandonment from material either blowing in, or eroding from, a contemporary soil. This 'silting' process is both gradual and incremental so that episodes of soil formation may take place during times of stability. **Tertiary fills** are often deliberately placed to level the feature to the present ground surface, or on the other hand could be the result of ploughing over the top, which ultimately has the same effect.

Although a site does not always become buried following its abandonment, most sites do. Burial often aids preservation of archaeological material. In other words buried archaeological sites are frequently better preserved than those that are not. Furthermore, the processes by which a site is buried will have a major impact on how well archaeological and biological remains survive (e.g. **15**).

14 *A sequence of ditch deposits from an Iron Age enclosure in the Quantock Hills, Somerset, southern England. Primary and secondary fills date to the Late Iron Age, and tertiary fills to the later Roman period*

There are two aspects that are key in this respect: the rapidity and energy of burial (**Table 4**). Generally speaking, material will be best preserved when burial is rapid and the energy levels by which the burying medium is deposited, low. For example, Roman archaeological materials from Pompeii have been preserved particularly well, firstly because the city was inundated almost instantaneously, and secondly because the material deposited was volcanic ash of a relatively fine grain size. If Pompeii had been located closer to Vesuvius during the AD 79 eruption, it might well have been impacted by pyroclastic 'bombs' rather than ash, in which case preservation would have been significantly worse. The converse of the rapidity/energy correlation is also in theory true: slow burial and high energies make for poor preservation. However, in many systems high depositional energy levels are associated with rapid deposition. For example the gravel layers illustrated in figure **11** most likely formed in high energy conditions (probably as the result of major storms) and are thought to have accumulated over a very short space of time. Biological preservation is therefore likely to be minimal.

Post-depositional processes

Even when a site has been buried, it is not necessarily protected from post-depositional change. Buried layers or features may be modified by a whole host of processes associated with variations in climate (be this on a seasonal level, or

Mechanism	Timescale and impact
Volcanism	Hours. Preservation potential will depend on the nature of the volcanic deposit that covers the site. Ash may provide good preservation, but lava is likely to cause widespread destruction (unless deposited on top of ash, as is the case at Bronze Age Santorini)
Glaciation	Years. The passage of a glacier over a site will probably lead to its destruction unless that site has been protected by other sediment in the meantime (e.g. alluvial or aeolian material)
Marine incursion	Variable from hours to years. The level of destruction will depend on exposure to tide, rapidity of burial and calibre of the sediment deposited
Alluviation	Normally years, but if resulting from a sudden change of stream course, hours. If a site is buried beneath floodplain sediments it is likely to be well preserved, if beneath channel sediments, destruction is probable
Lake level rises	Years. A low energy process meaning the preservation is ubiquitously good
Aeolian processes	Variable from hours to years. A low energy process that in theory means preservation is likely to be good
Colluviation	Variable from hours to years, depending on the nature of the triggering event. Preservation is also variable depending on the calibre and energy of material

Table 4 *Potential natural burial mechanisms and their likely rapidity*

15 *The Cromer forest bed (dark deposit in left foreground) on the north coast of Norfolk is the remains of a river more than 500,000 years old. No archaeological materials have been found on the site, but organic preservation is excellent. The deposits have been protected from the effects of a glacier that passed over the site about 400,000 years ago by thick layers of glacial outwash (fluvioglacial deposits – lighter-coloured deposits behind the speaker) derived from a previous glacier during seasonal melt*

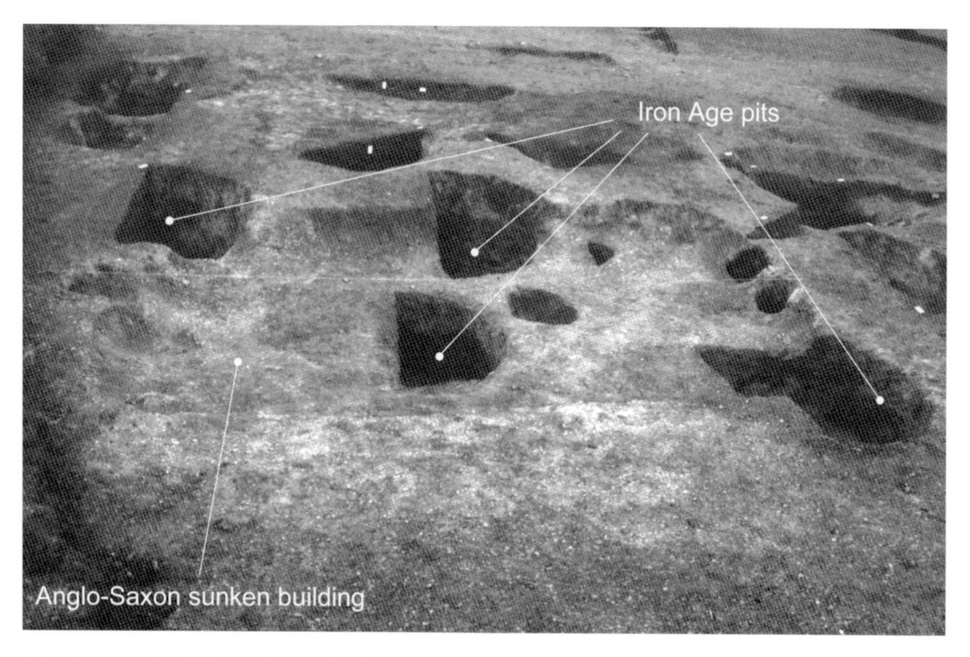

16 *An example of C-transforms, disturbance to existing Iron Age archaeological stratigraphy (pits) caused by the construction of an Anglo-Saxon sunken-featured building. Human action in the sunken-featured building could cause chemical change to the pit stratigraphy, for example, the input of phosphates*

relating to longer term processes affecting the entire Earth surface) and vegetation, as well as by renewed human activity. Schiffer has divided these processes into two categories, **N-transforms** and **C-transforms**. The latter are the result of human, or rather cultural, processes, through renewed activity on the site. Such activity may have a direct physical impact on the existing stratigraphy, such as the digging of pits, but it may also alter the chemical composition of this stratigraphy, for example by introducing phosphates (**16**). The classic example of a C-transform is the arable field, whereby humans create and manipulate an environment thereby influencing natural processes. N-transforms relate to natural processes that operate close to the Earth's surface and usually result in the physical disturbance of archaeological stratigraphy (**Table 5**).

In the previous paragraphs we have discussed how deposited biological assemblages may be removed from their original point of emplacement, have their integrity compromised through being mixed with other material, and how they can be physically damaged both during re-deposition and burial. Unfortunately, for the environmental archaeologist that is not the end of the story, as the chemistry of the soil or sediment in which biological remains are emplaced, in particular their exposure to oxygen, will also determine whether those remains survive or not (**Table 6**). In other words, even if biological materials were deposited in strata that are conducive to their survival, suitable preservation conditions need not persist until the present day. Examples of this include reductions in the water table that expose previously waterlogged remains to the air and hence cause their oxidation (this process is most obviously associated with wetland drainage schemes), or decalcification (a by-product of many soil-forming regimes) causing the dissolution of shell and bone.

Process	Explanation
Cryoturbation	Freeze and thaw processes causing finer soil/sediment particles to move towards the surface
Argilliturbation	Expansion (swelling) and contraction (cracking) of clays within soils/sediments as they become wet and dry respectively
Graviturbation	Slow movement downslope of deposits as a result of gravitation processes, allied to wetting and drying of the ground surface (also termed 'soil creep')
Deformation	Small scale faulting that causes a change in the orientation, shape or volume of a layer (e.g. through compression)
Floralturbation	Disruption of soil structure or sediment stratigraphy by the roots of plants, or by tree throw
Faunalturbation	Disruption of soil structure or sediment stratigraphy by animal burrows

Table 5 *N-transform mechanisms.* Modified from Michael Water's (1992) *Principles of geoarchaeology: a North American perspective*

Class of remain	Preservation conditions					
	Charred	Waterlogged	Desiccated	Mineralised	Calcic	Acidic
Plant macrofossils						
Terrestrial/aquatic plants	✓	✓	✓	✓		
Plant microfossils						
Pollen		✓				✓
Phytoliths		✓	✓	✓		✓
Diatoms		✓		✓		
Animal macrofossils						
Land/freshwater molluscs		✓	✓	✓	✓	
Marine molluscs		✓	✓	✓	✓	
Insects		✓	✓	✓		
Mammal and fish bone	✓	✓	✓	✓	✓	
Animal microfossils						
Ostracods		✓		✓	✓	
Foraminifera		✓		✓	✓	

Table 6 *Biological remains studied by environmental archaeologists and the conditions where they are commonly preserved. For explanation of classes of remains see* **Table 1** *and sections 2 and 3*

Summary

In this introductory section we have discussed what environmental archaeology is and how it has developed from nineteenth-century roots. We have outlined the materials studied by environmental archaeologists and the factors that have to be considered in the interpretation of those remains. We have discussed these important ideas, which underpin not only environmental archaeology, but also all application of natural science methods to the past, in a largely theoretical manner. However, having established these principles we can now move on to look at how preserved biological remains and geological materials are used in practice. This we will do in the following two sections, which look at palaeoenvironmental and palaeoeconomic reconstruction respectively.

SECTION 2

PALAEOENVIRONMENTS:

THE STUDY OF ARCHAEOLOGICAL LANDSCAPES

What does 'environment' mean?

At the beginning of the previous section we briefly discussed the term 'environment'. There, we describe it in relation to the physical landscapes occupied by past populations and the plants and animals (including those domesticated by people) that also inhabited it. The Oxford English Dictionary (second edition) defines environment as: 'that which environs (i.e. surrounds) or the region surrounding anything'. The OED describes landscape as: 'a tract of land with its distinguishing characteristics and features, especially considered as a product of modifying or shaping processes and agents (usually natural)'.

These definitions are neutral, as they should be, and represent our present-day view of what environment is, or rather, was, in the archaeological past. However, many archaeologists also consider past physical environments in terms of **natural** and **cultural** landscapes. Indeed, many archaeologists who subscribe to a post-processual viewpoint consider that any environment which contains people must de facto be a cultural landscape as people act as a modifying agent. Others suggest that the degree by which a landscape has been deliberately altered by people, and the purpose for which those changes were made determines the nature of that landscape.

The reason that we know that Stonehenge was surrounded by grassland in the Neolithic is because mollusc remains from ditch fills and buried soils beneath the various monuments indicate it was so. Significantly then, it is the sub-discipline of environmental archaeology which often provides the data on which these views of environment and landscape are based. This is despite the fact that reconstructions produced by environmental archaeologists are more often than not purely descriptive and do not specifically espouse any theoretical viewpoint. Indeed, another of the criticisms made by Julian Thomas of environmental archaeologists is that they do not make their interpretations interesting to the non-specialist reader, and do not incorporate their findings within any theoretical framework. Because the findings of environmental archaeological investigations are entirely open to interpretation by other archaeologists they can be claimed by anyone; post-processualists, processualists and culture historians (should any still exist) alike.

For the moment we will leave any further discussion regarding the relationship between environmental archaeology and archaeological theory to

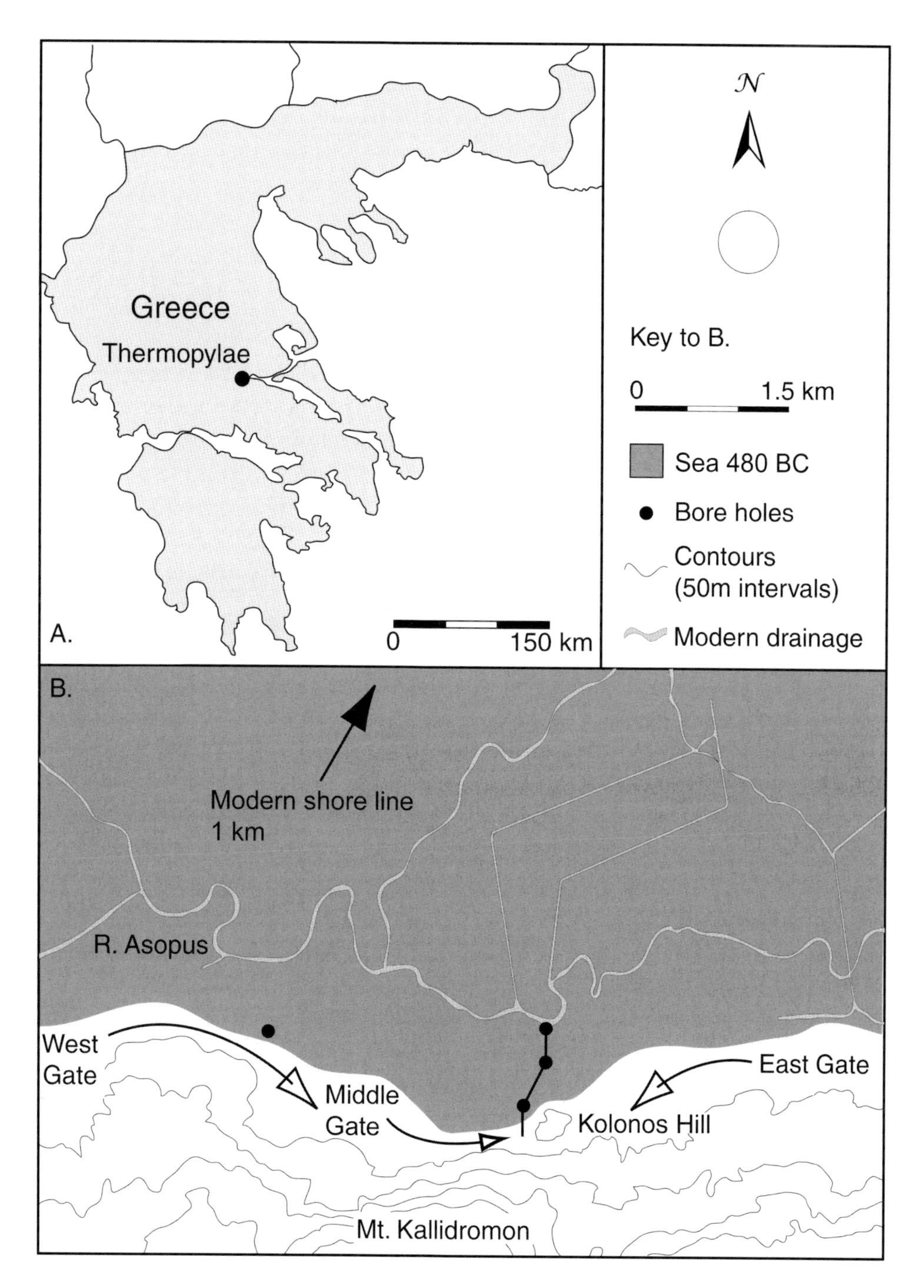

17 *The location of Thermopylae and the relationship of the 480 BC battle site to the modern topography. The battle was fought at the Middle Gate and, following the flanking manoeuvre by 5,000 immortals of Xerxes' army, the last stand of Leonidas and the 300 Spartans was on the Kolonos hill*

section 5, and return to the pragmatics of palaeoenvironmental reconstruction. As was outlined at the beginning of this section and in the last, environments can conveniently be divided into **topography** (the physical landscape) and **biota** (the living organisms that inhabit it).

Understanding past topographies

Prior to the mid-nineteenth century, topography of any area of the world as seen through contemporary eyes was perceived to be much the same as that which would have been observed by people in the past. This concept is termed **actualism**. In other words, in pre-mid-nineteenth century eyes, the course of rivers, the aspect of hills, and the position of coastlines had not altered. The discoveries of Lyell and a host of other nineteenth-century geologists changed this view in relation to the distant past, including that of the earliest humans, although of course uniformitarianism always provided a link to the present. However, it was still common well into the twentieth century for archaeologists studying later prehistory and the historic periods to think that the topographic changes were not relevant in their studies. Indeed, these views even persist to the present day. Many are the times that today's environmental archaeologists are taken to sites and have their location 'explained' in relation to present-day phenomena. Similarly it is common in the archaeological literature to include a distribution map of archaeological sites articulated by reference to modern topographic variables. However, the major lesson of earth science studies over the past 200 years is that the earth surface is a dynamic place even over very short time scales. To demonstrate this, it is useful to take a short look at a single example, the background to which will be familiar to most students of ancient history.

Changing topographies: the example of Thermopylae

The battle of Thermopylae in 480 BC, fought between the Greek city-state of Sparta and its allies, Thespiae and Thrace, and the Persian invaders of Greece under their Great King Xerxes, is well known thanks to the account of Herodotus. Thermopylae is perhaps most famous for the last stand made by King Leonidas and 300 Spartans on the hill of Kolonos in the final stage of the battle, and the death of all but two of them. However, the earlier stages of the battle are a subject of controversy among classical historians, largely due to the apparent disagreement between the movements of the respective armies outlined by Herodotus and the modern topographic situation. Thermopylae (literally 'hot gates', referring to thermal springs) lies on the eastern seaboard of northern Attica (**17**). Given the mountainous nature of the interior, the coastal plain on which Thermopylae is situated has been a natural invasion route from northern Greece and the Balkans into the south. Indeed, although the battle of 480 BC is the most famous, other actions have been fought on the same spot and for the same reasons in 279 BC and 191 BC – with very similar outcomes. In all cases Thermopylae had been selected as a position to defend by greatly outnumbered armies (e.g. Leonidas' allied Greek army is given by Herodotus as 7,500, as opposed to the Persian force of 300,000), because of the narrow front between the cliffs of Mount Kallidromon and the sea, that could be defended (**17**). The Persians of 480 BC and the Romans and their Macedonian allies in 191 BC both initially undertook costly frontal assaults and only achieved victory once they had outflanked their opponents by sending smaller forces through the mountains. However, far from being a narrow pass, the distance between the break of slope of Mount Kallidromon and the Gulf of Malia is over 5km at the present day. In order to tackle the apparent inconsistencies, John Kraft, George Rapp and colleagues undertook a geoarchaeological study of the region in the 1970s. As well as drilling their own bore holes to reconstruct the stratigraphy below the modern coastal plain, they also collected previous bore hole data from Greek government agencies, examined historic maps and considered sea-level curves of the eastern Mediterranean. Their results demonstrated that Herodotus' account is very likely to be an accurate description of the battle. In 480 BC the distance between the cliffs of Mount Kallidromon and the sea (where an Athenian ship was able to maintain contact with Leonidas' army) was between 20-100m wide, and could therefore easily be defended by an army of 7,500 men against any number of opponents. In the intervening 2,500 years the Sperchios river has deposited massive volumes of sediment in the Gulf of Malia, and therefore – despite the fact that the gulf has itself been lowered as a result of tectonic processes – the relative sea level has fallen, and the coastline regressed 5km. Kraft and Rapp's studies demonstrate that the 480 BC battlefield is likely to lie some 20m below the present ground surface as a result of material deposited by the Sperchios. These same processes have also reduced the slope aspect of the Kolonos hill as the intervening valleys have become infilled with sediment. Therefore, this feature would have been very much more impressive when Leonidas' 300 Spartans made their final stand at its summit.

The changing topographies outlined for Thermopylae are not unusual. River channels move across their floodplain, or even abandon a particular stretch altogether by breaking through their banks (**avulsing**) and flowing along a different course. This was the case for another of the plain of Malia's rivers, the Asopus, which seems to have been in a very different position in relation to the pass of Thermopylae in 480 BC than now. Relief may also change as ranges of hills and mountains are uplifted in relation to adjacent valleys by crustal processes operating within the Earth (**tectonics**). On the other hand, due to erosion in upland areas and the consequent infilling by sediment of valleys, relative relief may be reduced. It has been demonstrated that sea levels have risen and fallen in relation to the land in Attica, but these findings are equally applicable to other coastal areas. The key point of this discussion, which is worth emphasising again, is that all archaeologists must consider and attempt to understand the changing nature of topography of the area in which they are working, even if the period they are interested in is a relatively recent one. On most archaeological projects, it is the geoarchaeologist who considers how and why topographic changes have occurred and how humans fit into the picture. In what follows we examine three topographic situations that are common in temperate zones of the world by way of examples of topographic reconstruction: a coastal, riverine and hillside location. For each we discuss the techniques and approaches that are used to chart their changing behaviours. However, before we do so, it is first necessary to discuss the main proxy records employed in reconstructing the physical landscapes, namely sediments and soils – and the properties that define them.

Sediments and soils

We defined sediments and soils in section 1. To recapitulate briefly: sediments comprise material that has been eroded, transported and deposited, and usually imply instability in a landscape, while soils develop from a stable surface downwards, overprinting the underlying stratigraphy with soil-forming features. Therefore if we can determine whether individual layers in a stratigraphic section are sediments or soils it is immediately possible to suggest how stable the landscape was at any particular point. However, the physical and structural properties of sediments and soils are determined by the environment in which they formed, and so the sediment/soil record has a greater potential than simply determining whether a landscape was stable at various times in the past or not. If we are able to make the link between sediment/soil characteristics as seen in a piece of exposed stratigraphy and formation processes that led to its formation through the uniformitarian principles discussed in section 1, we are then able to reconstruct a wide range of variables relating to past physical environments. All this can be achieved simply by observations made in the field, which makes fieldwork the most important part of any geoarchaeological study. Laboratory analysis of samples taken by geoarchaeologists, in contrast to those studied by bioarchaeologists, is usually carried out to confirm, refute or refine hypotheses made during fieldwork, rather than to develop new ideas. As the standing section is the primary record, description of stratigraphy has to be conscientiously made, of high quality and most importantly, made to standard criteria. Only if this is the case can reliable comparisons with modern environments be made, and thereby the past situation interpreted.

Colour and grain size

The first aspect of any description of a sediment or a soil is the determination of colour. Ask a group of archaeologists to describe the colour of an archaeological context solely from what they see and you would get as many answers as the number of people asked. Clearly, such subjective descriptions are of limited use as one person's 'dark grey brown' is 'light yellowish grey' to someone else. To standardise colour descriptions, those studying sediments and soils use a **Munsell Soil Color Chart**. This is a book containing standard colour pallets with which stratigraphy can be directly compared in the field. Usually two colour determinations are made on the basis of the soil/sediment in a dry and moist state. In the Munsell Soil Color Chart colour is defined on the basis of hue, value and chroma (**Table 7**), while a descriptive term is also generated (e.g. 10YR 5/4 yellowish brown). As all geoarchaeologists use Munsell Soil Color Charts, reporting a colour on the basis of its Munsell number will confer immediate understanding to all those who read the description.

The colour of a sediment or a palaeosol often relates more to post-depositional modification processes than to the environment in which it formed. It therefore has limited interpretational value when reconstructing past environments. In contrast **grain size** (sometimes also called particle size), provides a good indication of depositional or soil-forming environments and therefore is an important element in the categorisation of both sediments and soils (**18**, **Table 8**). Grain size, as the name suggests, is a measurement of the size class of individual particles that comprise a sedimentary unit or soil and an estimate of which class predominates. Only in the case of very coarse particles (gravels – see **Table 8**) are individual grains actually measured; for intermediate-sized particles (sands) an estimate is made based on a visual key (**19**) or by 'feel' for the finest particles (silts and clays). Accurate determination of grain size by touch is not easy and requires long experience, but a good rule of thumb is that once wetted, a sediment/soil dominated by clay will readily smear when rubbed across the flat of one's hand, while that dominated by silt will not. Any grittiness detected whilst doing this means that sand is also present. In interpreting a depositional environment from sediment grain sizes it is usually the case that the finer the grain size, the lower the energy that deposited it, and vice versa. We will see some examples of this phenomenon

Hue		Value		Chroma	
2.5R	Red	0	Black	0	Neutral grey
5R	↑	1	↑	1	↑
7.5R		2		2	
2.5R		3		3	
5YR		4		4	
10YR		5		5	
.5R	↓	6		6	
5Y		7		7	
Y.5R	Yellow	8	↓	8	↓
		9		12	
		10	White	20	Absolute colour

Table 7 *Munsell Soil Color Chart terms and meaning*

when we look at riverine, coastal and hillslope environments later in this section. Grain size in soils obviously reflects the characteristics of the parent material, but because finer particles are preferentially moved by water through the soil profile (**eluviation**), there is a tendency for average grain size to decrease downwards, while clays collect immediately above the parent material (C horizon), forming an **illuvial** accumulation called a Bt horizon.

Sediments and soils rarely consist only of particles of a single size class, but rather a mixture of several size classes. **Sorting** is a term used to describe how

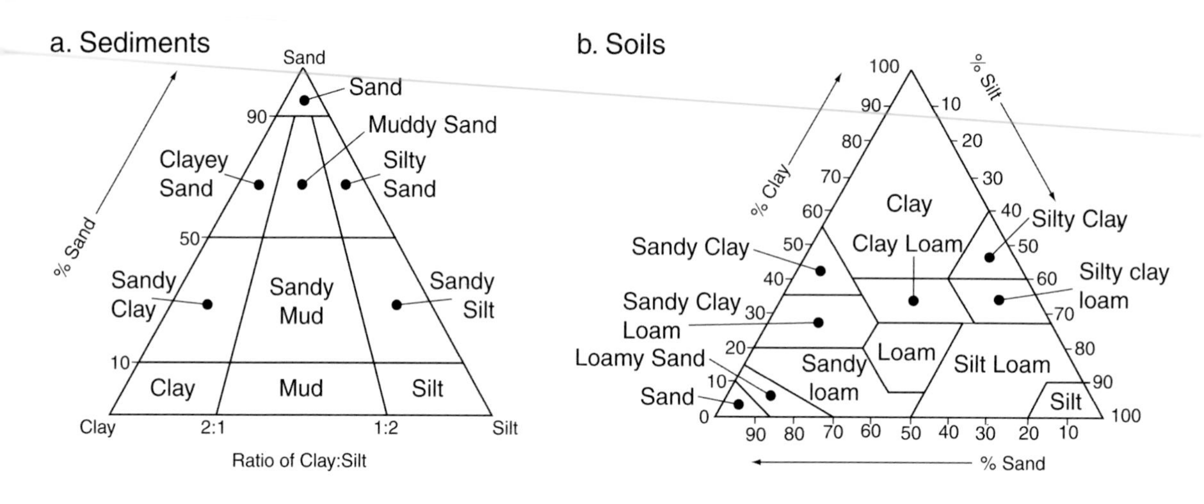

18 *Ternary (triangular) plots showing the classification of clay-sand size material for: a. sediments, and b. soils. These plots can be read by finding the proportion of material in each category (sand, silt and clay), determining where the intersection point is and reading off the description (in the larger font)*

Size (mm)	Category	Wentworth
>256	Gravel	Boulder
64-16		Cobble
16-4		Pebble
4-2		Granule
2-1	Sand	Very coarse sand
1-0.5		Coarse sand
0.5-0.25		Medium sand
0.25-0.125		Fine sand
0.125-0.0625		Very fine sand
0.0625-0.031	Silt	Coarse silt
0.031-0.0156		Medium silt
0.0156-0.0078		Fine silt
0.0078-0.0039		Very fine silt
<0.0039	Clay	Clay

Table 8 *The Wentworth system of sediment/soil size classes*

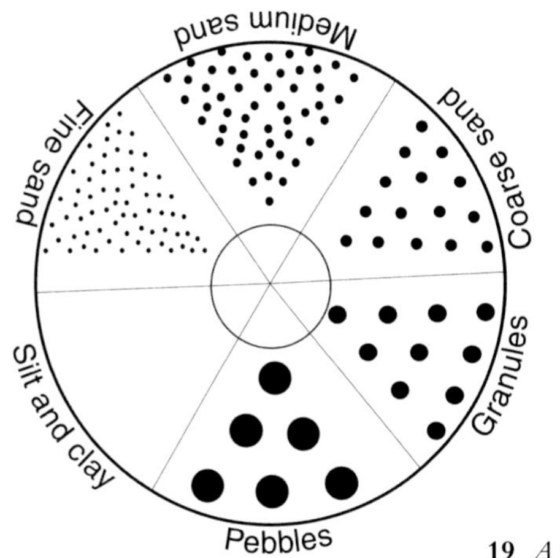

19 *A visual key for estimating grain size of sands*

far from the mean size class particles of a given sediment unit or soil group when grain size is measured. For example, loess, a windblown (**aeolian**) dust derived from the top of glaciers during cold climate events of the Pleistocene and which once carpeted most of northern Europe, is predominantly coarse silt. However, it frequently contains a lesser component of fine sand. Nevertheless, loess is invariably 'well sorted'; in other words, only a few particles differ from the predominant size class, and even then only by a single Wentworth size class (**Table 8**). We might contrast this situation with the colluvial layers seen in figure **7** where the predominant size class is also coarse silt (determined in this case in the laboratory). However, these layers contain secondary components of clays, very fine-medium silts, sands, granules and pebbles. As the secondary components are both numerous, and differ significantly in size from the predominant size class, this colluvium is poorly sorted. Sorting properties are almost impossible to determine in the field for silt and clay-dominated sediments and soils where there are no obvious coarser particles, but can be determined for sands and gravels using the visual key given in figure **20**.

It may not seem obvious from this description why sorting is an important sediment/soil characteristic. However, for sediments in particular it can be very useful in interpreting a depositional environment, and particularly whether post-depositional processes have modified the stratigraphy. In natural environments the majority of sediments are deposited in a well-sorted state. Only very high-energy conditions such as those generated by major storms or seismic events will produce poorly sorted debris flows. However, two other factors frequently transform well-sorted sediments into poorly sorted layers. The first is post-depositional disturbance as caused by natural phenomena (see **Table 5**), which leads to mixing of layers with different grain size distributions. The second is people – humans modify and thus mix stratigraphy by processes of re-deposition. For example, pits, ditches and post-holes are a common feature of many archaeological sites, but where does the extracted material go? Commonly it is 'dumped' to provide surfaces on which further construction takes place. As these dumped deposits will often comprise material from several different layers, the newly formed deposit will usually be poorly sorted. This is both because of the mixture of source material and also as a result of the relatively violent means of deposition (i.e. being thrown from a shovel or tipped from a cart or wheelbarrow), which together do not allow new sorting to take place. Particles of 'gravel' in the form of ceramic sherds, bones and other discarded debris are commonly introduced during this process, further altering the sorting properties. Therefore, when examining stratigraphy from an archaeological site, it is commonly possible to separate deposits that formed from natural landscape processes from those that are the result of human action (termed **archaeosediments** by Michael Waters), simply on the basis of sorting properties. Figure **21** illustrates this principle clearly. The stratigraphy photographed is the infill of an early Byzantine building within the Roman theatre of Sparta, Greece. The poorly sorted layers were both deliberately deposited by people to level the ground surface for construction, and the result of slope failure during major storms, while the fine, well-sorted layers formed during flooding of the theatre basin.

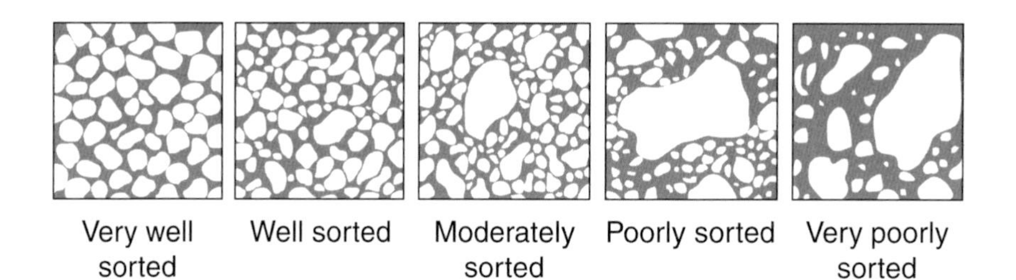

20 *A visual guide to estimating sorting properties of sand- and gravel-dominated soils and sediments. Elements shaded white are sand- and gravel-sized clasts and those in grey are finer-grained matrix*

21 *The stratigraphy associated with a Byzantine period (c. eighth-tenth century AD) building in the cavea of the Roman theatre of Sparta, Greece. Scale in 0.5m divisions*

Lithology and mineralogy

So far we have just focused upon various grain size properties, but another important element of any description and interpretation is the material that the individual grains are actually made of. This is termed lithology for gravels and some coarse sands and mineralogy for other sands and finer size grades. The importance of lithology/mineralogy is not so much in determining how a particular sediment/soil formed, but rather where the material from which it is comprised originated. Lithology and mineralogy both reflect the properties of their original parent material, so determining whether for example a gravel comprises clasts of flints and quartzites or limestones and sandstones is important as this information may help in reconstructing exactly where it was deriving its sediment from and hence the former course of the river. The mineralogy of a sand may help decide whether it was derived from the local river, or brought in from some distance for constructional purposes. The principles involved are of course very similar to the mineralogical study of clays to determine where pottery was manufactured. The mineralogy of the pottery itself is determined and then compared with that of known clay deposits. However, in the field it is only possible to determine the lithology/mineralogy of particles that can be clearly seen – frequently with the aid of a hand-lens – and so only gravels and coarse sands can be so described.

Structure

To the last category of description, (that relating to structure), whole books have been devoted, so only a brief summary is offered here. Structure concerns how particles are arranged in layers, and the relationship of one layer to the next. It is described in different ways for soils and sediments. In soils, structure develops on a temporal basis. That is to say, as a soil ages it first develops structures, which then continue maturing throughout its 'life'. Therefore, the degree of development of certain features also provides an indication of approximate soil age. In sediments on the other hand, structures are developed as the deposit is formed and are not subsequently altered, except by post-depositional disturbance. Sediment structures often relate to the moment (i.e. direction) of flow, be that a result of water, wind or gravity. In most cases flow is in a single direction: downstream in the case of a river, downwind in the case of aeolian environments and downhill for colluvial situations, but in the case of marine environments, flow may be bi-directional due to ingoing and outgoing tides. Thus sedimentary structures, and indeed the long axis of clasts, tend to align themselves with the direction of flow. The examination of these phenomena can, therefore, indicate in which direction this was. Flow processes produce characteristic cross-bedded structures in rivers, aeolian and some marine environments (e.g., see the basal sands in figure **9**), but these structures are much less clear in other terrestrial environments. Where there is no moment of flow, such as in lakes or in floodplain situations, sediments build up vertically in horizontal layers as particles fall from suspension. As the rate at which a particle falls through the water column is dependent upon its mass, and therefore size, these situations tend to produce so-called fining-up sequences. In other words: the predominant grain size decreases upwards through the sediment body. These same low-energy environments also commonly produce **laminated** sequences, where alteration in deposition occurs on a regular (e.g. seasonal or annual) basis causing the production of alternating layers or differing properties, each of a thickness less than 10mm.

Before we move on to look at soil structures it is worth considering one final aspect of sediment stratigraphy (or **lithostratigraphy** as it is more precisely known): boundaries. Boundaries, or **contacts** as they are called in the geological literature, are the imaginary lines that separate one layer from the next and may be of two types, sharp or gradual. These two types of boundary have very different meanings in the interpretation of stratigraphy. Gradual boundaries imply that conditions changed slowly between one layer and the next, through a series of intermediate stages. In other words, that deposition was part of a continuum. The geological term for this type of boundary is a **conformable** contact. We might conceive of such a situation in the case of an expanding lake, where rising water levels gradually cover the organic sediments of surrounding marshland, with inorganic mud. Sharp boundaries have a different meaning and are termed **unconformable** contacts (e.g. the boundary between the cross-bedded sands and overlying archaeological layer in figure **9**). They almost invariably indicate that a break in sedimentation occurred between the deposition of the two layers that they bound, although they do not indicate how long this period may have been. Boundaries that are sharp and straight indicate that a depositional environment changed suddenly, while the underlying surface may have been planed off in the process. Sharp boundaries that undulate almost always indicate that erosion has occurred between the deposition of two sediment layers.

Soil structure is more difficult to interpret than that of sediments. Several types of structure can form as a soil becomes more mature, but the rate at which it will develop these structures is entirely dependent on climate, interference, slope aspect, vegetation and parent material, although the first is by far the most important. Therefore, there are no hard and fast rules with regard to the rate of soil structure develops with respect to time. For example, it is impossible to say that because a soil has a Bt horizon, it must be 1,000 years old – too many other factors are involved. Perhaps the most significant structural aspect of soils is the development of **soil horizons** (**Table 9**). All soils, no matter how young, will develop A horizons characterised by a high humic content, but over time (measured at the order of thousands of years) one or a series of the various types

Horizon	Description
O	Dark-coloured organic remains at the surface in either un-decomposed (Oi), partially decomposed (Oe) or fully decomposed (Oa) state. Extremely rare in the archaeological record
A	Mineral horizon forming directly below the O horizon. Includes a mixture of humified organic and mineral particles, with the latter dominating. Usually darker than the underlying horizons. Where the horizon properties are determined by farming it is termed an Ap horizon.
E	Only present where clay has been washed through the profile. In these circumstances this horizon occurs immediately below an O or A horizon and is characterised by a light colour and a lack of clay and organic particles
B	The mineral horizon formed beneath an O, A or E horizon, but with little similarity to the properties of the parent material. The B horizon contains clays and minerals washed down from overlying horizons and will thicken with time. B horizons are subdivided as follows:
	Bh Concentration of organic matter Bhs Concentration of organic matter with iron and aluminium Bk Concentration of calcium carbonate Bo Residue of iron and aluminium Bq Concentration of silica Bt Concentration of clay Btn Concentration of clay together with sodium Bw B horizon of red colour but lacking illuvial clay By Concentration of gypsum Bz Concentration of compounds more soluble than gypsum
C	Parent material, which if weathered, is termed a Cr horizon

Table 9 *Soil horizon types according to the United States Department of Agriculture and outlined from the top to bottom of a soil profile.* Modified from Waters 1992, table 2.4)

of B horizon will develop depending upon the above variables. As the soil matures still further the B horizon will thicken, a distinct clay-rich Bt horizon will develop as clay is washed through the profile, and in semi-arid and arid conditions carbonate will collect to form a Bk horizon. The formation of these illuvial (Bt) and **calcic** (Bk) horizons is frequently combined by the evolution of a depletion zone of a light colour beneath the A horizon, called an eluvial, or E horizon. Alongside the development of horizons, many soils develop aggregate structures called **peds** which are separated from one another by spaces called **voids**. Ped structure, or rather the shape and extent of the resultant voids, is an extremely important part of the soil-forming process as it determines the rate at which clay minerals and carbonate can pass down through the soil profile. Ped structures are determined by the same variables as the development of soil horizons, although the grain size of the parent material plays an important role. There are four basic types of ped structure, granular, prismatic, blocky and platy, which can each be divided into size categories from very fine to very coarse, and maturity, using the terms weak, moderate and strong. Soils that have no peds have either a single grain structure, where each grain is separate from one another, or a massive structure, where particles form a single cohesive mass.

Now that we know something of soil and sediment properties and how these can be used to reconstruct past environments, we will briefly look at three of those environments and see how, with our newfound knowledge, we can interpret the evolution of past landscapes: coastlines, rivers and inland hill and mountain ranges.

Coastlines

Ever since the Palaeolithic, people have preferentially exploited coastlines over other topographic zones. The director of the Boxgrove excavations in Sussex, England, Mark Roberts, has argued that the hominids who exploited the Boxgrove flint sources some 500,000 years ago came to the site by moving along the southern English coast. Then, as later in prehistory, the coast would have been less densely vegetated than the interior, making it easier to travel along, particularly following the development of boats. Coastal situations would, therefore, have been the first to be settled by newly arriving populations. Coastlines are also of importance to humans for the resources they offer. They are by definition ecotonal environments, bordering terrestrial and marine ecosystems. This has also made them a popular area of exploitation throughout human time. Land mammals and flint tools from terrestrial sources are found in Mesolithic shell (marine) middens together with fish and marine mammal bones, just as the Romans developed salterns, and constructed fish traps in the intertidal zone, whilst growing cereals and herding cows and sheep on fully dry land. Finally with the advent of marine trade in the Bronze Age (if not before), coastlines took on new economic importance as the hub at which imported goods arrived, and products for export left.

The rapidity with which coastlines can evolve through erosion of coastal features and sea-level change is already clear from the discussion on Thermopylae. Sea-level change in particular has been significant over archaeological timescales – during the course of the last 150,000 years world sea levels have been between 125m lower, and 5m higher than present. In periods of low sea level, as for example occurred towards the end of the Weichselian cold stage, some 18-25,000 years ago, large areas of continental shelf that are now submerged beneath the world's sea would have been land, and the coastlines of the time were tens if not hundreds of kilometres away from those of the present day (**22**). This means that archaeological sites from the late Pleistocene, and even the early Holocene, which relate to human activity in terrestrial environments, are now submerged. Indeed Palaeolithic artefacts are commonly dredged from the sea beds surrounding Europe, while the discovery of Mesolithic finds recovered from fishermen's nets in a part of the North Sea called the Dogger Bank, indicates that the area was exploited by hunter-gatherers of that time (**22**). The lesson of all of this for environmental archaeology is one that was discussed in section 1: we cannot assume that the landscape setting of the past was the same as that of today. An absence of early Mesolithic coastal-type sites on the modern English east coast is only to be expected – the coastal sites of the time were several tens of kilometres to the north-east.

Sea-level change

A complex of different factors causes variation in sea level, but the most important are **eustatic** and **isostatic** processes. Eustatic sea level variations are a product of the absolute quantity of water that is in the Earth's oceans at any

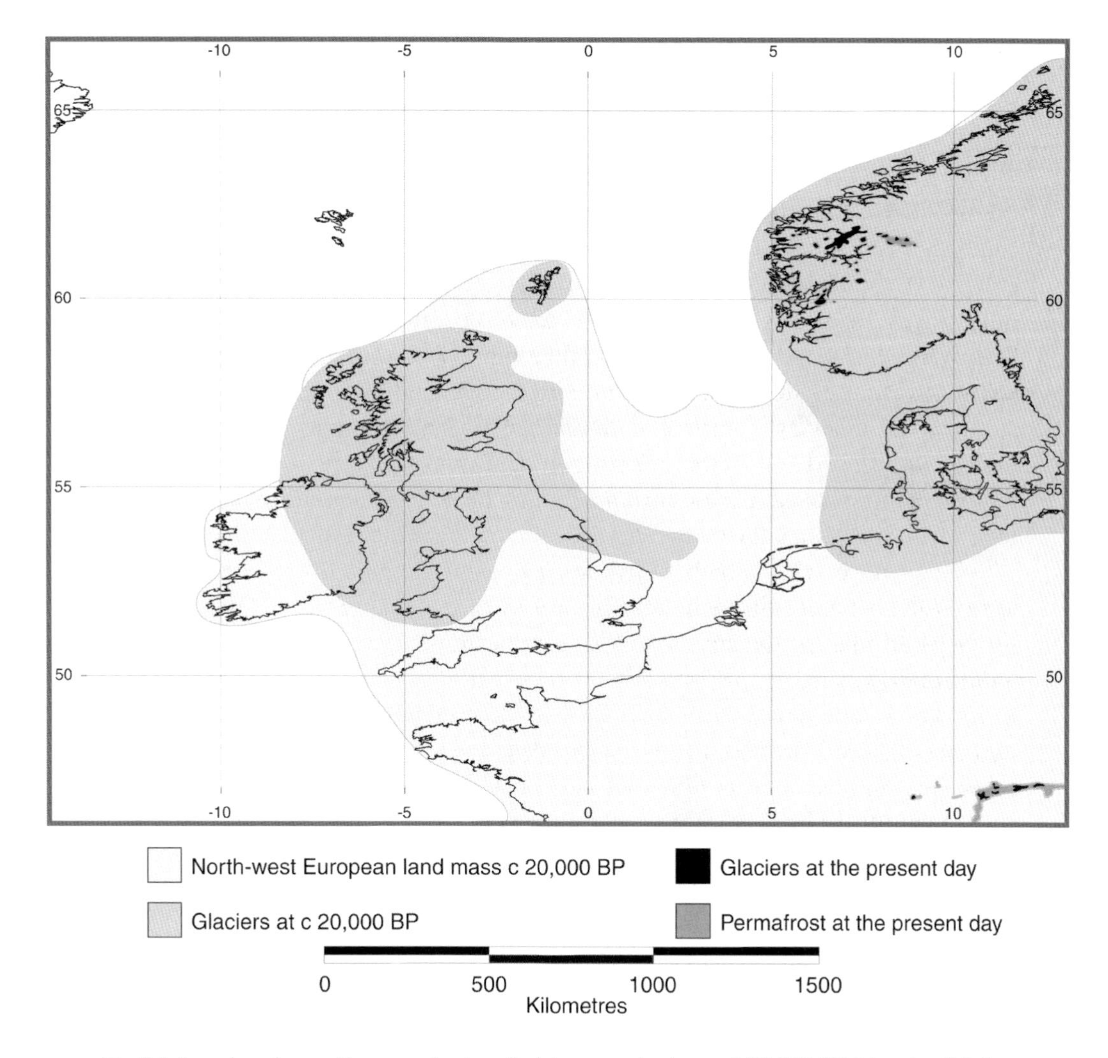

22 *Britain and north-west Europe at the time of minimum sea level around 20,000 BP. Note that Britain is joined to continental Europe and a large plain exists in what is now the North Sea, which has been termed 'Doggerland' by Bryony Coles*

particular time. When the amount of water in the ocean increases, sea levels rise, and when it decreases, sea levels fall. During the Quaternary period the amount of water in the world's oceans has been largely dependent on the extent of glaciers developing in the Arctic and Antarctic. In other words, the world only contains a finite amount of water, and the more of this that is locked up as glacier ice, the less that is present in the sea. As glacier extent is a product of climate, eustatic sea levels are too. Isostatic sea level variations in the Quaternary are also largely the result of glaciers, but this time are caused by crustal movements occurring near the Earth's surface and resulting from the weight of glacier ice. During the **stadials** (the coldest parts of 'cold' geological stages such as the Weichselian) glacier ice developed over many parts of northern Europe (**22**), North America and northern Asia, often to a thickness of several kilometres. Given that every cubic metre of ice has a mass of 0.92 tonnes, this adds up to a considerable downward pressure exerted on the underlying rock strata. Contrary to popular belief, rock, even granites and basalts, is not unyielding – when subject to the enormous pressures exerted by the overlying glacier, the rock is compressed. However, this compression is counterbalanced by a 'swelling' or uplift of rocks immediately beyond the glacier margin (a so-called **forebulge**). Given that sea levels are low during glaciation the immediate effect is minimal. However, as temperatures warm and the glacier retreats, the sea moves in to those areas compressed by the ice, and which are below the new sea level. Conversely those areas in the forebulge remain high and dry. Nevertheless, following glacier retreat over time the Earth's crust re-establishes its pre-glacial equilibrium, so that compressed rocks once beneath ice sheets expand, and the forebulge subsides. These re-adjustment processes also have sea level implications, with levels falling in those areas once compressed, and rising in the area of the forebulge. A good example of these processes is the east coast of Scotland, an area which had been glaciated at the very end of the Weichselian about 14-12,000 BP. In the St Andrews area raised shorelines are found dating from the earliest part of the Holocene, which are now 25-30m above current sea level, but which because of ice loading were at sea level at that time. As eustatic sea levels were about 25m below their current levels at the time, this represents a crustal rebound of more than 50m. Returning to the coastal Mesolithic of Britain, these crustal processes explain the reason for the preservation of shell middens along the coast of Scotland and its islands, when they are largely absent from England.

Given the importance of coastal situations for past populations and the dynamic nature of these environments, an important task for geoarchaeology in such contexts is to establish how past sea levels relate to archaeological activity. In other words: did people exploit the intertidal zone, and if so which part; or were they instead active only in fully terrestrial environments adjacent to the coast; or alternatively was human activity entirely unrelated to the position of the coast? A reasonably clear picture of these relationships can quickly be built up from examining the stratigraphy of coastal deposits exposed in sections (e.g. in ditches, sea cliffs, channels etc) or in bore holes drilled for the purpose, and relating this stratigraphy to the archaeological site being examined. Marine deposits in northern European situations commonly consist of grey silts, clays and sands (or in some circumstances gravels), while terrestrial deposits often take the form of peats and of silts and clays of a brownish hue. Archaeosediments, or artefactual material found within a sequence, indicate human activity in the marine environment. Successions of conformable sediment beds changing upwards in a core or other vertical sequence, from wood peats (forming in fen carr), to *Phragmites* peats (forming in a saltmarsh), to grey clays (marine margin), and to sands (fully marine), would be indicative of sea level rise (**transgression**), while the reverse would indicate either a decrease in the rate of sea level rise (i.e. where deposition of sediment outstrips rise in sea level) or an absolute fall in sea-level (**regression**). Such sea-level events can be set in a chronological framework by dating peats that are conformably overlain or underlain by marine clays using the ^{14}C technique. The dates become **sea-level index points** that either indicate a **positive sea-level tendency** (i.e. rise in sea level) if the dated peat is overlain by marine deposits, or a **negative sea-level tendency** (i.e. a reduction in the rate of sea level rise, or a fall), if they are overlain by terrestrial deposits. These index points can be plotted as a graph of altitude versus date to produce a sea-level curve, showing how sea levels have changed with time (**23**).

Reconstructing sea levels on the basis of stratigraphy is a useful preliminary step, but such studies can never elucidate the intricacies of the changes. For example, grey silts and clays form across the intertidal zone and cannot be differentiated with any certainty. Another potential problem is the presence of unconformities in the sequence that are unrecognisable in the stratigraphic record. Trying to establish – albeit unintentionally – sea-level index points above or below such unconformities can lead to misleading reconstructions of past sea-level movement, and hence relationships between people and the coastline. For example dating the top of a peat that is separated by an unconformity from a marine clay, will merely provide a date for some unspecified period of peat growth, not the last episode of peat growth before the salt marsh in which it was developing was inundated. However, three classes of biological remain can help in addressing such problems, and at the same time provide high resolution sea level data. Species of **diatoms** (single-celled algae), **Foraminifera** (single-celled animals possessing a calcareous 'shell') and **ostracods** (small bivalve crustaceans) are all distributed according to the concentration of salt in water and water depth. Given that salinity and water depth are strongly related to sea level, remains of these organisms can provide a proxy for past sea levels. Furthermore all three classes of organism are sufficiently small that they can be found in profusion in the small samples that are retrieved in cores, as well as those taken from upstanding vertical sections. The

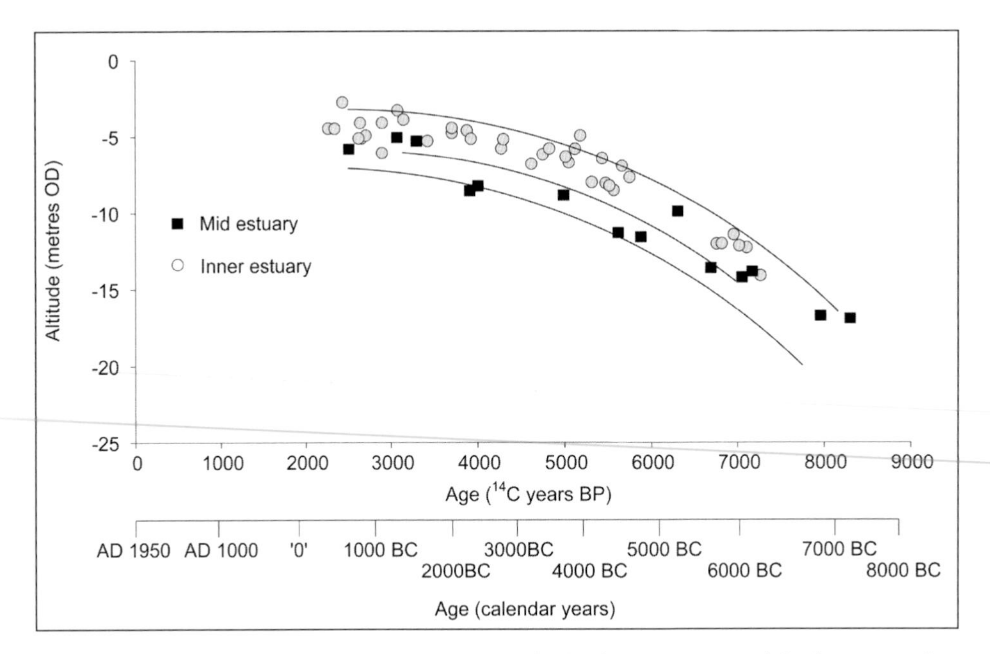

23 *A sea level curve for the Thames estuary, southern England. The inner estuary is defined as areas to the west of the town of Tilbury (including London) and the middle estuary, as areas to the east*

procedure with all three is to extract the sub-fossil remains, identify the specimens recovered to as high a taxonomic level as possible and then to assess which part of the coastal zone the identified assemblage is likely to have come from in light of the present day ecology of the species identified.

DIATOMS

Diatoms survive in the archaeological record as 'shells', or **frustules**. They are made of silica and are preserved in waterlogged conditions. Frustules, which can vary in size between 5 µm and 2 mm, are shaped in different ways (e.g. circular, elliptical or rod-like), and have varying types of microsculpture depending on species. Therefore high level taxonomic determinations can be usually be made. Being algae, diatoms are plants, so they need sunlight to photosynthesise. For this reason they do not live in water depths exceeding 200m. Nevertheless different diatom species occupy a wide range of **benthic** (sea-, lake-, riverbeds) and **planktonic** (i.e. floating freely within the water column) habitats. As well as their use in sea level studies, diatoms have also been studied from lake sequences where they provide information on past water depth, and for recent periods, on the impact of acid rain. Diatom frustules are extracted from enclosing sediments by soaking a sub-sample in hydrogen peroxide to remove organic matter. Frustules are then removed by flotation in a 'heavy liquid' (i.e. in a liquid that is denser than water; lighter organic remains tend to float, whereas denser mineral components will remain at the bottom). The flots produced by this procedure are mounted on microscope slides and the diatoms identified under high power microscopy. At least 500 frustules must be counted to produce statistically meaningful results.

FORAMINIFERA

Foraminifera only inhabit saline water bodies. They are preserved in the archaeological records as **tests**, which are the hard, shell-like parts that surround the soft body of the living organism. Two forms of tests exist, one termed **calcareous** where the test is built up of calcium carbonate and relatively smooth, and **agglunitating** forms, where the test consists of a mass of particles of various minerals that have been accumulated by the animal and bonded together. Being of mineral composition, foraminiferal tests do not need waterlogged conditions to survive, but do require the surrounding sediments to be calcareous. Tests are extracted from the sediment matrix by soaking a sub-sample in hydrogen peroxide to remove organic material and then passing the mixture through a 63 µm sieve. If the residue retained does not contain large quantities of sand grains it can then be sorted under a high-power binocular microscope. If large quantities of sand are present, heavy liquid can be used to separate the foraminiferal tests. Foraminifera, like diatoms, can be benthic, or planktonic, dwelling, but as they have no light requirements species may live in exceedingly deep water. Foraminiferal species distribution is highly dependent on sea surface temperature and they have therefore been used to reconstruct past marine climate. Foraminifera are also used as source material for oxygen isotope studies to reconstruct world ocean volume variations during the Quaternary, thereby producing isotope curves of the type illustrated in figure **4**.

OSTRACODS

At first sight, an ostracod looks like one half of a minute monkey nut. Ostracods vary in size between 0.6-2mm and are preserved in the archaeological record as a shell – called a **carapace** – shaped like a hospital kidney dish. Each animal lived inside two, hinged carapaces during its life, but on death the soft body parts deteriorate and the carapaces separate. Such remains are preserved in the same conditions as Foraminifera, and identification is again based on the shape and microsculpture preserved on the surviving part. The vast majority of ostracod species are benthic, although both salt and fresh water habitats are occupied, while species distribution is largely dependent on water temperature and salinity. Ostracods are separated from sediment by the same procedure as outlined for Foraminifera, except that a 125µm mesh is used for sieving.

AN EXAMPLE: GOLDCLIFF, GWENT LEVELS, WALES

Having looked at the importance of coastal environments and means of identifying past sea level, we can now turn to an example to illustrate how the techniques and approaches work in practice. The Severn estuary of south-west

Britain has been intensively studied over the past two decades, during which time a considerable number of Mesolithic to Romano-British sites have been located in intertidal stratigraphy. The area is of particular interest given the fact that the coastline now surrounding the Severn has the second largest tidal reach (after Tierra del Fuego at the southernmost tip of South America) anywhere in the world. Given the number of sites found, the most obvious questions asked were whether they occurred on their contemporary coastline, and if so how the sites related to past sea level. Following on from that the use people made of marine and intertidal resources was also investigated. Undoubtedly the site where the most detailed studies have been carried out is that of Goldcliff, near Newport on the Welsh side of the Severn. Here, investigations directed by Martin Bell between 1991 and 1998 led to the discovery and subsequent excavation of a number of later prehistoric structures associated with intertidal muds, peats and an early Holocene palaeosol. Most activity in the Goldcliff area is associated with an island of raised bedrock, which at certain times in prehistory projected above the saltmarsh and at others was home to vegetation communities different from those in lower lying areas. The initial evidence of human activity is formed by a series of Mesolithic flint scatters associated with animal bones and charcoal, and dating to around 5500 BC. These activities took place on both the dry land of the island and a peat that fringed it, both of which were occupied by woodland, according to the pollen evidence. However, it would seem that these Mesolithic people also exploited lower zones of the foreshore as is shown by human footprints in the surrounding intertidal muds. The presence of associated deer footprints crossing those of the humans may provide a hint of Mesolithic use of the area.

From about 4000 BC a slow-down in sea level rise, recognised around much of the southern British coastline (**23**), caused the formation of peat over the whole Goldcliff area. This peat went through a number of **seral** stages from reed swamp to fen marsh and finally raised bog, during which time the landscape was predominantly wooded according to a series of pollen diagrams. However, humans were present, as is demonstrated by a number of woodland clearance episodes from 2000 BC onwards, while from around 1500 BC the vegetation consisted predominantly of grasses and herbs. More direct evidence of human activity is found from around 1100 BC. By this time rising sea levels were reasserting an influence and a tidal channel had formed between the island and other parts of the shore. To ease passage to the island during ebb tide events, the Bronze Age population built a trackway across the channel, at least one timber of which had been previously used in a boat. The island itself would appear to have been used, at least in part, for mortuary practices as two skulls of a similar date were found in pits dug on it.

The period between *c*.600 and 300 BC saw an oscillation of negative and positive sea-level tendencies represented by alternating deposits of peat and mineral sediment formation. It was in this changing environment that the most extensive later prehistoric human activity at Goldcliff took place. This began around 360 BC when a series of rectangular buildings was constructed on the surface of a shelving peat, connected to one another by brushwood trackways. Wood from these buildings has been both dendrochronologically and ^{14}C dated and suggests that the last building may have been constructed as late at 150 BC. Most were constructed of wood that according to dendrochronology was cut in winter months, while the wood of one (building 6), could be more precisely ascribed to a cutting date of April/May 273 BC (**24**). Little occupation debris was found associated with any of the buildings, but floor deposits were located and sampled. Mites in these suggested that cattle had occupied the structures, a finding confirmed by numerous cattle footprints found between the individual structures. Plant macro-remains and diatoms associated with the buildings suggested they were built just as sea levels were beginning to rise, while there are indications in the stratigraphic record in the form of thin mineral layers in the floor deposits, that the buildings were on occasion inundated by rising tides. The most obvious question of this Iron Age activity, given that no marine resources were being exploited, is: why bother exploiting an area that was becoming increasingly inundated by tidal waters from the Severn? Bell has suggested that rising sea levels are associated with changing vegetation, and in particular low saltmarsh herbs and grasses. These are ideal

24 *An Iron Age rectangular building (number 6) in the intertidal zone at Goldcliff.* Photograph courtesy of Martin Bell

for grazing and would be a useful resource for a pastoral community, particularly if exploited on a seasonal basis.

At Goldcliff, then, we have a dynamic coastal environment exploited from the Mesolithic to the Iron Age at varying intensities. Exploitation was of explicitly marine resources (in the Mesolithic), but also of the indirect products of sea level rise, namely herb-grassland pasture (in the Iron Age). Ultimately, however, nature was to make the Goldcliff intertidal zone unusable from the Late Iron Age onwards. The Romans had a very different approach to managing the coastal zone, as we shall see in section 6. Their sea wall, as represented by a marker stone recording work by a century of the second legion commanded by Statorius Maximus, was found in 1878, and is 300m landward of the Iron Age buildings. It was from behind this sea defence that the Romans exploited the intertidal zone.

Rivers

Like the sea, rivers have functioned as transport arteries and zones of diverse and concentrated resources for the entire period of human life on Earth. As far back as the Lower Palaeolithic, some 300-500,000 years ago, it is thought that Neanderthals and their immediate predecessors, *Homo heidelbergensis*, used river valleys to penetrate into the middle of Europe, even as far north as Britain. During much of the Holsteinian and Cromerian interglacials (**4**), northern Europe was covered by thick woodland occupied by dangerous predators such as lions and panthers. Therefore, the lesser vegetation of the river, margins offered an easier and less dangerous route. In later periods rivers became a focus for settlement – a quick look at the distribution of towns of the late Roman Empire as listed on the Antonine Itinerary, plotted against rivers provides convincing evidence. In the case of arid and semi-arid zones, they become a focus for agriculture too. The best known example of the latter is the River Nile, whose floodplain, which is enriched by an annual input of silt, has been cultivated since the early Neolithic, and in the Roman period was the 'breadbasket of the Empire'. In such climates water is the key to soil fertility – therefore in Egypt the only areas that can be cultivated are those with a ready supply. Of these the Nile valley is by far the largest. Rivers were also effectively the highways of the ancient world and have been vital for trade from perhaps as long ago as the Mesolithic. Before the invention of steam power and the internal combustion engine there were only three ways of transporting produce, raw materials and trade goods: by pack animal, cart and by boat. Given the limited weight that could be either pulled by, or loaded onto a horse or ox, the boat was preferable, so the course of rivers determined many trade routes.

The importance of rivers in the subsistence, transport and trade of pre-industrial societies means that river valleys are some of the richest archaeological landscapes in the world, so it is important that we understand how river systems operate in order to understand these aspects of past societies. However, because the vast majority of rivers are dynamic even over short periods of time we also need to understand how rivers can change. Such changes can occur on a number of scales for example; flow through a river alters on a seasonal basis, river channels move across their floodplains at a scale of decades, centuries or millennia and may even be 'captured' by another river, in which case the original river valley will be abandoned. At even longer timescales, climate change may cause more fundamental changes to the river, causing it to change from for example a slow flowing, predictable and seasonally consistent stream[1], occupying only part of the valley, to a raging torrent covering the whole valley, albeit that the latter only occurred during a small fraction of the year. In the following pages we will examine two of the more common types of rivers. We will look at the types of deposits they produce and assess where within such rivers archaeological sites are likely to be preserved, and how the remains might be interpreted.

What are rivers?

Of all the world's environments, rivers have certainly been the most thoroughly studied by geomorphologists, so that we know in great detail how, where and in what micro-environment every morphological feature and each type of deposit formed. Huge treatises have been written explaining what diagnostic characteristics are displayed by **fluvial** (i.e. relating to moving water) deposits and landforms forming under particular conditions. Even the most basic geomorphological text books will contain several chapters on rivers. Rather than attempting to summarise all this work, we will instead concentrate on river environments found at northern latitudes and the opportunities these provide for environmental archaeologists. Readers who wish to know more of the minutiae of the products of fluvial processes are referred to the 'Further Reading', where several texts are recommended.

Rivers are natural conduits that transport precipitation falling on land back to the sea. They are, therefore, a vital part of the world's 'operating system', fulfilling an important role in the carbon cycle by transporting carbonate, in the form of weathered rock to the sea, as well as constantly diluting the oceans with fresh water. Rivers flow according to gravity and along the line of least resistance. Consequently, the headwaters of rivers are in mountains or hills and their channels always run downhill along the weakest geological strata to the sea. If sea levels drop for the reasons previously discussed, rivers will re-adjust to the new situation by steepening the slope – termed **gradient** by geomorphologists – from their headwaters to their outflow. In other words, they will cut down through their existing bed, leaving remnants of their previous bed and floodplains perched above the new level as a **terrace**. The same is also true of tectonically active areas where the continental crust is undergoing uplift; rivers simply readjust to the new situation by downcutting. Therefore, over time the interaction of changing sea levels and tectonics can cause the formation of many terraces at various levels above that of the present river (a so-called terrace staircase), where the oldest river beds and floodplains occur at the highest elevations.

The world's rivers can be divided into four basic types; straight, meandering, braided and anastomosing (**25**). However, many geomorphologists dispute the existence of straight channel forms, seeing them instead as a sub-category of meandering rivers; in other words a meandering river without pronounced meanders, or in technical parlance: a low **sinuosity** river. Rivers may display different **bedform** properties along different parts of their course (**stretch**). This is because the type of bedform found directly relates to the gradient of the stream, the nature of sediment supply to the stream and the amount of water in the stream (**discharge**). So for example, a river may be characterised by a braided bedform in the upper part of its catchment and meandering in its lower reaches. Anastomosing rivers, while common in North America, are much rarer in Europe, where they are seen as transitional forms between braided and meandering streams. For example many of Europe's rivers, including the Thames, Meuse and Vistula, are characterised at the end of the Weichselian late glacial period by anastomosing bedforms, although all are meandering at the present day. Anastomosing rivers consist of deep, multiple channels occupying only a restricted part of the floodplain (channel belt), interspersed by dry land, or seasonably submerged islands, while extensive floodplains and even marshland occur beyond the channel margins (**25**D.). Whereas the channels and islands are inherently unstable in braided rivers, they are extremely stable in anastomosing systems. Islands within the river exist for long periods of time and consequently may contain evidence of multiple phases of occupation.

Braided rivers

As the name suggests, braided rivers consist of a mass of relatively shallow channels spread out across the floodplain (**25**C). These channels have a short life and may be created, filled and re-cut along a slightly different alignment during subsequent episodes of flow. Channels mostly run parallel to the direction of flow, but frequently intersect with one another. Braided rivers occur where the supply of coarse-grained sediment and discharge is high. At the present day these requirements, combined with a steep gradient, are only found in mountainous reaches within the temperate zone. The world's mountainous regions are largely comprised of hard rock geologies, while weathering processes are most intense at high elevations. Therefore, within mountain ranges there is a ready supply of large calibre particles that have been liberated from the bedrock by frost action, slope failure or faulting, and which is readily

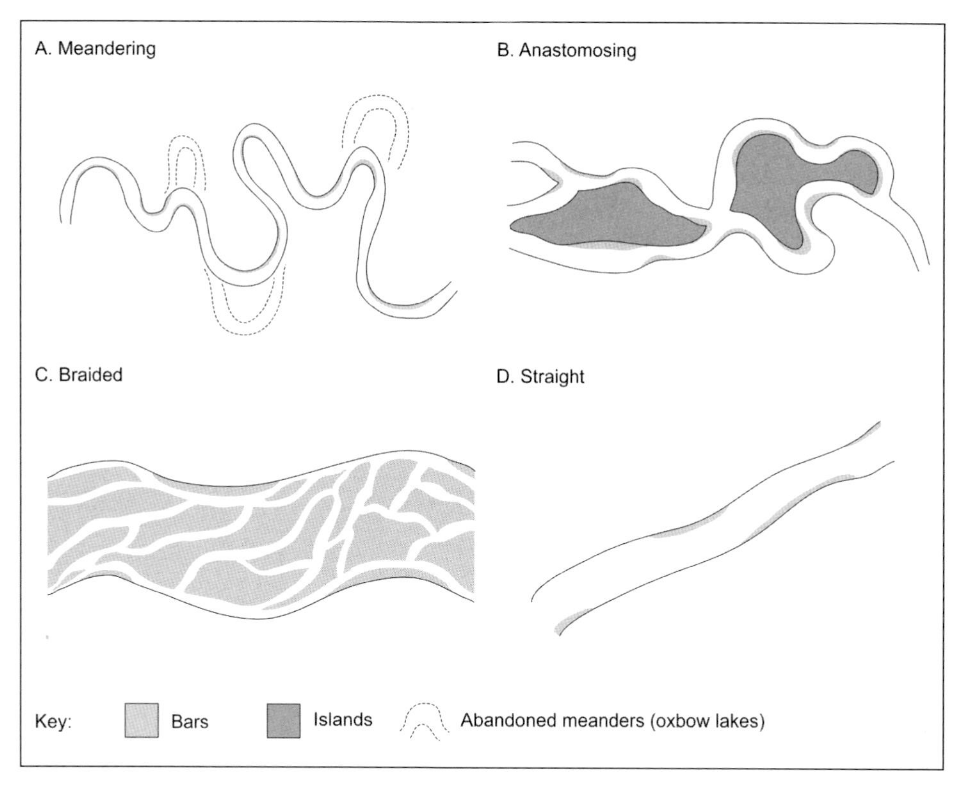

25 *River forms*

available for transport. In many mountainous areas rainfall is concentrated in late autumn, winter and spring, and for much of the winter at least, falls as snow. The combination of this precipitation pattern, together with the retention of water as either snow or ice during the winter months, means that discharge is concentrated during the spring, producing a seasonal – or flashy – pattern of flow. High spring discharge combined with a steep steam gradient makes for high energy streams, which can easily transport the available sediment particles. Braided rivers are therefore the most dynamic of bedform types and are characterised by seasonal flow: much of the river valley is covered by surging water during the spring, but by autumn only a fraction of the many channels on the floodplain are occupied. In the past, braided rivers were not restricted to mountainous zones. In northern Europe for instance, the cold climates of Weichselian stadials caused many rivers to adopt braided bedforms. Permafrost conditions caused extreme seasonality in discharge while the cold, periglacial climates increased weathering of the geological substrate, even in the lowland zone.

The vast majority of sediment movement in braided rivers occurs at flood stages, i.e. peak discharge. Sediment particles are moved along channels by flowing water and deposited when water energy drops to a level where it is no longer capable of moving a particle of that size. By this process, bars are formed, parallel or at right angles to flow within the channels. Such bars are characterised by a decrease in average grain size from the upstream end of the bar to the downstream side. This characteristic is caused by the way that particles are deposited, each being forced over mounds of previously deposited particles and simply falling down the far side of the file to form so-called **lateral accretion** deposits. In plan, bars can resemble tear drops with the 'tail' facing upstream, while individual particles align themselves either with, or at right angles to, flow. During periods of lesser flow the tops of bars form land surfaces. However, in the majority of cases the succeeding flood events either erode the bar by driving a new channel through it, or deposit another bar on top of the original, thereby building up the channel bed. It is this characteristic that causes the remnants of former braided river environments to comprise sediment stratigraphies several metres thick. Given the dynamic nature of braided environments it is hardly surprising that evidence of *in situ* human activity within the channels is exceedingly rare. Nevertheless, archaeological material does survive in braided river deposits, although the material is reworked and therefore older than the river sediments in which it was found. For example over 80 per cent of England's Lower Palaeolithic record comprises re-worked finds of stone handaxes and flakes found by gravel diggers while extracting aggregates from former braided river beds now perched in terraces. In rarer circumstances evidence of human activity on bar surfaces is found in a near *in situ* state, where a gentler depositing medium led to the rapid coverage of the bar surface by finer sediment. Such is the case at the Middle Palaeolithic site at Lion Pit, West Thurrock, Essex, where some 11 Levallois flint cores and about 150 flakes were found at the top of a braided river gravel attributed to the Saalian period. The mint condition of over 90 per cent of the artefacts, when considered alongside the fact that many of the flakes could be refitted to the cores, suggests that very limited movement of the artefacts had taken place. It would appear that the hominids here were making use of the gravel exposed on the bar surface as raw material and it is lucky for us that the site was inundated by intertidal sands rather than subject to further braided river processes.

Meandering rivers

Meandering rivers contrast markedly with those of a braided type. They consist of a single winding (**high sinuosity**) channel, a reasonably constant flow and a low gradient (**8**). Meandering rivers tend to be characteristic of lowlands, temperate climates and non-flashy discharge, in other words much of Europe and North America at the present day. The low flow energies of meandering rivers mean that only along the base of channel bed itself are coarse-grain deposits, comprising coarse sands and gravels, transported (**27**). Higher up in the water column fine sands, silts and clays, derived from soils bordering the river, are moved in suspension. The latter are frequently transported onto

26 *Gravels originally deposited in a braided river of the Weichselian period in the valley of the river Huecha in northern Spain. Scale bar divisions in 0.1m intervals*

natural **levees** that border the river (in the case of sands and silts), or onto the surrounding floodplain (silts and clays) during peak flow events that occur either annually or biannually. Over time, levees build up into topographically distinct features, which may dominate the floodplain. The floodplains accumulate by vertical accretion of mineral particles, so that relatively thick laminated deposits can form. However, because flood events are comparatively rare the floodplain is commonly vegetated, while soil formation also commonly takes place within the floodplain deposits. Therefore, floodplain sequences commonly appear as either homogeneous silts and clays, pedogenesis and rooting having destroyed the lamina structure, or more rarely as a series of horizontally bedded muds and palaeosols. The latter occurs where flood events are rare, but of high magnitude and transport high concentrations of fine sediment. The relatively gentle mechanism of floodplain deposition combined with surfaces that are stable over several years means that archaeological remains are both common and well preserved in many floodplain situations. For example, whole Iron Age and Romano-British farming landscapes have been preserved in the upper Thames valley of southern England. Not only do the ditches that once divided fields survive, often with organic-rich lower fills preserved by waterlogging, but also settlement sites. Aerial photographic surveys carried out in the 1970s have been able to trace these features over huge areas, suggesting that the floodplain of the Thames and its tributary, the Windrush, were intensively used in the last few centuries BC, and the first four

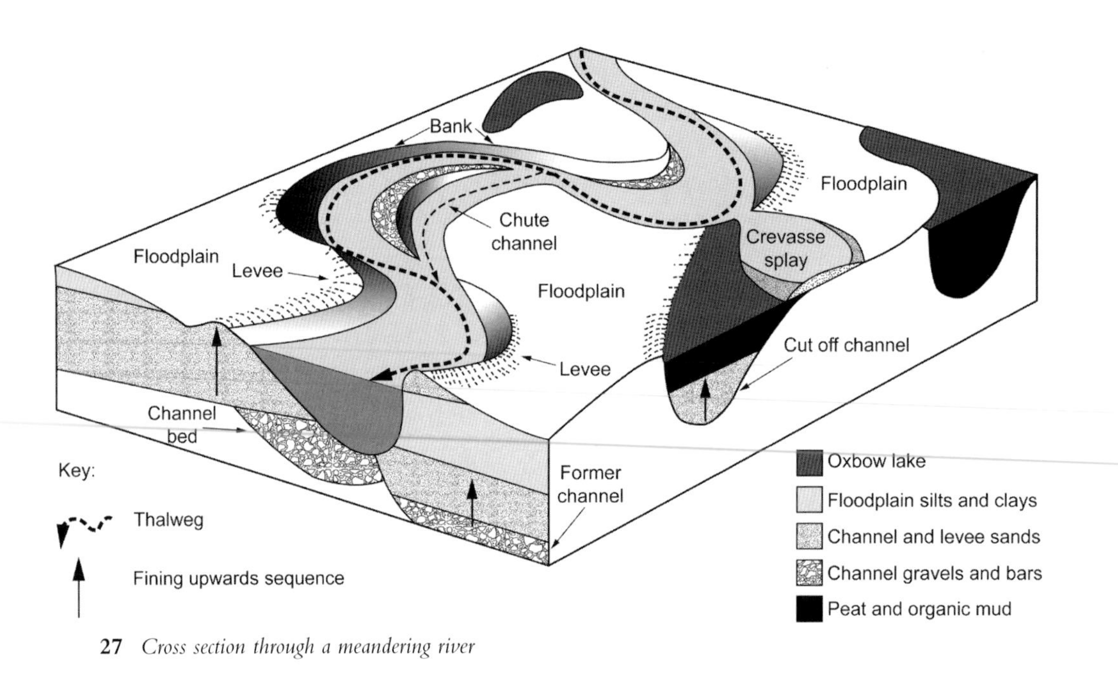

27 *Cross section through a meandering river*

centuries AD. At Mingies Ditch near Stanton Harcourt, for example, an Iron Age farmstead dating to 300-110 BC and consisting of five roundhouses (not all contemporary) was found to survive in an excellent state of preservation, complete with waterlogged plant remains and numerous bones. It would appear that this was a settlement of pioneering pastoralists specialising in horses, but that flooding led to the abandonment of the settlement within 50-100 years of its founding.

Processes operating in the channels of meandering rivers are somewhat different to those on the floodplain. Flow is concentrated across the channel from one outer meander bend to the next (termed **thalweg**, see **27**). This has the effect of eroding the bend of the outer meander and then transporting the eroded material to the next inner bend where it is deposited as a **point bar**. By this process, meanders tend to become increasingly exaggerated with time. Unsurprisingly people have not occupied sites or indeed farmed the channels of meandering rivers, but they have used them for other purposes. We have already discussed their use as transport routes, but many rivers also have been seen as sacred places by many prehistoric groups. It is as a result of either accidental losses during trading trips or ritual offerings that the richest Neolithic find spot in London is in fact the Thames – a meandering, non-tidal stretch of the river until the later Bronze Age. Countless polished stone axes have been dredged from the river in the last few decades as the river is kept navigable for present vessels, together with undoubted offerings from later prehistoric periods, such as the famed Battersea Shield. Channels of meandering rivers were also of course obstacles. Such is the case at Testwood in Hampshire, where a palaeochannel of the River Test was spanned by a bridge constructed in the middle Bronze Age (**28**). Surprisingly, given the relatively high energies of flow within the channel, the wooden supports of the bridge have been preserved beneath the gravel and sand bedload of the former river channel.

Although meandering rivers do not have the dynamism of braided rivers, they should not be assumed to be stable environments. The process by which meanders become extended also leads to parts of adjacent outer meanders eroding towards each other, eventually leading to a breach. This process results in the abandonment of the former meander bend which then becomes an **oxbow lake**. Oxbow lakes gradually fill with fine-grained sediment by means of vertical accretion, but like other low-lying areas of the floodplain they tend to contain semi-permanent water and are, therefore, highly vegetated. As a result, sediment accumulating in oxbow lakes and other floodplain hollows frequently comprises organic muds and peats. Oxbow lakes can also be formed by a process called **chute cut off**. This occurs during episodes of high flow when the river drives a channel across the point bar of the inner bend of a meander, effectively cutting off the outer meander bend (**27**). Should this process of oxbow lake creation be extensive, the river will eventually be constrained to a relatively narrow channel belt. This is because the fine-grained fills of the oxbows form cohesive banks in which the river cannot readily create new meanders. When this situation is combined with further deposition

28 *The remains of a Bronze Age bridge across a former channel of the River Test at Testwood, Hampshire. The timber piles supporting the bridge are under plastic covers*

within the channel, the whole channel belt is raised, creating a very unstable situation. In such circumstances a weakness will be found through the levee and between oxbow lakes, and the river will change course rapidly (avulse) to flow through the lowest parts of the surrounding floodplain. This process has played a major role in the development of the landscape of the central Netherlands. Both the Rivers Rhine and Meuse, which flow through this part of the country, have changed course over what is essentially flat terrain (**29**). This fact, combined with the high sediment loads carried by these major rivers, means that many sites of later prehistoric date are buried beneath several metres of floodplain deposits. Nevertheless, in spite of what would appear to be difficult circumstances, the Dutch have developed a reliable method for prospecting for such sites. Discoveries of later prehistoric sites in the central Netherlands 'rivers' area have always been made on levees or 'islands' of windblown sand left over from the late Weichselian cold stage. Furthermore Iron Age and later sites are associated with artificial soils called **plaggen**. These were formed by people transferring their organic refuse to fields surrounding their settlement to improve fertility, but a by-product was the creation of distinctive organic deposits in the stratigraphic record. The approach to prospection, therefore, is to carry out auger surveys to actively look for levee deposits (characterised by coarse silts and sands), loess islands or plaggen soils. Where these occur trenches can be sunk with a reasonable chance of locating prehistoric archaeological sites.

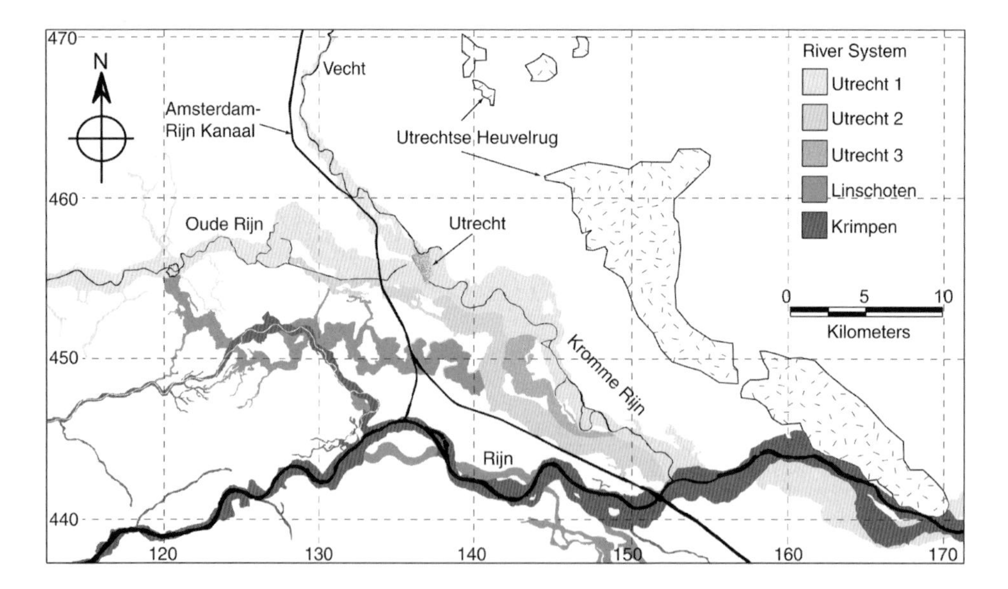

29 *Courses and former courses of the River Rhine in the Utrecht area of the Netherlands. Modified from Berendsen and Stouthamer,* Palaeogeographic development of the Rhine-Meuse delta, the Netherlands (2001)

Inland hill and mountain ranges

A large proportion of continental land mass is made up of ranges of hills and mountains, cut through by the sort of river valleys that we have just discussed. As we have seen, human activity gravitated towards riverine areas, but the surrounding hills and mountains were not totally neglected. This is especially true of areas of northern Europe such as the Jurassic limestone and chalk geologies of southern England, northern France and south-west Belgium, where the landscape is gently undulating, rather than truly mountainous. Some of the most important later prehistoric sites in southern Britain for example, including such famous locations as Stonehenge, Avebury and Maiden Castle, are located on these gently sloping chalk hills, more commonly termed downs. Springs and streams fed from them, and perched water tables within the limestone and chalk bedrocks offered a ready water supply, while the soils developed on such geologies are easily tilled, free-draining and relatively fertile, enabling arable agriculture to be both straightforward and productive. From the Neolithic period onwards, uncultivated areas have been used as grassland pasture, so successfully in fact that in the medieval period sheep grazing the limestone and chalk hills of southern England provided wool for much of northern and western Europe.

Chalk downland and the Jurassic limestone hills of north-western Europe are today characterised by thin soils of at maximum 0.5m thick, and which are termed **rendzinas**. These commonly have thin A horizons and relatively stony, calcareous B horizons. They support a vegetation of grassland, or scrub woodland, where the land is not cultivated. However, these environments are entirely the product of human interference in the landscape and document the intensity of agricultural use by later prehistoric populations. Hints as to the situation predating the development of the rendzina soils can be found in palaeosols buried by Neolithic monuments such as South Street longbarrow near Avebury and from more extensive palaeosols buried in dry valley sequences throughout southern England. Rather than rendzinas, these palaeosols are Brown Earths, developed in windblown loess which, as we have already discussed, had been deposited to depths of 1-2m and more across northern Europe towards the end of the Weichselian cold stage. Loess, unlike chalk and limestone, is a soft substrate and, therefore, soils can rapidly develop to a considerable depth. As has previously been discussed, Brown Earths are characterised by the development of deciduous woodland as the end, or **climax** stage, of vegetation colonisation in northern Europe. In southern England the Brown Earth soils would have supported oak and broad-leaved lime forests on both chalk downland and Jurassic limestone hills. Therefore it is environments of this nature that the builders of Neolithic monuments would have encountered. Having developed in loess, Brown Earth soils are mainly comprised of coarse silts and have very low stone contents. This combination, as well as being a major contributing factor to their fertility, also means that such soils are highly susceptible to erosion following the removal of the covering vegetation. After all, material that was once windblown is readily transported by wind again once the binding provided by plant roots is released. Silt particles can also be more easily moved by water action than any other size grade (clays tend to bind together, while greater energies are required to move larger sand- and gravel-sized particles). In other words the Brown Earths of northern Europe provided a double-edged sword for early agriculturalists: high fertility and ease of cultivation, but a time-limited resource that would rapidly disappear through erosion.

So far we have just considered hills – an equally important part of both downland and limestone upland is the valleys that separate the hills. The larger valleys contain rivers, which we have already examined. The smaller features are often dry, and are therefore termed **dry valleys**. It is thought that these features were formed by streams of water that exploited the weakest part of the chalk and limestone bedrock. The streams orginated from seasonal melting of snow and upper portions of permafrost on the surrounding downland during colder episodes of the Pleistocene. Palaeosol evidence suggests that during the early part of the Holocene prior to the Neolithic/Bronze Age the same Brown Earth soils that existed on the surrounding upland were also found in dry valleys (e.g. **7**). However, following the removal of woodland on the surrounding downland in order to provide land for agriculture in the Neolithic and Bronze Age the character of the dry valley soils changed. We have already seen that forest clearance on Brown Earth soils causes erosion. Much of the eroded material is windblown and can be transported long distances, certainly beyond downland areas, but still more material is transported by water and therefore moves to the lowest point in the landscape, the dry valleys. Therefore, as soil thickness decreased on the hills, the deposits in the dry valleys became thicker. Archaeological material on the hills was transferred to the valleys, while sites in the valleys were buried by the reworked soils, or **colluvium**, as sediment accumulating by gravitational processes in dry land situations is known. There is, then, little doubt that downland and Jurassic limestone environments are important in the study of northern European prehistory, and indeed in the Roman (the majority of Roman villas in southern Britain are located in such environments) and medieval periods. A primary task for geoarchaeology is, therefore, to determine when soil erosion/valley deposition occurred, what caused transfer of soils (e.g. forest clearance, agricultural practice, pastoralism) down slope, the mechanisms by which this transfer occurred, whether regional patterns of erosion/deposition can be seen and lastly whether as a result of the erosion/deposition the archaeological record of these areas is biased.

Except for palaeosols beneath Neolithic and Bronze Age monuments, the prime evidence (i.e. the downland soils) for later prehistoric and historic landscape processes in hilly regions has been removed. The absence of suitable

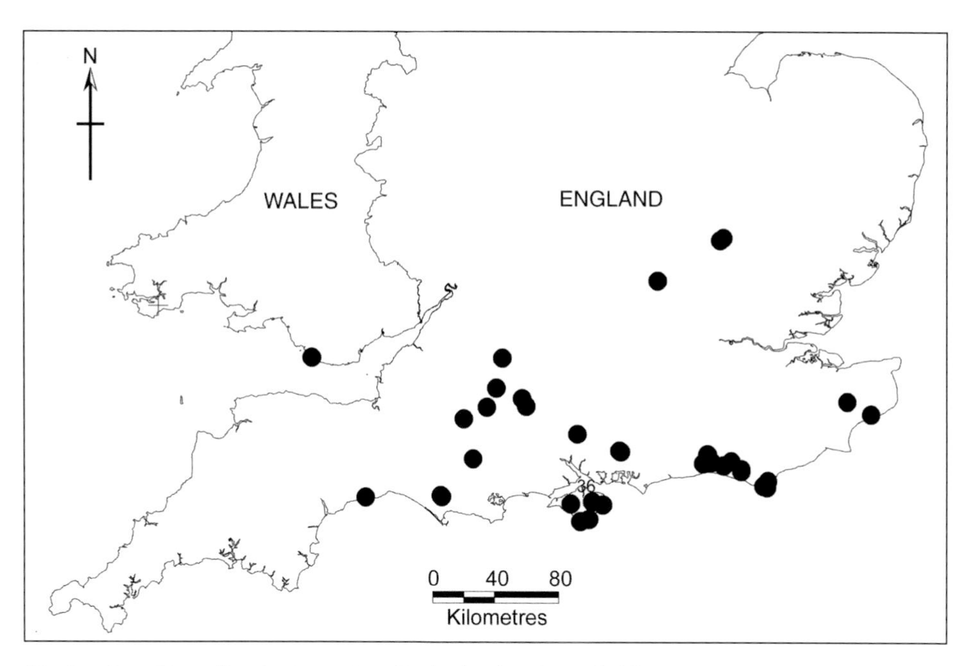

30 *Location of dry valley sites in southern England and Wales studied by environmental archaeologists*

study material on slopes means that an indirect approach to studying erosion history has to be taken. That used in southern England has been to examine sequences of dry valley sediments. Studies in Wessex and Sussex undertaken by Martin Bell in the 1970s and Mike Allen and others in the 1980s involved cutting sections through dry valley deposits using a mechanical excavator and then dating the stratigraphy from associated artefacts collected from controlled excavations of an adjacent strip, as well as by ^{14}C dating of charcoal (**7**). The stratigraphy was described in the field and sediment samples analysed for their micromorphological and particle size properties in the laboratory, while snails (see below) were also examined from continuous samples taken through the deposits to provide an indication of the environment in which erosion occurred. These studies clearly indicated that it was not forest clearance *per se* that was responsible for soil erosion from the surrounding slopes, but rather agriculture and in particular the use of intensive agricultural practices from the late Bronze Age onwards. The evidence suggests that the initiation of dry valley sequences in Wessex and Sussex was not synchronous, primarily because agriculture was carried out at different intensities, in different places and at different times, entirely as the result of local cultural and environmental factors. Therefore, whereas some dry valleys such as Redcliff on the Isle of Wight began to fill with reworked soil in the late Neolithic period, others, such as Chalton in Hampshire did not accumulate material until the medieval period. The exact mechanism by which this erosion and transport cycle occurred has been the subject of some debate in the archaeological and soil science literature, even following Bell and Allen's investigations. Prior to the 1990s it was thought that **overland flow**, in other words transport of soil particles down slope by rivulets, or even sheets of water passing over the ground surface during rainfall events, was the main route to dry valley sediment accumulation. Such a model would suggest that because high rainfall has been a constant feature of the northern European climate in the Holocene, sequences should be continuous, and that in dry valley fills we therefore have an unbroken record of land-use around each valley from the point at which intense agriculture began on the surrounding slopes. However, this view was challenged by John Boardman in the 1990s who suggested, based on the observation of present day fields, that erosion on cultivated fields was most significant during extremely high rainfall events. Most soil was eroded as a result of the development of shallow channels called **rills**, or during particularly violent storms, by much deeper features called **gullies**. As rills remove the entire soil profile, the transported material includes not only silts and clays, but also granules of bedrock that are incorporated in the B horizon of the eroded soil. As the eroded soil is transported down the rill by what is essentially a small stream, particles of different sizes separate and are deposited together in much the same way that has been indicated for river environments. The same is true of gullies, but because these features penetrate the bedrock, gravel-sized particles are eroded and re-deposited in the dry valley. Here then is the explanation for the presence of lenses of chalk limestone granules and sheets of flints in dry valley sequences (e.g. **31**). If such deposits are the result of intense storm activity affecting cultivated fields on dry valley slopes, the implication is that the sequences in the dry valleys have not formed continuously, but rather as a series of short-lived, but intense episodes. Unfortunately, this makes the deposits rather less useful for reconstruction of land-use through time, given that a large part of the soil profile on the valley sides is removed in each erosion episode. Therefore, sediments accumulating in the valley bottom in each erosion event may relate to the entire history of soil development on the slope.

Recently scientists mapping soil transport across modern arable fields in England have developed a new technique: examining spatial concentrations of the caesium isotope ^{137}Cs. This isotope is emitted during atomic bomb tests, which have been carried out since the mid 1940s and by 'nuclear accidents' such as Chernobyl in 1986, and builds up within a soil on a time accumulative basis. Therefore, where erosion has occurred, ^{137}Cs levels will be low (because soil material containing ^{137}Cs has been removed) and where accumulation has occurred it will be high. Where a situation can be found in a nearby area where neither soil erosion or deposition has occurred since the mid-twentieth century, a control can be established of how much ^{137}Cs should be expected at different depths in the soil profile given no interference. By comparing the control with ^{137}Cs measurements taken from a series of sub-sampled cores (**32**), or profiles exposed in soil pits taken/dug across the field being studied, soil movement can

31 *Sequence through a dry valley at Sweetpatch, Brighton, southern England showing sediments accumulating as a result of rill and gully processes. Scale bar in 0.5m divisions*

be quantified. Data of ^{137}Cs collected in this way, together with microtopographic mapping data collected in separate studies by the Universities of St Andrews and Exeter, have been interpreted as meaning that it is the action of the plough itself that is the main cause of soil transport. The idea here is that mouldboard ploughs always turn the tilled soil in a single direction with a tendency for this to be downhill. Over several years of ploughing soil is transferred across the field in a downhill direction as shown by a build-up of ^{137}Cs at the downslope side of the field. Eventually this reworked soil is transported by the same process across more fields to arrive at the base of the dry valley. If the applicability of these modern observations could be confirmed to the archaeological situation in the centuries and millennia prior to 1945 – which of course is difficult given the absence of atomic bombs or nuclear accidents prior to the mid-twentieth century – the pre-1990 view of dry valley fills accumulating gradually and by increments would prevail once more.

32 *Core samples being taken from a site in the Quantock Hills, Somerset, south-west England for ^{137}Cs study to reconstruct patterns of soil erosion.* Photograph courtesy of Michelle Collings

Although the examples we have considered relate to southern England, the results are equally applicable to other regions, even those beyond the margins of northern Europe. For example much of Mediterranean Europe is characterised by mountainous terrain divided by valley systems. Here a similar transfer of soil material to valleys has occurred, but at a much higher magnitude than in northern Europe. In the only study of the entire Mediterranean region to date, Claudio Vita-Finzi has suggested that the deposits that accumulated in the Mediterranean valleys occurred during two more or less discrete episodes. One of these, termed the **Older Fill** by Vita-Finzi was commonly of a distinctive red colour and formed during the late Pleistocene. The second, the **Younger Fill**, was of a buff or grey colour and was dated to the late Roman period on the basis of artefacts found within it. As the events occurred at the same time across the whole Mediterranean area, a regional phenomenon had to be the cause. Climate change to cooler and wetter conditions – a **pluvial** in the case of the late Pleistocene event – was suggested as being this triggering factor.

During the 1980s and 1990s the idea that all deposits forming the Younger Fill were deposited at the same time was exploded as more sequences were examined and dating of the sequences improved. The same was later found to be true of the Older Fill. Instead it was observed that there was no pan-Mediterranean pattern of erosion with time. Therefore, a more unpredictable agent must have been responsible, in the caes of the Younger Fill, the most obvious being people. Studies carried out in Greece and Italy in particular suggested that agriculture was just as much the root cause of the Younger Fill in the Mediterranean as it was of dry valley fills in northern Europe. However, the differing climate and consequently farming methods employed in the

Mediterranean region meant that the erosion occurred as a result of different mechanisms than those seen in northern Europe. The mountainous terrain of much of Italy, Greece and Spain cannot be cultivated by simply creating fields that disregards the contours. Such a strategy is impractical, in that the farmer and plough team would have to walk up and down very steep slopes. It also risks drastic soil loss as winter and spring rain would simply wash precious soil off the field and onto that of a neighbour, or indeed into the valley below. To get around this problem many Mediterranean cultures artificially created flat agricultural land by building terraces that follow the contours of mountainous foothills (**33**). Such terraces require constant maintenance to keep them from collapsing and thereby causing the loss of the retained soil. This is likely to have been carried out during times when the crops grown on the terraces were in demand by nearby populations. However, in times when economies deteriorate and areas become depopulated, the demand for food falls and there is consequently no perceived need to maintain terraces. It is in these periods of economic stress that Tjeerd van Andel and colleagues see major phases of erosion occurring, based on data from valley fills in southern and central Greece. For example the Upper Flamboura deposits exposed in several valleys of the southern Argolid date from the late Roman period, just the time when the area is thought to have been suffering severe depopulation. More recently still this view has been challenged by John Bintliff who has suggested that not all erosion phases correlate with economic downturns and that valley deposition is most likely to occur where climate changes to more arid conditions combine with intense agriculture. Whatever the correct explanation for the Younger Fill – if indeed there is a single process which can explain all – this one subject has dominated Mediterranean geoarchaeology over the last 20 years to the exclusion of almost everything else.

Reconstructing past biota

There are two obvious components of the biota: plants and animals. However, as we saw when discussing ecology in section 1, each part of an ecosystem – which is what an environmental archaeologist is trying to reconstruct for a past situation – is interdependent in a manner which accords with its trophic status, i.e. its place within the ecosystem hierarchy. Ignoring for a moment factors associated with colonisation and climate, the animals of an ecosystem are only members of that system because of the characteristic plants that are also members. Thus the nature of the animal population, both in terms of its species composition and trophic structure, depends to a large extent on plants. The plants in turn are present as a result of variables associated with climate and geology, as well as factors associated with colonisation routes. In other words the plants of an ecosystem (the lowest trophic levels) provide food for

33 *Active agricultural terraces in the Parnon foothills, Peloponnese, Greece*

herbivorous animals, which in turn are the prey of carnivores. Above these, many ecologists place humans (the highest trophic level). What all this means for environmental archaeology is that if the lowest trophic levels can be reconstructed, in other words the plant communities that formed part of a past ecosystem, then using uniformitarian principles with analogies of present day ecosystems, it is possible to populate a reconstructed ecosystem with some of its animal inhabitants. The first step in reconstructing past biota therefore is to determine the nature of past vegetation (**palaeovegetation**).

Palaeovegetation studies

Assuming geoarchaeological studies of the type previously described have been carried out, one element of an ecosystem on which past plant communities will have been dependent may have been determined; namely the geology. Geological substrate is the main determining factor of the soils in which plants grow. The vagaries of a further element affecting ecosystems, climate, are known from the last few tens of millennia, but rarely with the exactitude that would help in the reconstruction of individual ecosystems. Indeed as climate is often reconstructed using the remains of plants and animals that survive in the geological record, there is a dangerous potential for circularity of argument in using the same proxy climate data in helping to suggest what biotic groupings were present in past ecosystems!

Reconstruction of past vegetation in environmental archaeology is undertaken using a variety of proxies. Historically the first type of plant data used in palaeoenvironmental reconstruction was the most obvious, namely the seeds, fruits and wood of plants. We will look at plant macrofossil analysis later in this section. However, by far the most important technique, in terms of both the quantity of work that has been undertaken, and the usefulness of the data for reconstructing vegetation at a variety of scales, is **palynology** (pollen analysis).

Pollen analysis

Pollen grains are produced by all plants that develop seeds (the **angiosperms** and **gymnosperms**) as part of their reproduction cycle. Pollen fulfils the male role in the process and is dispersed from the host plant in the hope that at least some grains will reach the stigma of the female flower where fertilisation can take place. As the pollination process is a very uncertain one it is in the plants' reproductive interest to produce large quantities of pollen grains to maximise the chances of fertilisation taking place. The amount of pollen produced by a plant depends on the strategy it uses for dispersal of the grains. Plants using an **anemophilous** (wind) means of dispersal, such as pine (*Pinus*) and spruce (*Picea*) produce the greatest number of grains, as there is the greatest uncertainty that individual grains will reach female flowers. On the other hand plants employing an **entomophilous** (insect) strategy such as lime (*Tilia*) produce much smaller quantities of pollen as there is a much greater likelihood of the grains reaching their destination. No matter which strategy is employed, however, the final resting place for the vast majority of the pollen grains is not the female flower, but the Earth's surface. Where this surface is a water body, such as a lake, or a marshy or boggy place, there is a high chance that the pollen grains will be preserved in the archaeological record, as long as that environment has remained waterlogged to the present day. Even in such beneficial preservation conditions, where oxygen is excluded, not all elements of the pollen grain survive. A pollen grain consists of three elements: the living cell (which actually carries out the fertilisation process), which is surrounded by a cellulose covering, called the **intine**. This is in turn further coated in a highly durable substance called sporopollenin. Only the latter, termed the **exine**, survives for more than a few years (**34**).

As well as pollen grains, **palynologists** also frequently examine **spores**, which are preserved in exactly the same conditions as pollen. Spores are produced by evolutionarily primitive plants that do not flower, such as ferns (**Pteridophyta**) and mosses (**Bryophyta**). However, in the same way as pollen grains spores fulfil a role in the reproductive cycle, being dispersed by the wind to locations where a second stage of growth can occur. This location need not always be on another plant.

The environments where pollen and spores are most likely to be preserved (e.g. lakes, backwaters of river and bogs), are also characterised by high sediment deposition rates. Over time therefore the pollen rain falling on their surface will become incorporated in different sediment layers. If these layers are separately sampled it is consequently possible to look at changes in the pollen rain, and therefore vegetation, over time. The job of the palynologist is to

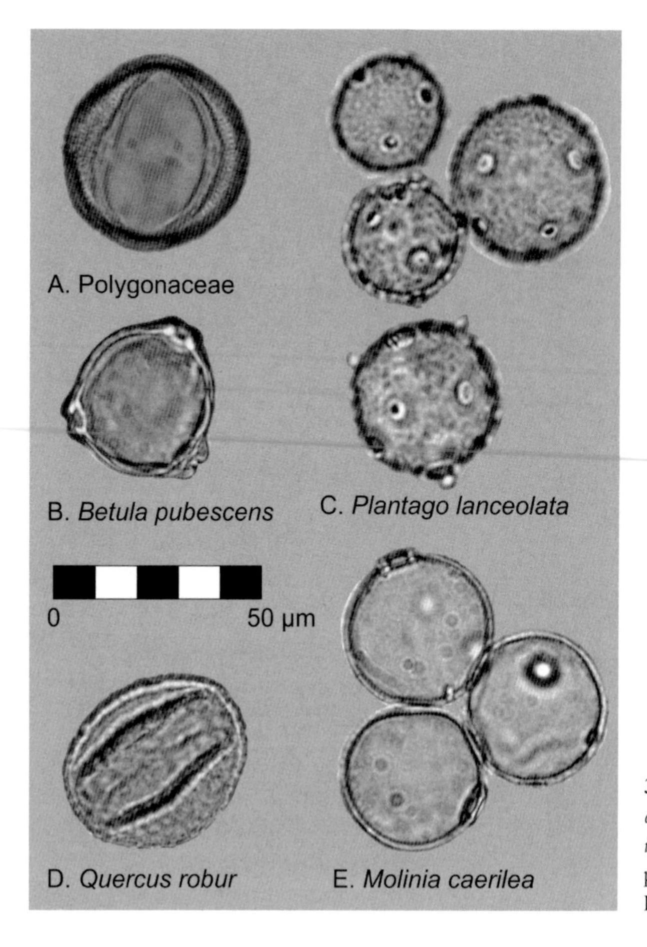

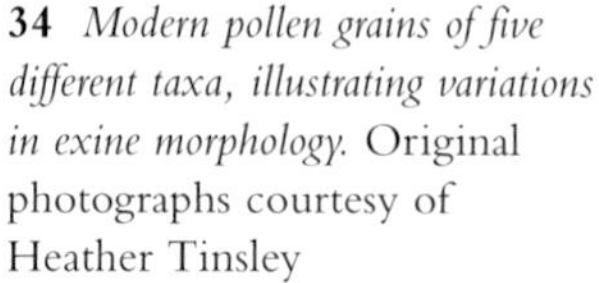
34 *Modern pollen grains of five different taxa, illustrating variations in exine morphology.* Original photographs courtesy of Heather Tinsley

identify pollen grains from such samples (e.g. **35**) and to reconstruct the past vegetation represented by each sample using the ideas that we reviewed in section 1. In addition to locations such as lakes and bogs, palynologists have also studied pollen preserved in soils buried beneath prehistoric monuments, in the gut contents of preserved bodies and in organic residues surviving in pots.

The potential for preservation of pollen and spores in the archaeological record is, however, just one factor in their archaeological study. If the grains or spores could not be identified to a relatively high taxonomic level (e.g. beyond family) there would be little potential value to the technique of palynology. Luckily, evolution has played us a useful card in that most genera of plants have developed characteristic, and most importantly, different, exine morphologies (**34**). These differences are in part related to the means of dispersal. For example, pollen grains of anemophilous plants tend to be small and smooth, while air sacks have been developed by some to enable grains to stay airborne and travel long distances. In contrast exines of entomophilous

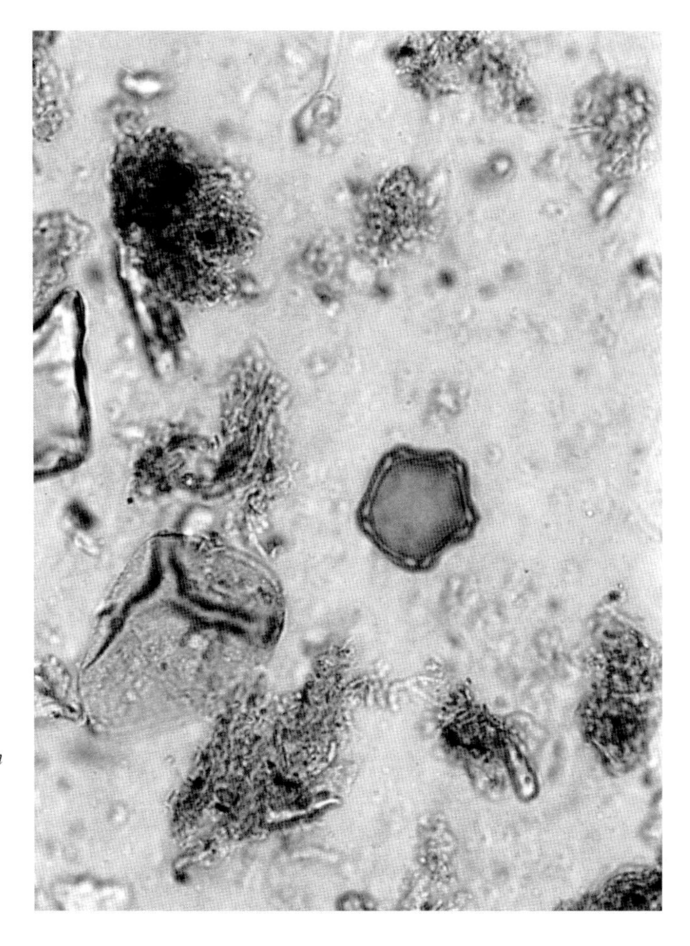

35 *Pollen grains of Alnus (alder) and Cyperaceae from an archaeological site at Salcombe, Devon. Note the differences in preservation from* **34**. Photograph courtesy of Heather Tinsley

plants are relatively large, highly armoured and frequently include spines. These protect the living part inside the intine from mechanical damage during transport, while at the same time enabling the grain to adhere to an insect. Thus pollen grains and spores can almost always be identified to the family taxonomic level, frequently to genus and occasionally, with particularly characteristic exines, to species.

SAMPLING FOR POLLEN ANALYSIS

As pollen grains commonly vary between 0.08 and 0.02mm in size they obviously cannot be collected individually. Sample collection methods, however, vary depending upon the context from which the researcher wishes to obtain information. The most common situation on an archaeological site is that a section of suitably waterlogged sediment is made available for sampling. Monolith tins (stainless steel boxes measuring around 50cm in length, with widths and depths of 5-10cm) are hammered into the cleaned section face. Placing a piece of wood on top of the tin and using a block or similar broad-headed hammer is the most efficient method and also has the advantage of not causing any damage to the monolith tin. Normally, several monolith samples are collected from a sequence, making sure that each tin overlaps its neighbours, so that the whole stratigraphy is preserved in the sampled record (**36**). Once all the tins have been inserted and marked with reference and location details (it is particularly important to label the top and bottom of the tin as it relates to the section, so that the vegetation history of the site is not reconstructed back to front!), the position of the tins added to the section drawing and photographs taken of the samples *in situ*, the samples can then be extracted. Extraction has to be carried out carefully by cutting around the back of the tin so as to leave at least 2cm of sediment protruding above its top. Each tin is then separately wrapped in polythene film and placed in cold storage as swiftly as possible once it reaches the laboratory where it is to be studied. The wrapping and cold storage are necessary measures to prevent the sample from drying and the included pollen grains oxidising, as well as preventing contamination by modern pollen.

Off-site locations such as marshes and lakes rarely have vertical exposures suitable for palynological sampling. Where such exposures exist on, for example, upland peat moors, it is highly likely that deterioration of pollen

36 *Monolith samples for palynological study being taken from a site in Glastonbury, Somerset.* Photograph courtesy of Vanessa Straker

grains near the exposed surface will have occurred. Where exposed sections do not occur samples must be obtained from below the ground surface with minimum contamination. The most common means of achieving this is by drilling bore holes – either using powered apparatus, or in relatively soft sediment, manually powered – and recovering samples using a closed chamber auger head. Typically closed chamber auger heads, such as the Russian types, have a flange against which a half-tube is turned on reaching the depth at which a sample is required. Sample chambers of mechanically powered augers commonly consist of an open tube into which a plastic sample tube is placed. This is then filled with sediment as the auger head is drilled into the ground. Assuming that continuous samples are required throughout the sequence, normal practice is to drill two bore holes side by side. Assuming a sample chamber of 50cm length, core samples from the first bore hole are taken at depths of 0m, 0.5m, 1.0m, 1.5m etc., and those from the second at 0.25m, 0.75m, 1.25m, 1.75m etc. Thus two suites of overlapping core samples are produced. The reason for taking what would at first sight appear to be twice as many samples as are required, is the cutting head at the basal end of the auger chamber. The cutting head is required in all closed chamber augers to penetrate consolidated sediments, but at the same time it may disturb sediment between successive samples. Therefore, having a parallel set of samples at different depth intervals means that in theory a combined, undisturbed suite of samples for the entire stratigraphy exists. As with monolith samples, those collected in closed chamber augers are usually wrapped and placed in cold storage as soon as possible.

Collecting a representative sample of stratigraphy is just the first step of the pollen analytical processes. The next step is to sub-sample the monoliths or cores. Because of the very small size of pollen grain and spores, and as they are often found in such profusion, it is common practice to collect sub-samples of only 1cm^3. These are collected in sealable plastic bags and placed in cold storage, for the same reasons as already reviewed, until required for analysis. The number of sub-samples taken from each monolith will vary depending upon a number of factors such as the purpose for which the analysis is being carried out, the inferred deposition rate, the complexity of the stratigraphy (and of course the amount of money and/or time available to the palynologist). So for example a greater number of sub-samples would be taken from each core of a slowly accumulating bog sequence, with the intention of determining whether early Neolithic people were creating localised and temporary woodland clearances, than from rapidly accumulating lake deposits, where the idea was to reconstruct broad vegetation changes spanning the whole Holocene.

LABORATORY ANALYSIS OF POLLEN SUB-SAMPLES

In the laboratory the sub-samples taken from the monoliths or cores are treated to extract the pollen grains and spores. Firstly the sub-sample is chemically treated to break down bonds that hold individual sediment grains together. This procedure may be achieved in a number of ways, for example by boiling in a 6 per cent solution of potassium hydroxide (KOH). The sample is then sieved through a 100µm mesh to remove material too coarse to be pollen grains. At this stage the sample consists of individual grains of material <100 µm. However, as well as pollen grains/spores this residue also contains mineral grains and other organic matter. These are removed by the addition of a number of chemicals and/or separation using a heavy liquid. Organic matter such as cellulose is removed by the addition of an oxidising agent and then acetolysis, while the mineral residue is removed using either hydrofluoric acid (HF) or by use of a centrifuge and heavy liquid (mineral grains are denser than pollen grains and will therefore tend to pass through the heavy liquid to the base of the processing tube). Carbonate minerals can be removed by less strong acids such as hydrochloric acid (HCl). At this stage of the treatment procedure the sub-sample consists largely of pollen grains/spores. The next stage in the procedure is to identify these and count the number of each type in the sample. However, first the sub-sample is usually stained with safranin, which enhances the sculpture on the individual grains/spores, thereby easing identification. Then the sub-sample is mounted on a microscope slide.

Microscopes used for examining pollen slides are relatively high-powered epi-illuminating types. Magnifications of between 100 and 1,000 times are required to identify all sizes of grains and spores. Identifications are initially made using keys, photographs of the grains of known pollen types or from reference slides of modern pollen/spores. However, after working on a number of sequences palynologists will only need such aids for rare or deformed pollen/spore types, and will be familiar with the appearance of the common exines for the geographic area in which they work. In the normal course of events no palynologist will attempt to identify all the pollen grains and spores on a single slide. To do so would often take an extraordinarily long time, while the end result would have little greater statistical reliability than counting just a few hundred grains/spores from the sample. Therefore at the identification stage the palynologist will make yet another sampling decision. Two approaches are commonly taken. The most common is to decide on the number of pollen grains that will be identified from each sub-sample, and then to systematically traverse the slide, keeping a tally of all grains/spores identified, until that number, termed the **pollen sum**, is reached. Typically between 200-500 grains/spores are identified and counted depending upon the context in which the work is being carried out. For example as part of a trial 'assessment' of deposits that are potentially rich in pollen, only 200-300 grains may be counted, while for detailed research into whether there is evidence for forest clearance in an early Neolithic sequence, 500 or more grains may be counted.

A second approach, which is used in order to counter the problems associated with use of percentages discussed later in this section, is to introduce a so-called exotic pollen type, in a known concentration, to the sub-sample before it is stained with safranin and mounted on a microscope slide. The exotic pollen must not be part of the natural flora of the region from which the pollen samples have been taken, otherwise its addition would merely contaminate and thereby bias the sample. Therefore, in Europe and North America *Lycopodium* spores and *Eucalyptus* are commonly used exotics. Both are native plants of Australasia and have the further advantage that their exines cannot be mistaken for anything else. When counting a slide from a sub-sample where an exotic has been added, identifications are made in the normal way by traversing the slide as previously described. However, the palynologist will have decided to continue identifying only until a certain number of exotic grains have been counted. This quantity will in turn depend on the same factors as discussed for standard counts as above, but will be also affected by the concentration of exotic pollen used and the perceived concentration of non-exotic grains/spores after a couple of slides have been worked upon. Although the second approach would appear to be more complex, it does have the advantage of enabling **pollen/spore concentrations** to be calculated. This is possible because the exotic pollen was introduced to the sub-sample in a known concentration and, therefore, the counts of non-exotic pollen/spores can be directly compared to that concentration. In this way changes in pollen deposition and preservation through a sequence can be easily determined.

POLLEN DIAGRAMS

Once pollen/spores from all the sub-samples in a sequence have been extracted, identified and counted, the results can then be presented and interpreted. By far the most common way of representing stratified pollen data is as a **percentage frequency histogram** (**37**). These diagrams are produced by calculating the proportion made up by each taxon of the **total land pollen** (TLP) and then plotting these percentages against a depth scale. The result is a diagram where the Y axis consists of the depth scale, while there are many X axes, each one representing a different plant taxon. Plant taxa are usually grouped along the top of the histogram by type into categories such as 'trees', 'shrubs', 'grasses', 'herbs' and 'aquatics'. Spores are not usually included within these groups and plotted separately. In addition to the percentage variation of individual taxa it is common practice to plot summary data on the right hand side of the diagram. This consists of the percentage importance of each plant group and is calculated by adding the percentages of all constituent plants in a sample. So, for example, in figure **37**, the percentage figure for 'trees, shrubs and climbers' is the sum of the separate percentages for pine, birch, oak, alder etc. The advantage of presenting summary data of this type is that major changes in vegetation through a sequence, for example woodland clearance or reforestation, can be seen at a glance. Examination of these summary plots is usually the first stage of vegetation reconstruction. There are other categories of data that are presented on percentage pollen frequency histograms that do not relate to pollen at all. For example, it is standard practice to represent the stratigraphy from which the pollen came in a lithology column to the immediate right of the depth scale (e.g. **37**). This is an important element of the diagram as it allows the analyst to see whether changes in the pollen record relate to changes in stratigraphy. For example the interpretation of what a fall in the percentage of alder might represent is likely to be very different if it occurred within a peat deposit, than if it occurred across a peat-inorganic mud boundary. In the latter situation the most likely scenario is that alder was simply killed off by rising water levels, but in the former, a climatic or even a human cause would be more likely. A further common feature of modern percentage pollen histograms are counts of microscopic charcoal, which are usually plotted at absolute counts for each sample to the right of the spore data (**37**). Charcoal greater than 0.04mm is counted during traverses of the microscope slides at the same time as pollen and spores are identified. Peaks in frequency of such charcoal logically relate to the frequency and magnitude of burning events in the surrounding landscape. Recent research suggests this was the result of campfires rather than forest burning. Therefore, peaks in frequency of microscopic charcoal seen in a histogram mainly relate to the extent of human activity within a pollen catchment and may therefore allow more detailed interpretation of a diagram. For example, a fall in the percentage of lime pollen (but not other trees) at the same time as the frequency of microscopic charcoal increases (a common phenomenon in pollen diagrams from southern Britain dating to the Later Neolithic and Bronze Age), may suggest human clearance of lime wood for agricultural purposes.

Some readers having read the previous paragraph may have noticed an apparent sleight of hand. We stated that the percentage contribution of each plant taxon was calculated in relation to total land pollen, in other words as a proportion of fully terrestrial plants. The reason that this approach is taken is that terrestrial habitat is what the palynologist is really interested in, for it is on land that people lived, hunted and farmed, and here also that changes in the Earth's climates most clearly manifested themselves. However, aquatic taxa are of some importance too so the question remains as how to represent them. The usual approach is to calculate the relative proportion of each aquatic taxon as a percentage of the total land pollen plus the total aquatic pollen, or TLP+A as it is sometimes abbreviated. Presenting the data in such a way obviously reduces the apparent importance of aquatic plants within a histogram as the percentage calculation is on the basis of a larger population. Nevertheless, this approach allows interpretation to be focused on the terrestrial elements while still allowing consideration of what was occurring in wetland habitats. Spores are portrayed in the same manner as aquatic pollen, but as a percentage of total land pollen plus the sum of the spores in each sample (TLP+S). Many percentage pollen histograms constructed in the 1970s and earlier were

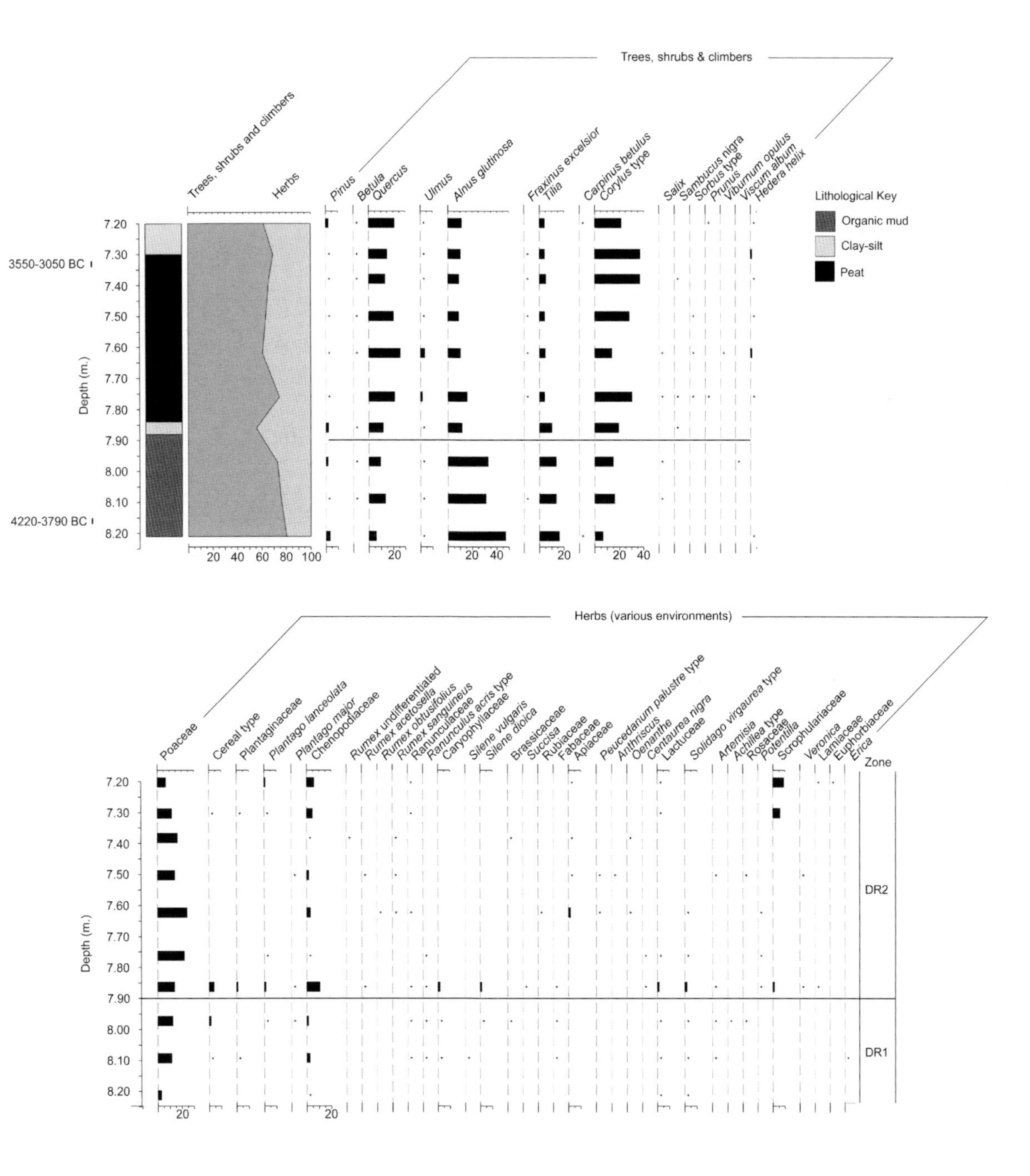

37 *Percentage pollen diagram (trees and herbs only) from Deanery Road, a site in central Bristol, England. The analyst was Heather Tinsley*

calculated on the basis of the proportion of individual taxa as a percentage of total tree pollen (TTP). This approach is consistent with the focus on tree populations that was the mainstay research of most palynologists of this period. However, diagrams of this type are now seen as being outmoded and play down the importance of herb and grass taxa that are of particular importance in archaeology, not least because many were deliberately grown by people.

Not all pollen histograms are constructed on the basis of percentage frequency, and various forms of **absolute frequency histogram** are used as well. Perhaps the most useful is the pollen concentration diagram, which can be produced if an exotic taxon has been introduced (see above). If it has, it is possible to calculate the concentration of each plant taxon in relation to the known concentration in which the exotic pollen was produced. If the sediment sequence from which the pollen samples have been taken is well dated it is possible to produce a pollen influx diagram, where the amount of pollen accumulating in each cm^3 of the sampled site on a yearly basis is calculated and plotted. The advantage of absolute pollen diagrams of this type is that the variation displayed by each taxon is entirely independent and not related to fluctuations in other plant taxa to which it is not related. The disadvantages are that the diagrams are difficult to interpret, and because terrestrial and aquatic taxa are given equal weighting, the latter tend to dominate the diagram because aquatic habitats are the most commonly sampled. This makes what might be ecologically or archaeologically important variations in the terrestrial pollen record difficult to see. Given that there are so many types of pollen histogram it is important when first encountering a new example, that the reader determines what type of pollen data it is representing. Only then is it possible to suggest what the variations of pollen frequency in the diagram might mean.

INTERPRETATION OF POLLEN DIAGRAMS

We have already seen that the first stage in interpreting the meaning of changes in pollen frequency through a sediment sequence is determining whether gross changes in vegetation can be seen in the summary data on the right hand side of a histogram. The next stage is to further define any changes that might appear in terms of the variation in individual taxa that occur in the main body of the diagram. For example, the summary part of the diagram may suggest that forest clearance has occurred because there is a fall in frequency of tree pollen. However, only by examining the variation in the individual taxa can it be determined whether certain trees were preferentially cleared, and if clearance occurred in dry or wetland forest. The changing representation of plant taxa may also indicate changes that are not seen in the summary record, for example whether a decrease in one species of herb is compensated for by an increase in another, or where a taxon that is present as a low proportion of the assemblage declines as a percentage still further. The latter situation is relatively common and is typical for instance of the palynological representation of the relatively well-known Neolithic elm decline. Elm is relatively poorly represented in the palynological record for reasons explained below, but its decline in the forests of the fifth and sixth millennia BC have long provoked arguments amongst palynologists and archaeologists, which we will discuss further in the text on insects below. The elm decline has also been used as a sort of stratigraphic marker (albeit one that relates to a period of perhaps 800 years), that links pollen diagrams of the Neolithic. One common means that palynologists use to represent changes in vegetation seen in a pollen histogram is to divide it into zones (frequently called **pollen assemblage zones**). This involves drawing horizontal lines across the diagram, which demarcate samples with a broadly similar pollen composition. The explanatory text that accompanies the pollen histogram is usually based on the zonation.

From what has been stated above, it may appear that interpretation of palynological data is a relatively straightforward task that can be carried out by simply following a few guidelines. However, this is not the case; a vast shelf space of articles and books has been devoted to the many pitfalls of interpreting pollen assemblages and various techniques to get around them. It is clearly impossible to adequately cover this huge literature in a book such as this, but we can make readers aware of some of the problems and constraints of using palynological data. Perhaps the most obvious one is that different plant taxa produce varying amounts of pollen, depending upon which strategy of pollen dispersal they have evolved and other related factors. Thus, even though 50 per cent of the pollen in a sample may come from a single taxon, that fact alone does not mean that half the vegetation consisted of that plant, particularly if that plant is anemophilos. Conversely if 10 per cent of the total land pollen was of an entomophilous taxon such as lime then it is likely that the plant was an extremely important part of the local vegetation and may have comprised over half the vegetation. Attempts have been made to circumnavigate the problem of differential representation by using correction factors known as **R-values** (**38**). The corrected percentages are then plotted as a histogram – so here is yet another type of pollen histogram that the reader should be aware of. Figure **38** shows an example of this approach, and indeed Anderson's analysis of Eldrup Forest Denmark is a classic in the study of the Neolithic-Bronze Age lime decline. Both lime (*Tilia*) and beech (*Fagus*) pollen is under-represented in percentage pollen histograms, despite the latter being aerophilous. However, when the data are converted to R-values the true nature of the vegetation change is revealed: a huge lime decline occurred from around 4200 BC onwards caused by human forest clearance, and the former lime forest was replaced by one of beech and oak. Despite the obvious advantages of the R-value approach, relatively few pollen diagrams of this type have been produced, mainly because R-values have only been calculated for tree taxa. Also, pollen productivity of the same taxa varies in different climate zones.

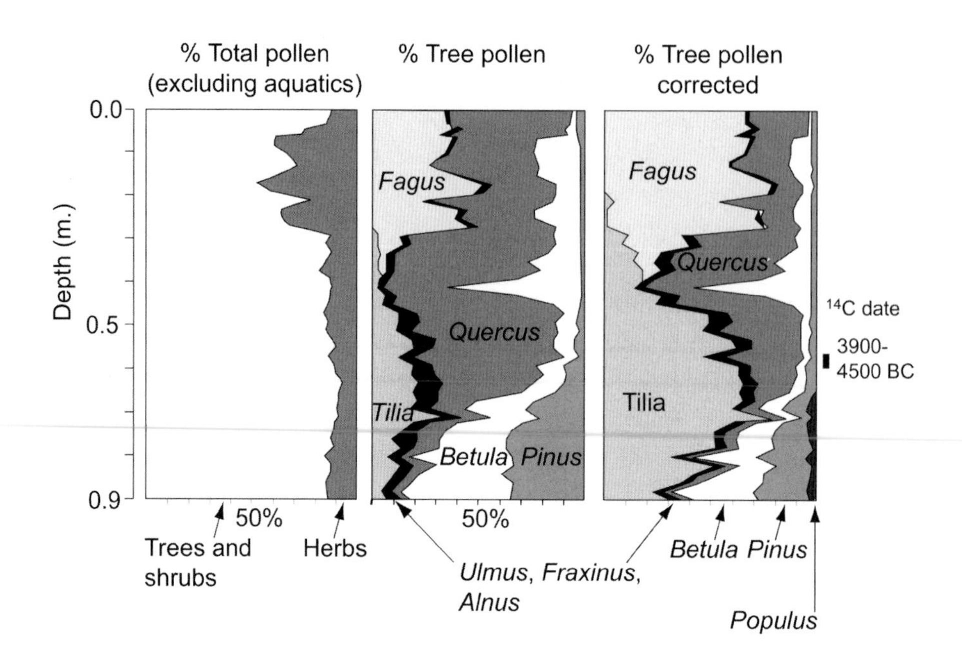

38 *Pollen histogram from the Eldrup forest, Denmark, using the R-value 'corrected' approach. Note the lime decline from the late Neolithic onwards.* Redrawn from Anderson (1973)

Representation of plant taxa in a pollen histogram does not only relate to the quantity of pollen that is produced, but also to how well that pollen survives in the archaeological record. As was discussed in section 1, this is the study of taphonomy. Although pollen will survive in almost all anaerobic environments, it is preserved better in some than in others. Where the conditions of preservation are in any way marginal, for example in an organic mud at the margins of a river that may be subject to occasional dry episodes, or a peat sequence that has been exposed by cutting, or even when samples taken for analysis are stored for long periods of time, deterioration of the pollen exine occurs. Damage to the exine can thus be caused by both chemical (e.g. exposure to oxygen) and mechanical (e.g. abrasion by sand in a river environment) means. If this damage occurs in a uniform manner for all taxa the problem would be a minor irritant, but unfortunately the pollen of some taxa deteriorates at a faster rate than others, leaving a palynological record that is no longer representative of the surrounding vegetation. It is extremely difficult to allow for such losses to the pollen record despite the fact that it is known that pollen grains of certain taxa such as beech are more susceptible to damage than others, for example lime. It has been noted that the armouring of a pollen grain that protects it during insect or bird pollination also promotes preservation in the archaeological record. In other words, although entomophilous pollen grains are produced in lower frequencies they have a better chance of surviving in the archaeological record than anemophilous types. Qualitative allowances can be made in interpreting pollen assemblages that are thought to have suffered from differential representation, if it is observed that the majority of pollen that is counted has a durable exine, or if the exines of the majority of grains appear deformed or corroded.

Before proceeding we might take a step back and consider where the sampled pollen has actually come from. In other words, where exactly did the reconstructed vegetation communities live? The pollination process inherently implies that transport of pollen grains has occurred to the sampled site. The question is how far this transport has been. This in turn relates once more to the pollination strategy used by each taxon as well as factors such as wind direction, nature of the vegetation and size of the sample basin. Plants that reproduce by anemophilous means rely on long-distance transport of pollen grains to reproduce, while pollen grains of certain taxa such as pine have developed features that allow them to travel hundreds of kilometres. Indeed pine pollen has been found in ice cores from Greenland, over 1,000km from the nearest pine tree. Conversely, pollen transported by entomophilous means may be moved much shorter distances, while self-pollinating plants such as cereals and vines drop pollen grains within a radius of less than 10m of the flowering plant. The bottom line is that pollens from different taxa have a varying capacity for being transported.

A further consideration is secondary transport. Pollen landing on a body of water is likely to be moved by it, both horizontally downstream and vertically through the water column at different rates depending on speed of flow and pollen grain morphology. Therefore, there is a much greater potential for transport than the factors relating to the pollination strategy would suggest. Extensive secondary transport frequently causes mechanical damage to the exine, so some allowance must be made for these factors when interpreting the assemblage data. However, there is no way of estimating the distance of primary transport by examining the pollen grains themselves. Instead other means have to be employed.

Research over the last 20 years has demonstrated a relatively straightforward relationship between the size of the sample basin, be it lake, peat bog, pond or whatever, and the distance from which pollen grains incorporated within it have been transported (**39**). The larger the basin size the greater the proportion of long-travelled pollen and the smaller the percentage of locally derived pollen. This is a finding of great importance as it enables palynologists to focus their attention on different types of sample site depending upon the questions they want to ask. Sediment cores from the middle of large lakes for example are the most useful for reconstructing vegetation on a regional scale, but may say very little about the impact on the local vegetation of a single settlement. To examine the latter, sediment from a smaller catchment such as a pond or a hollow, would have to be sampled. The simplicity of this relationship is appealing, but it is not the only determinant of the degree of transport. Prevalent wind obviously has an impact: the vast majority of pollen deposited in a basin will have come from up-wind, therefore, the environment that is reconstructed will relate to the landscape in that direction.

The nature of the vegetation itself is also important. Long distance transport is greatly restricted in a forested environment where the branches of trees and leaves interfere with the movement of pollen grains, but is correspondingly unrestricted in open conditions where no such obstruction exists. Unfortunately, to allow for this factor in past situations would mean a certain circularity of argument. The local vegetation would have to be known before the proportion of grains transported a long distance could be calculated, which would then enable the nature of the local vegetation to be calculated!

It is hardly surprising given the nature of the problems previously outlined that palynologists commonly only work in that one area of environmental archaeology. However, despite these problems palynology offers the environmental archaeologist perhaps the most reliable means of palaeoenvironmental reconstruction, the only widely used method of regional vegetation reconstruction, and certainly the methodology best supported by theory due to the long history of the technique and the large number of palynologists. In the next pages we will look at other means of palaeoenvironmental reconstruction that can be used to examine local environments.

Plant macrofossil analysis

By far the greatest use of plant macrofossils in environmental archaeology has been for the study of past economies, so we give extended coverage to studies

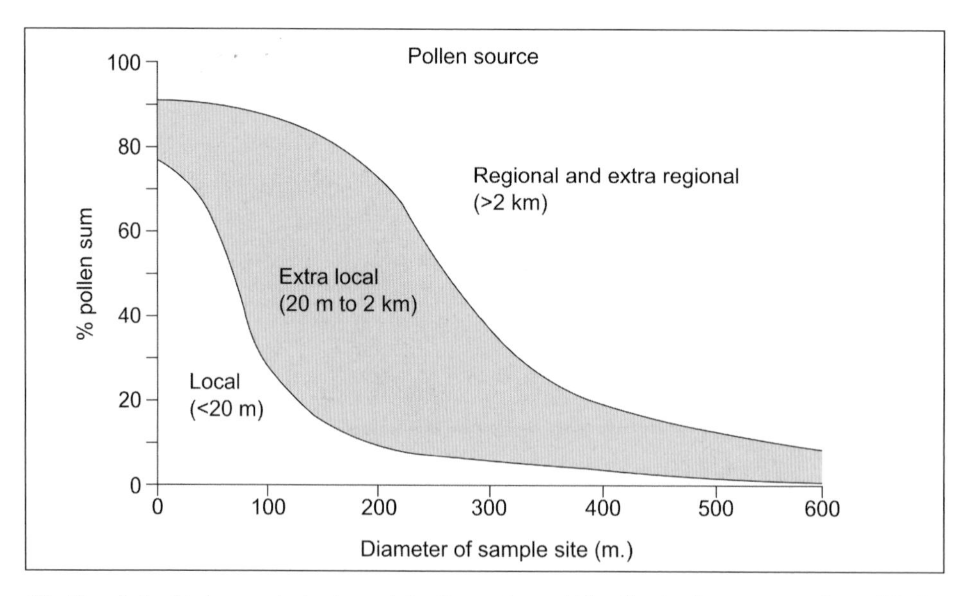

39 *The relationship between basin size and the distance from which pollen has been transported.* Modified from Jacobsen and Bradshaw (1981)

of this nature in section 3. Here we deal with the use of plant macrofossils, either from locations away from archaeological sites, or alternatively remains that are incorporated in archaeological deposits by chance processes, and their use in examining past environments. The visibility of plant remains in such contexts has meant that they have a long history of study, extending to well before that of pollen. The first landmark study – for Britain at least – of plant macro remains for the information they provide on past environments was carried out by Clement Reid in the last years of the nineteenth century, and published as *The origin of the British flora* in 1899. Elsewhere Ferdinand Keller had already published studies of exceptionally well-preserved plant remains found stratified in Swiss lakes.

Plant macrofossils found in off-site contexts can range from seeds of flowering plants, buds of trees, fruits, moss remains, leaves, wood and even vegetative tissue (**parenchyma**). Thus, it is quite possible that a single plant can be represented in the archaeological record by a number of macrofossil categories. For example, acorns (fruits), leaves and wood are all potentially preserved from an oak tree. Nevertheless, preservation is rarely uniform, and such remains as leaves, flowers and vegetative remains are comparatively rare even in the best preservational contexts. Plant macro remains are most commonly preserved in waterlogged conditions such as lake sediments, features on river floodplains and in peat bogs, although the exact opposite condition, desiccation, can also preserve the same types of remains. The plant remains that are preserved are usually a mixture of autochthonous species, actually living in the depositional feature that is being sampled, and allochthonous species, whose macrofossils have blown, fallen or slipped in. In all situations except peat bogs the latter usually predominate and therefore, the information potentially provided by plant macrofossil remains concerns the vegetation component of past environments beyond the sampled feature, albeit – unlike pollen – of a local nature. As for peat bogs, because they contain the remains of plants growing *in situ*, plant macro remain studies of their contents are particularly useful for constructing the past bog environment, especially given that up to 80 per cent of the peat commonly comprises plant remains. The plant remains can provide information on the resources available to past human communities in the bog environment and can help 'calibrate' pollen diagrams as we shall see later. In upland bogs, they can indicate something of past water levels in the bog, which has in turn been used as an indicator of rainfall. Past precipitation is normally reconstructed by studying remains of *Sphagnum* moss. There are many species of *Sphagnum*, each of which inhabit different ecosystem niches, depending on the surface water depth in a bog. So, by comparing the relative importance of each species in the stratigraphic record a reconstruction of water depth variation with time and hence rainfall can be determined. Studies of this type have been carried out by Keith Barber in bogs in north-west England. These studies

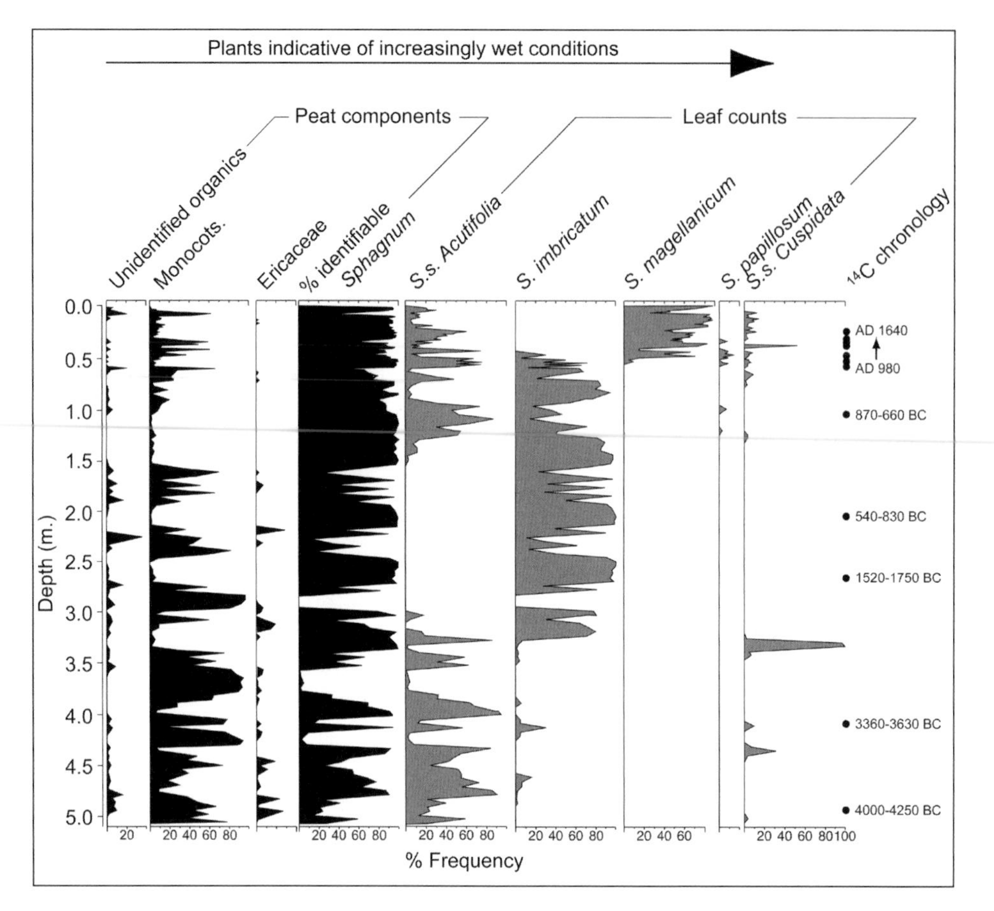

40 *Reconstruction of bog surface water depth, a proxy for past rainfall, from the site of Bolton Fell Moss, north-west England, based on plant macro remain studies.* Modified from Barber *et al.* (1994)

show that over the last 7,000 years rainfall variations have operated on an 800 year cycle (of dry to wetter conditions and back to dry again), with a general tendency for wetter conditions with time (**40**).

SAMPLING AND PROCESSING METHODOLOGIES

Methods for sampling and then processing samples to recover plant macro-remains vary considerably depending upon the nature of the preserving medium (i.e. waterlogged versus desiccated conditions), the visible concentration of plant remains and the preferences of the individual researcher. Ideally samples should be taken as continuous blocks through an exposed section (only rarely are plant remains present in sufficient concentration to be sampled in a core), with the block thickness determining resolution, as we discussed in section 1. For a typical northern European waterlogged situation sample sizes should vary between 250g and 2kg. Smaller samples can be used to modify sample desiccated contexts given that the absence of moisture means that individual plant remains weigh only a fraction of their waterlogged counterparts (most of the weight of a plant is water). Sample processing carried out to extract the plant remains relies on obtaining representative specimens greater than 250µm without damaging the individual remains. Samples from waterlogged contexts are usually soaked in a weak solution of sodium hydroxide (for acidic or neutral sediments) or nitric acid (for basic sediments), to separate the plant remains from their matrix. They are then gently washed through a 250µm mesh to remove the matrix, and the resultant residue stored in ethanol or water (containing a fungicidal agent) prior to identification. Plant macro-remains, whether waterlogged or desiccated, are identified under a low-power binocular microscope and using a modern comparative reference collection. Wood remains and parenchyma are identified by making a thin section of the specimen, examining it under a high-powered microscope and comparing it against a photographic key, the most common ones used being by Schweingrüber for woods and Hather for parenchyma. Once all specimens in a sample have been identified they can then be quantified and tabulated prior to interpretation. Some researchers have attempted to produce pollen-style histograms from plant macrofossil data and then interpret them in the same way. However, such an approach is fraught with danger as species from aquatic habitats will be (for a waterlogged context) hugely over-represented.

APPLICATION OF PLANT MACRO-REMAIN STUDIES FROM OFF-SITE LOCATIONS

As plant macro-remains and pollen commonly survive in the same location, one of the main uses to which off-site plant macrofossils have been put is in helping to interpret pollen diagrams. For example (as we mentioned when discussing pollen data), we cannot know for certain whether the pollen from which our diagram was constructed was of local or regional derivation. As plant macro-remains are mainly from the immediate locality of the sampled feature, they can provide that information. In other words, a list of plant macro-remains recovered from a particular stratigraphic level indicates which components of a pollen sample from the same level originate from the immediate vicinity of the sample site. Such an understanding is particularly important for pollen types such as pine, which is both a large pollen producer and has grains adapted for long distance travel. If pine needles (which are highly durable) are not found in the macro-fossil record it is very unlikely that pine trees were growing locally.

Another aspect of pollen diagrams that plant macro-remains can help interpret is in differentiating between two or more species within a genus. As we have already seen, pollen can rarely be identified to a taxonomic level higher than genus, so for example *Quercus robur* cannot be differentiated from *Quercus pubescens* (both are deciduous oaks). However, macro-fossils can be identified to species level, and hence could determine whether the pollen recorded in the diagram was from *Q. robur*, or *Q. pubescens*, or indeed both.

Finally, there are certain plants whose flowers never open; because no pollen is released they are unrepresented in pollen diagrams. Therefore macro-fossil records are the only way of recognising such plants in the archaeological record. Such is the case with the violet (*Viola* sp.). Specimens of this genus were identified during analysis of plant macrofossils from a later Mesolithic shell midden and flint working area at Westward Ho!, south-west England, but were not found in an accompanying pollen diagram. The analyses carried out of these 5500–4550 BC deposits by Robert Scaife and Dominique de Moulins demonstrates the advantages of examining both classes of evidence together. Pollen data suggested that the midden began to form in a fen carr, where oak and hazel were dominant, although there was a change towards the top of the deposit where willow, birch and ivy became more important. The macrofossil record fleshes out that from the pollen in suggesting that the willow found was either *Salix caprea*, *Salix cinerea* or a hybrid (sallow). On the basis of abundant fruits, dogwood (*Cornus sanguinea*) would also seem to have been an important part of the woodland community, despite the fact that it is represented by only a couple of grains in the pollen record. The plant macro-remains also indicated something of the ground flora, which is poorly defined in the pollen diagram. This would seem to have included woody nightshade (*Solanum dulcamara*), hempagrimony (*Eupatorium cannabinum*), blackberry (*Rubus fruticosus*), as well as meadowsweet (*Filipendula ulmaria*) and sedge (*Carex* sp.). The hazel nuts were also of interest as they were both smaller than today and misshapen. It has been suggested that this was caused by environmental stress, where the hazel trees were growing at the extreme of their tolerance to standing water, or a result of disease. Neither could be observed from the pollen records.

Palaeontology

Bizarre as it may sound, the remains of fossil or sub-fossil animals can often provide information not only of past fauna, but of flora too. The reasons are linked to the ecosystem dynamics that we looked at in section 1. As herbivorous animals are entirely dependent on plants for food, they will only inhabit areas where particular types of plant grow. While mammalian herbivores will often feed upon a wide range of plant species, smaller organisms may feed only upon a single plant genus or even species. If we move one step up the food chain the same applies. Most carnivorous mammals will prey upon a range of other creatures, but certain carnivorous mammals and many invertebrates, will feed upon only a restricted range of prey species. As a result of this interaction between plants, herbivorous and carnivorous animals it is therefore often possible to obtain indirect information on the nature of past vegetation from animal remains. Being indirect, this type of information may not be as useful or reliable as that provided by plant remains themselves, but as palaeontological material frequently survives in locations where palaeobotanical remains do not, there is often little choice. On the other hand, in a few cases palaeontological remains may provide greater details of past plant communities than if remains of the original plants had survived in perfect conditions. For example, the beetle species *Mecinus pyraster* eats only ribwort plantain (*Plantago lanceolata*), a plant that is thought of today as a weed of grassland habitats.

Of course palaeontological study also provides us with information on the composition of past animal communities. Indeed most of those working in zooarchaeology would say that this is where their data have their greatest value.

Insects

Insects are a group of invertebrate animals, characterised (as everyone who has ever studied biology knows), by having six legs and a segmented body. Insects, or Insecta as they are termed by taxonomists, are also by far the most diverse class of life on earth, representing over 75 per cent of the world's animal species and easily outnumbering the number of plant species. One other aspect of the insects makes them particularly unusual amongst the world's inhabitants: the longevity of individual species in the fossil record. Whereas most species of mammal have an evolutionary history that can be measured in hundreds of thousands of years – for example an earlier species of our own genus in Europe and Western Asia, the Neanderthals, began to evolve from around 600,000 years ago and was extinct by 28,000 BP – many insect taxa have remained unchanged since the Tertiary period, some 2 million years ago. Scott Elias quotes the example of the aquatic leaf beetle *Plateumaris nitida*, which, as well as being relatively common at the present day, has also been identified in Oligocene period shales from Colorado in the western United States, making this species at least 30 million years old. The diversity of the insects is both the reason for, and direct result of, their success. Insects inhabit all of the world's

Orders	Common name
Hemiptera	Bugs
Homoptera	Bugs
Diptera	Two-winged flies
Trichoptera	Caddis flies
Hymenoptera	Ichneumons
Odonata	Dragonflies
Coleoptera	Beetles
Families	Common name
Pentatomidae	Stink bugs
Lygaeidae	Seed bugs
Cicadelidae	Leaf hoppers
Chironomidae	Non-biting midges
Gerridae	Water striders
Saldidae	Shore bugs
Corixidae	Water boatmen
Notonectidae	Backswimmers
Formicidae	Ants

Table 10 *Orders and families of insects found in archaeological deposits*

terrestrial and aquatic ecosystems, as well as many marine marginal settings. Insect species occupy the most specialised niches within ecosystems that other orders of animals are simply incapable of exploiting. It is precisely this speciality and therefore their restricted tolerance to factors such as climate and vegetation, that makes insects so useful in reconstructing past environments. However, not all insects are **stenotopic**, and occupy restricted niches, other **eurytopic** species live a much more generalised existence, and because of this tend to be much less useful to the **entomologist** (a specialist in insect studies) working on archaeological material.

TYPES OF INSECTS STUDIED

There are many different orders and families of insects that have been found in archaeological deposits (**Table 10**). Of these, by far the most attention has been focussed upon the **Coleoptera** (beetles). There are a number of reasons for this. Firstly beetles are the most diverse of the insects. Over 300,000 species are known at the present day with an average of 1500 new species being named each year. These figures mean that beetles alone account for over 25 per cent of all living species, and more importantly from the point of view of the **archaeoentomologist**, that beetles occupy a huge diversity of ecosystem niches. This extreme specialisation of many beetle species means that palaeoenvironmental reconstruction using sub-fossil beetle data is potentially more detailed than any other biological proxy. Nevertheless, the pioneering work in using beetles for archaeological purposes was carried out in the United Kingdom, precisely because Britain's beetle fauna is comparatively impoverished, by Russell Coope, Peter Osborne, Harry Kenward and Paul Buckland. The 3,800 species known from Britain (cf. 30,000 species in North America) represented a manageable number with which to both establish identification criteria and to become familiar with the modern ecology.

A second reason that beetles are a particular focus of study is because remains of this order are more readily preserved in archaeological deposits than those of the other orders and (no less importantly) can be more easily identified to a high taxonomic level. Beetle **exoskeletons** have evolved to protect the soft internal organs necessary for the organism to survive and hence have taken on armour-like properties. The exoskeleton is comprised of three different parts (**sclerites**): the **head capsule** (which as the name suggests surrounds the beetle's head), the **pronotum** (protecting the upper body) and the **elytra** (which protect the wings) (**41**). Each schlerite has morphological characteristics, such as shape, microsculpture, and on occasion even colour, that enable species to be differentiated. These features are retained in sub-fossil material that has been preserved in waterlogged or mineralised conditions, as the schlerites are made from a nitrogen-containing protein called **chitin** that is insoluble in water, alcohol, most acids and bases (an exterior waxy layer that coats the exoskeleton and which prevents water loss rapidly decomposes on death). If on the other hand, chitin is

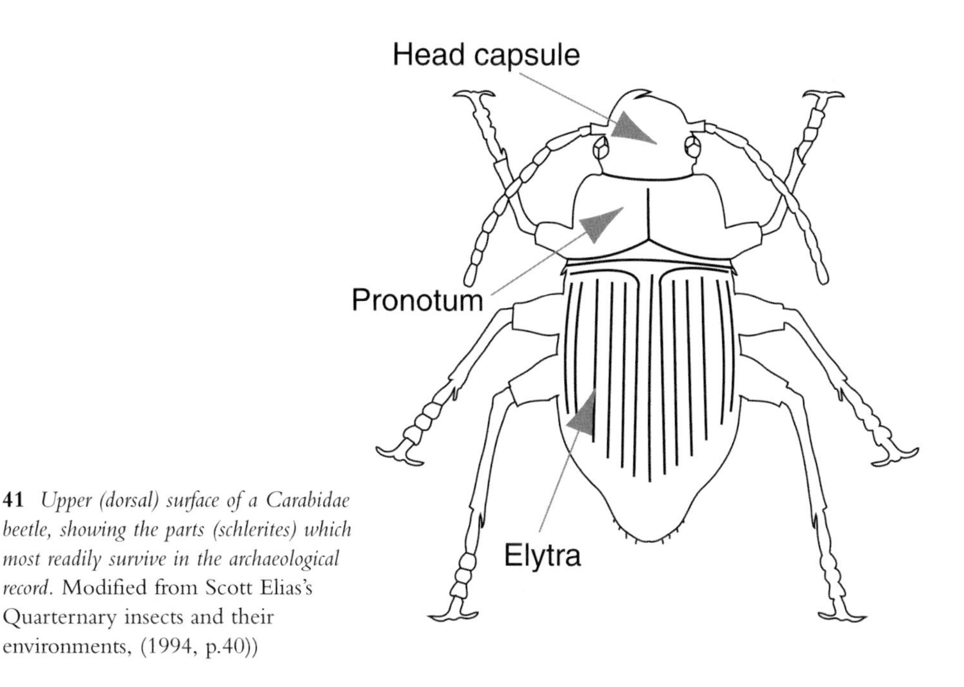

41 *Upper (dorsal) surface of a Carabidae beetle, showing the parts (schlerites) which most readily survive in the archaeological record.* Modified from Scott Elias's Quarternary insects and their environments, (1994, p.40))

exposed to the air for any length of time it oxidises and decomposes. It is for this reason that most archaeoentomological studies are of fully waterlogged sediments such as peats, well-fills, pits dug below the watertable, or latrines.

The final reason why beetles are studied by archaeoentomologists is that thanks to the work of naturalists over the last two centuries, the modern ecology of beetles is better known than any other insect order. This of course means that past environments can be reliably inferred from modern analogues on the basis of sub-fossil beetle remains.

Despite what we have said of beetles, it is becoming increasingly common to study one other insect family, the **Chironomidae**, for the information they can provide on past environments, although rarely as part of archaeological projects. Chironomids are a family of the Diptera (true flies), more commonly known as midges (although not of the biting variety), that inhabit aquatic environments. As well as water pH and salinity, the distribution of different Chironomid species is most heavily dependent on surface water temperature, a factor that makes them eminently suitable for palaeoclimate reconstructions. Chironomids go through a life cycle that starts with a larval stage living at the base of an aquatic body, and progresses through a maggot-like stage which again lives only in the water, to a fully adult stage that resembles a mosquito (although once again it does not bite). It is the highly scleritised head capsule of the maggot stage that is preserved in aquatic sediments, and is that which can be most easily identified to a high taxonomic level.

SAMPLING FOR INSECTS

As hundreds of Chironomid head capsules may survive in 1cm^3 of sediment, so it is possible to sample sediments at a much finer interval than for beetles, and therefore provide a palaeoclimate reconstruction at a higher resolution. Chironomid head capsules are extracted from sediment samples cut from cores or monoliths taken in the same way as for pollen analysis, by soaking in solutions of either potassium hydroxide (for organic samples) or hydrochloric acid (for calcareous samples) and sieving the resultant slurry through a 90μm mesh. Identification is carried out using a high-power binocular microscope and using published keys.

In contrast to Chironomids, samples for beetle analysis are taken from sections exposed in waterlogged features. Samples are simply cut from a thoroughly cleaned section face (contamination with modern insects is a very real possibility given the sheer quantity of insects in the environment) at depth intervals depending upon the nature of the sediment unit sampled and the questions asked (see under non-marine analysis below). There are no hard and fast rules on how large a sample should be, but the intention should be to obtain about 1 kg of organic 'flot' after initial processing. In other words, the fewer organic remains visible in a sampled sediment, the larger the sample will need to be. So perhaps 10kg of a highly organic well-fill will be sufficient, while 100-200kg of organic mud from an alluvial situation will need to be processed. Given that such assessments become more reliable with experience, they can easily be catastrophically badly estimated by the uninitiated, so samples are ideally taken by the archaeoentomologist. However taken, the first stage of sample processing is to wash the whole sample through a 250μm mesh – a task that is relatively easy for silt- and sand-dominated sediments, but altogether more problematic for clays (**42**). Clay-rich samples can, however, be disaggregated by leaving the sample to soak in a solution of 'calgon' (sodium hexametaphosphate – better known as dishwasher powder!) in a covered bucket for a few days prior to sieving. Likewise, highly organic samples can be similarly soaked in a dilute solution of potassium hydroxide to promote the separation of individual organic particles. Once sieved the residue retained will predominantly comprise organic material, which must be stored in water prior to further processing. The next stage of processing is to place the residue in a bowl or bucket with a spout at the top and add water so that the sample is completely covered to a depth of several centimetres. A lightweight oil such as paraffin or kerosene is then added so that the whole surface of the water is covered. The oil is then worked into the residue by hand mixing for several minutes. The oil will coat the impermeable surface of the insect schlerites, but not the permeable plant remains. Cold water is then added gradually to the bottom of the bucket by means of a hose. As oil and water do not mix, the former, together with the coated insect remains, stays on the surface and as the water levels rise, both oil and insects pass through the spout and a

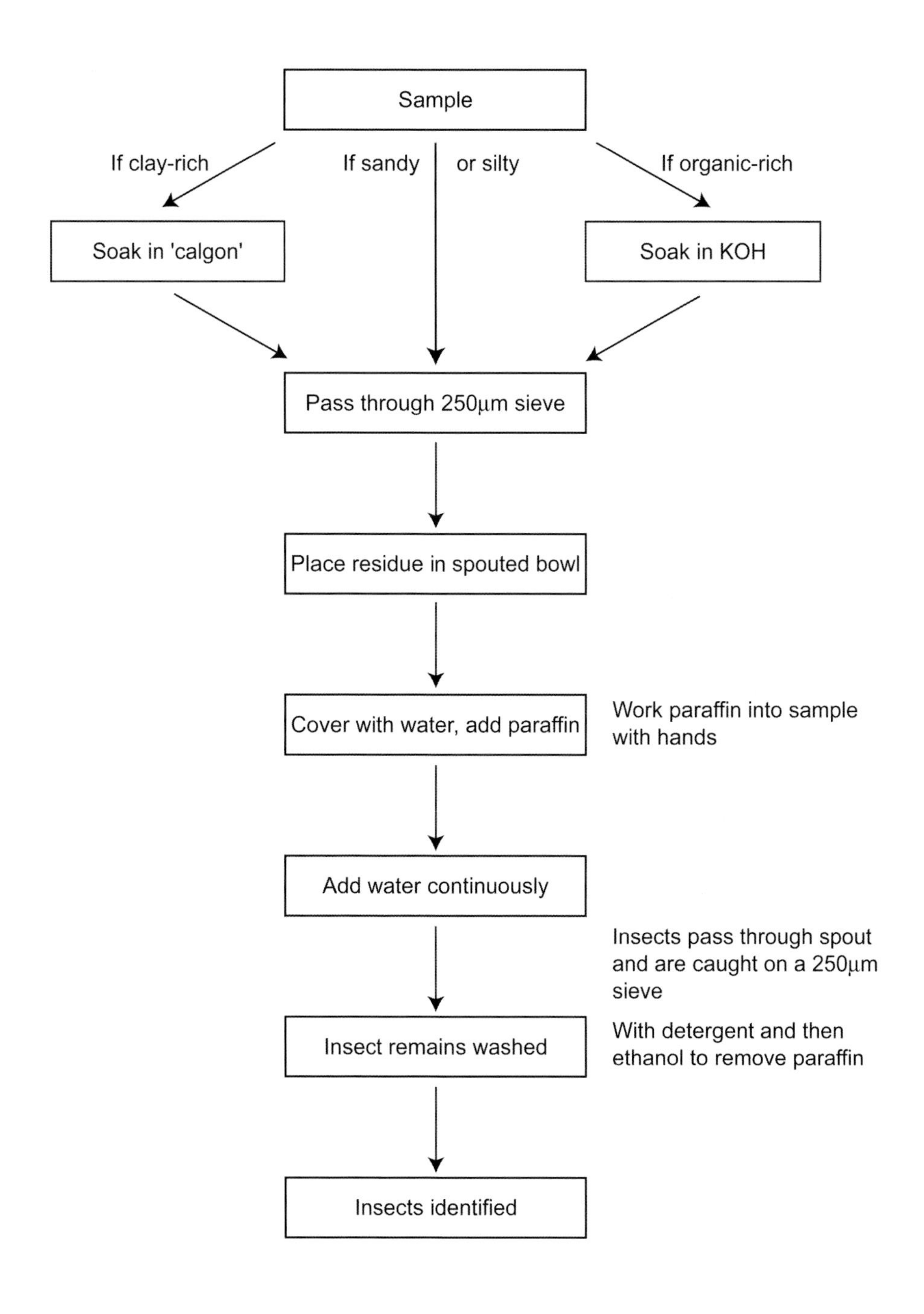

42 *Methodology for extracting beetle schlerites from archaeological sediments*

250μm sieve that has been placed underneath. It is on the latter that the insect remains are caught. The water supply is switched off as the last oil passes through the spout thereby allowing the oil that has passed through the sieve to be re-used. The oil-soaked insect remains retained on the sieve must be cleaned prior to microscopic sorting and to that end the flot is washed with firstly detergent and then ethanol. Commonly, the organic remains retained in the bucket are also wet sieved through a 250μm mesh and sorted as a control (e.g. head capsules are commonly filled with silt and clay particles and do not float).

APPLICATIONS OF BEETLE ANALYSIS

Beetles have been used for two main elements of palaeoenvironmental reconstruction: climate and habitat. Of these the former has a longer history, dating from Carl Lindroth's studies in the 1940s of beetles from interstadials (warm episodes) of the last (Weichselian) cold stage at Frösön, Härnön, Hälsingland, Pilgrimstad and Angermanland in southern Sweden. These suggested that summer temperatures had been warmer than those of the present day. However, the real father of modern beetle climatology is Russell Coope, who from 1955 pioneered quantitative approaches of using sub-fossil beetles from off site locations (e.g. lake and river deposits, and peats). Initially Coope's studies used a technique based on **range overlap**. The present spatial distribution of all predatory and scavenger beetles recovered from an ancient deposit were plotted. The geographical area where the ranges of all of these species overlapped was then identified and the most recent weather station records from within that area were sought and applied to the past situation. Using this approach Coope was able to suggest that summer temperatures in southern England during the last (Eemian) interglacial of 128-120,000 BP were some 3°C warmer than at present. In other words an Eemian London would have basked in temperatures comparable with modern Lyon, in south-east France.

Despite the apparent success of Coope's methods there was one major problem: in every diagram there were examples of species whose modern geographic ranges did not overlap with any of the others. This was particularly true of insect assemblages recovered from the colder parts of the Weichselian cold stage. For example the upper samples from Upton Warren, in the English Midlands, contained the dung beetle *Aphodius holderi*, which is only presently known from the Tibetan plateau and has a present day range that consequently does not overlap with any of the Scandinavian and alpine species in the assemblage. In short, many species had to be excluded from such range overlap reconstructions because the insect communities of the last cold stage have no modern analogues. A solution to the problem was found in the 1980s with the development by Coope, together with Timothy Atkinson and Keith Briffa, of the **mutual climatic range** method (MCR) (**43**). MCR relies on knowing two modern climate components for each species present within an ancient

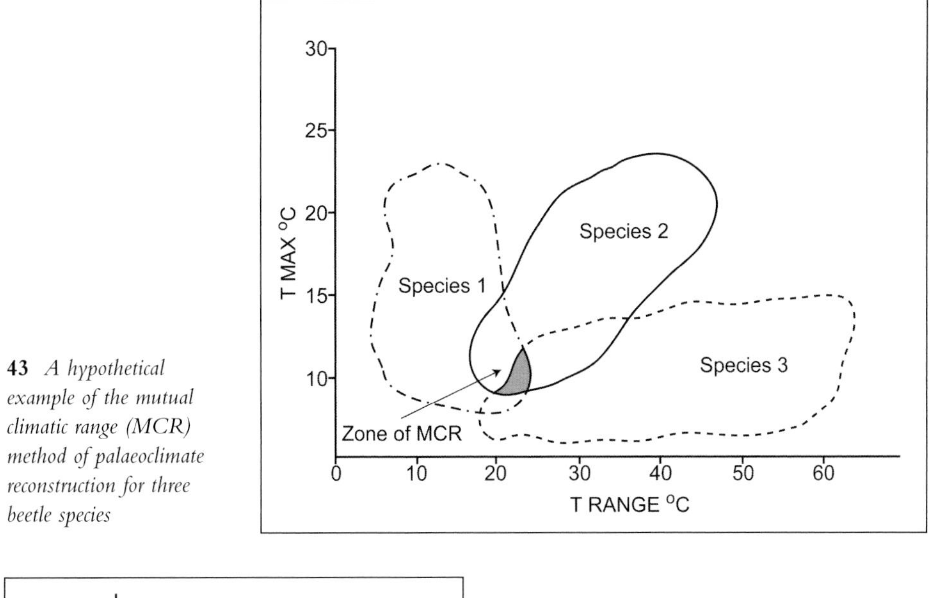

43 *A hypothetical example of the mutual climatic range (MCR) method of palaeoclimate reconstruction for three beetle species*

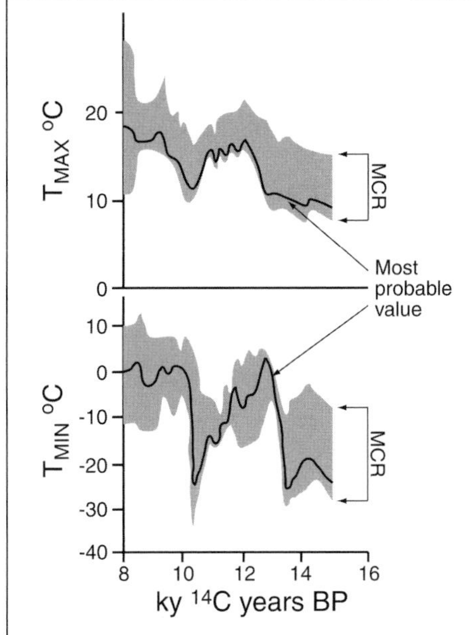

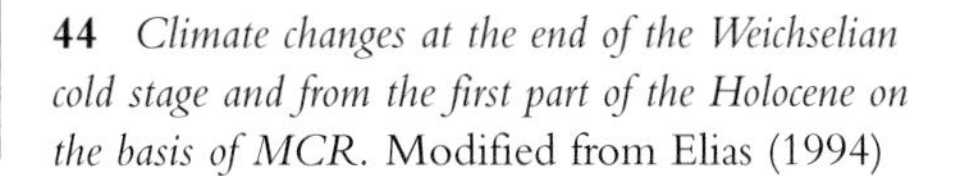
44 *Climate changes at the end of the Weichselian cold stage and from the first part of the Holocene on the basis of MCR.* Modified from Elias (1994)

assemblage: summer warmth (termed T_{MAX}) and the temperature difference between the warmest summer, and coldest winter month (T_{RANGE}), which relates to seasonality. These data are obtained by averaging weather station records within the geographic range occupied by the beetle. A computer is then used to compare T_{MAX} and T_{RANGE} for every predaceous and scavenging beetle species in the assemblage and to calculate the area of overlap – the so-called '**climatic envelope**' – on a biplot of the two variables. The overlap area can be read from the biplot to indicate the past T_{MAX} and T_{RANGE} (the minimum winter temperature [T_{MIN}] can also be calculated by subtracting T_{RANGE} from T_{MAX}) (**43**). Using this powerful method Coope has been able to reconstruct climate changes in the British Isles for the end of the Weichselian cold stage and the first part of the present (Holocene) interglacial (**44**). The results show that the harsh polar winters (averaging –23°C) of the Weichselian Late Glacial Maximum BP were replaced by warmer winters of 0°C, from around 15,000 BP within a period of less than a century. These were followed by a subsequent gradual decrease in temperature to around 11,700 BP, and a further rapid rise into the Holocene. The significant aspect of these discoveries were that they appeared to show that climate changes were both more rapid, and occurred at an earlier date than those indicated by using pollen as a climate proxy. The explanation of these differences is that during the extreme cold of the Weichselian, trees retreated to southern Europe, and because warming occurred so rapidly there was a long lag period before they could colonise northwards from these far refugia, whereas insects respond much more quickly to climate change.

Reconstructing climate for the middle Holocene and later using beetle evidence has proven to be a lot more difficult, due to just one factor: people. Human manipulation of natural ecosystems following Neolithic forest clearance has opened up environments for a mixture of insects that previously inhabited both warm and cold habitats. Instead the focus of Holocene archaeoentomology has been habitat reconstruction. In Britain, the home of archaeological entomology, well over 100 sites where insect faunas are preserved by local waterlogged conditions have been investigated. These are mainly of later prehistoric (Neolithic to Iron Age) and historic date and have focused on features such as pits, ditches, wells, latrines, occupation layers, stored food products, middens, and gut contents of preserved bodies. By later prehistory, not only were insect faunas changing as a result of forest clearance for agriculture, but many species that live in extremely close association with people – so-called **synanthropic** taxa – became more common. For example parasites such as the human flea (*Pulex irritans*) cannot live without a person as a host, while pests of stored products, such as the grain weevil (*Sitophilus granaricus*), exploit a humanly created context. The extremely narrow nature of the habitats occupied by some insect species makes archaeoentomology particularly attractive for reconstructing human activity areas. The beetle *Tipnus unicolor* for example lives only amongst rotting straw mixed with human cess and appears to have been common in urban environments of the medieval period (but is rare today because of changes in methods of sewerage disposal). However, as well as the synanthropic species that are present in archaeological insect assemblages, a **background fauna** of insects will also be found that is indicative of conditions in the wider environment. These background species become incorporated in archaeological deposits through chance events, for example being blown into an open privy – or from the insect's point of view,

45 *Sources of insect remains in human dwellings according to Harry Kenward. The building shown is a reconstruction of a Neolithic 'longhouse' in the Archeon, Alphen aan de Rijn, the Netherlands*

being in the wrong place at the wrong time. The sheer complexity of micro-environments that are associated with human activity has led Harry Kenward to suggest that only samples containing more than 50 insect taxa can provide reliable evidence of the intricacy of past human environments. He has also emphasised the importance of determining the origin of all insect taxa present within a sample for environmental reconstruction (**45**). For example, are the species found pests of stored products, parasites (human or animal), from occupation litter, or are they background fauna that have blown in? If the spatial distribution can be resolved then the study of sub-fossil insects from archaeological contexts is a very powerful tool for reconstructing habitation, subsistence and craft activities.

There have been numerous examples where insect studies have aided archaeological interpretations in situations where other forms of bioarchaeological data provide no answers. Coleopteran data collected from the West Heath Spa site, Hampstead Heath, London, enabled Maureen Girling and James Greig to suggest that the well-known decline of the elm tree in southern Britain during the early Neolithic was the result of Dutch Elm disease. The appearance of the elm bark beetle (*Scolytus scolytus*) occurred in stratigraphy just below that in which a sudden drop in elm pollen was seen in an accompanying pollen diagram. This beetle is the carrier of a fungus called *Ceratocystis ulmi*, which is the cause of the disease. Previous to their investigations it had been thought that the primary elm decline, recognised in many southern English pollen diagrams spanning 4200-3800 BC, was a direct result of humans deliberately removing elm trees. Recently, Mark Robinson has pointed out that *Scolytus scolytus* is present in Mesolithic Britain, so current data suggest the Neolithic Elm decline was the result of severed coinciding factors.

Analysis of beetles from a Bronze Age shaft at Wilsford, Wiltshire supported a conclusion that although surrounded by numerous barrows and being located less than 2km from Stonehenge, this 24m deep hole may have fulfilled the more mundane function of a well. Wilsford Shaft is drilled through chalk, an environment in which insects from archaeological deposits are usually extremely rare, but were preserved here by waterlogging. The beetle assemblage is almost entirely of background fauna, and dominated by dung beetles such as genera *Geotrupes* and *Aphodius*. This indicates that in the Bronze Age, as now, the area surrounding Wilsford Shaft was cattle and/or sheep-grazed pasture. A diverse assemblage of plant-eating beetles (collectively termed **phytophages**) were also found indicating that the grassland contained poppies, thistles, nettles and clover. Peter Osborne suggests that the shaft was a formal water source around which animals were herded to drink, but were prevented from falling in by wooden barriers as indicated by the furniture beetle, *Anobium punctatum*. Paul Buckland's well-known study on beetles from both Iceland and Greenland shows that these were only colonised by Europeans in the early Medieval period (Greenland had been the home of Inuit since at least 2000 BC), around AD 900 (and in the case of Greenland, abandoned again by AD 1350). The study demonstrates how humans enriched the existing beetle fauna by introducing synanthropic species such as the dung beetle *Aphodius lapponum*, although following the abandonment of Greenland most of these died out.

Despite the success of insect studies of such rural archaeological sites it is in the reconstruction of urban archaeological environments that insect data have proved most impressive. The Roman, Anglo-Scandinavian and medieval town of York has been studied in the most detail. Here analysis over three decades by Harry Kenward and Paul Buckland has enabled the squalor that was York in the Middle Ages to be reconstructed in unprecedented detail (and which can be witnessed first hand by a visit to the Yorvik Centre). It would appear that the typical flooring materials of Anglo-Scandinavian York, rushes and reeds, were rarely changed and hence beetle assemblages normally associated with rotting straw and compost were found within every dwelling. Indeed, it has been suggested by Russell Coope on the basis of similar evidence from Viking-period Dublin, that such rotting flooring was a sort of central heating, in other words the fermenting plants gave off enough heat to warm a house. Bedding in York was typically of heathland plants, but in at least one case was infested with biting ants, while the very structures occupied by people were infested with deathwatch and powderpost beetles. If that were not enough, much of the food that was stored for any length of time seems to have been infested with weevils, while the people themselves were rife with fleas and lice. Only insects enable living conditions to be reconstructed in such detail and they provide a conflicting impression from the impressive artefacts, which are indicative of a prosperous society comprised of craftspeople.

Non-marine molluscs

Unlike insects, non-marine molluscs can never be used to provide detailed information on what plants grew in a past situation, as no snail species is specific to a particular type of plant. However, assemblages of snails from archaeological sites can be used to infer the general nature of the vegetation, for example 'woodland', 'long grassland', 'short grassland', 'arable' in the case of land molluscs and 'moving water' (river), 'lake', 'canal/ditch' and 'marsh' in the case of fresh water snails. While at first sight such broad interpretations may appear less than satisfactory when compared to those possible from pollen or insect analysis, mollusc shells will frequently survive in just the oxidising conditions where the other two categories of remain will not. As most archaeological sites in both Europe and North America are on dry land where such conditions prevail, analysis of non-marine Mollusca will often provide the only means of palaeoenvironmental reconstruction.

Technically speaking the Mollusca are a group (or a phylum to be taxonomically correct) of invertebrate animals, where the soft body parts of the organism are enclosed in a shell. It is an extremely large group of animals, but only two classes contain non-marine taxa: **Gastropoda** and **Bivalva**. If we look at the gastropods first, some breathe by gills and are members of the sub-class **Prosobranchia** (not all of this category are aquatic!) and others breathe by a primitive lung – the sub-class **Pulmonata**. It should be noted that slugs of the families Arionidae and Limacidae are members of the Pulmonata and are therefore molluscs. Evolution in the slugs has caused an extreme reduction in shell size to the extent where it is only found as a plate-like structure within the body. The Bivalva are members of two subclasses, the **Palaeoheterodonta** and the **Heterodonta**. During their life, non-marine molluscs show a marked preference for habitats which contain an abundance of lime which they need to build their shells. Therefore they are found in profusion on calcareous geologies such as oolitic limestones, chalk and some types of windblown sand (e.g. beach dunes and loess).

Land and freshwater Mollusca have been studied in archaeology from early in the twentieth century. Excepting bones of the vertebrates, snail shells are some of the most obvious animal-related finds on many archaeological sites. It was, therefore, hardly surprising that they should have been studied from such an early stage. Some of the earliest mollusc-related studies carried out in archaeology were those undertaken by A.S. Kennard from excavations of early Neolithic causewayed enclosures in Sussex, southern England, conducted by E.C. Curwen in the 1920s and 1930s. Kennard interpreted these snail assemblages – which were collected mainly by picking visible shells out of sections or spoil – mainly in terms of climate, while also using the data he collected as a means of relative dating. The early Neolithic was interpreted from snails recovered from enclosures at Whitehawk and The Trundle as being both a warmer and a more humid period than at present due to the presence of large forms of *Arianta arbosturum* and *Trichia hispida*, together with the profusion of shells of species such as *Carychium minimum*, *Discus rotundatus* and the Zonitidae in great profusion. Both methods and interpretations appear 'primitive' when we look back, but at the time the work was pioneering. Indeed, analyses of molluscs from The Trundle and four other early Neolithic causewayed enclosures in Sussex was carried out during their re-excavation in the 1970s. The results led Ken Thomas to conclude that the reason for the presence of *Carychium*, *Discus rotundatus*, the Zonitidae and the large-shelled *Arianta arbusturum* and *Trichia. hispida* and hence the inferred humidity, was that the causewayed camps were built in woodland clearings. Woodland is the favoured habitat of most of these taxa at the present day.

After the 1930s the use of land snails to reconstruct past environments pretty much died out. This was largely as a result of the expanding use of pollen analysis, which could tackle all the questions posed by Kennard of his snail assemblages, but in a much more reliable way. It was only in the 1960s that the use of land snail analysis in archaeology went through a resurgence, largely through the work of John Evans. Most commentators would date the advent of modern analysis of non-marine Mollusca to the production of Evans' PhD thesis in 1967, and its adoption in 'mainstream' archaeology from the publication of his later book, *Land snails in archaeology*, in 1972. However, from the late 1950s, similar techniques of molluscan analysis, this time on Pleistocene geological deposits, had been pioneered by Brian Sparks at Cambridge. Sparks' techniques were more systematic than those of Kennard. They involved taking samples of sediment from an exposed section, processing those samples to remove all shells greater than 0.5mm (i.e. those sizes that could be readily identified with a low-power binocular microscope), identifying and quantifying them, and producing a histogram, much the same as those we have already discussed for pollen analysis. Evans refined Sparks' methodology and applied it to the study of prehistoric monuments in Wessex that were then being examined by teams from the Ministry of Public Building and Works led by Geoffrey Wainwright. Evans' work demonstrated that, contrary to then held views, most of the well-known Neolithic monuments such as the Avebury, Durrington Walls and Stonehenge henges were built in an open landscape, and that the countryside was mainly given over to pasture at the time of monument construction. Therefore, these late third millennium structures confirmed a long and relatively continuous history of human interference with this southern English landscape that had begun with the construction of causewayed enclosures such as Windmill Hill and long barrows, such as West Kennet, some 1,500 years earlier.

Studies of land snails similar to those conducted by Evans in Wessex are now routinely carried out as part of most excavations of prehistoric sites on chalk and limestone geologies, and even occasionally of sites of the historic period. Studies have also been carried out of freshwater molluscs associated with archaeological sites located in or near water. These are of course exactly the situations where waterlogged deposits are likely to be found and therefore where the remains of other classes of biological remains can be recovered. Nevertheless fresh water molluscs can provide details relating to the water body in which they once lived that are not obtainable from the remains of other organisms. For instance shells recovered from samples associated with the Dover Bronze Age boat (*c.*1600 BC) indicated that it had been abandoned in a freshwater channel despite lying less than 100m from the present seafront, as only non-marine snails were found (**46**). The molluscan evidence suggested that this channel was relatively shallow, but contained fast-moving water (*Valvata* sp., *Ancylus fluviatilis*), had a stony bed (*Ancylus fluviatilis*), and lacked marginal, muddy areas (a lack of marsh and 'slum' species – see below) in turn suggesting a steep bank. Beyond lay open grasslands (*Pupilla muscorum*, *Vallonia costata*), but with stands of longer vegetation (*Discus rotundatus*, Zonitidae, Clausiliadae). Following abandonment of the boat a pioneer fauna dominated by *Lymnaea peregra* colonised shallow pools in the boat's hull, but the boat rapidly became buried by tufa carried by the river. Later on the channel in which the boat lay was marginalised from the river, water flowed less fast and became muddier, more vegetated conditions developed (shown by increases in 'freshwater catholic', 'slum and marsh' species). These results from molluscan studies tied in well with analysis carried out on the same sediments of diatoms, plant macro remains, ostracods and insects, but contrasted with the structural, mineral and other biological evidence, which had suggested that the boat had been used in marine conditions.

Other than studies of snails from archaeological sites, attempts have also been made to analyse mollusc samples from off-site situations in order to reconstruct wider environments. These studies have largely been focused on dry valleys, and the majority of work was carried out in southern England (**30**). As we saw earlier, dry valley fills potentially contain evidence of environmental changes over a long period of time, and as they are situated in areas of

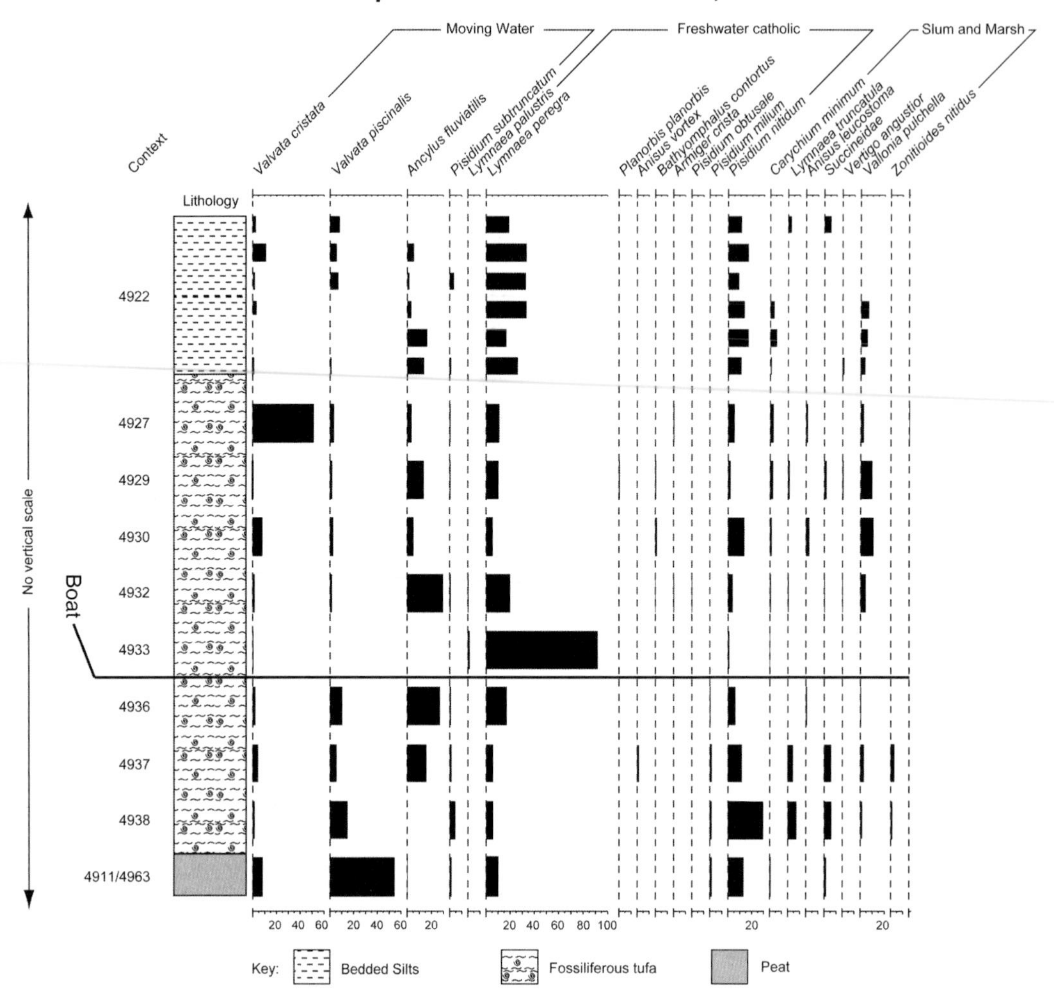

46 *Mollusc histogram of sediments associated with the Dover Bronze Age boat (continued on next page)*

calcareous geologies, often include well-preserved assemblages of land snails. Through molluscan studies of three dry valley sequences in Sussex and Hampshire, Martin Bell was able to demonstrate that despite Evans' evidence showing Neolithic clearance of the primeval woodland in Wessex, in Sussex, woodland clearance that caused erosion only occurred in the Bronze Age. Mollusc assemblages from dry valley fills provide information on the environments on the slopes surrounding the valley bottom and can therefore be used to infer past land use. So for example mollusc assemblages from the dry valley of Toadeshole Bottom East, Brighton provide a record of Late Weichselian glacial, early Holocene, later prehistoric and historic period environments (**7** & **47**). The base of the sequence is a marl, which formed in the Weichselian

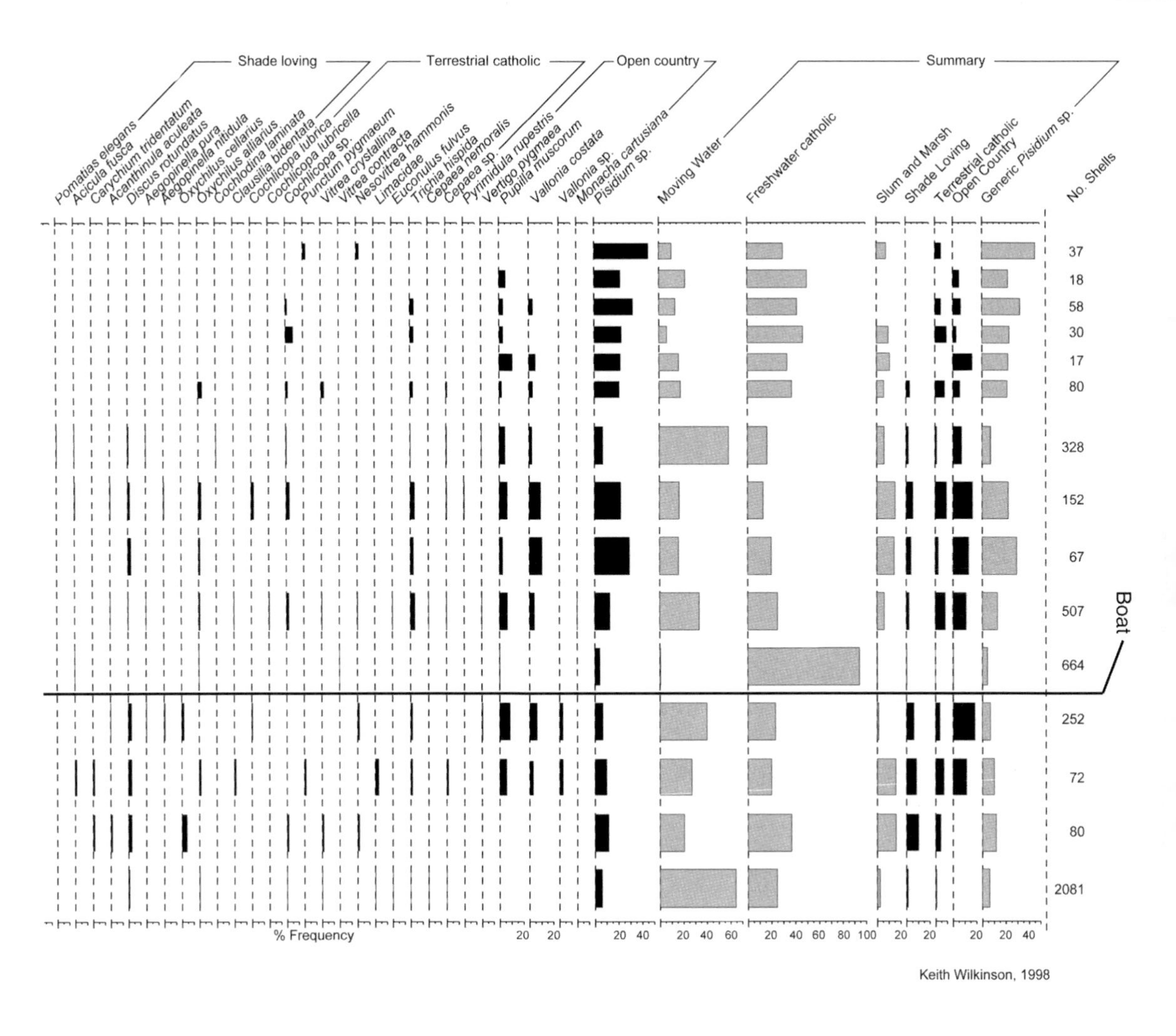

Late Glacial period (cf. the mollusc assemblage with **Table 3**), and is characterised by a large form of *Pupilla muscorum*, as well as *Vallonia costata* and *Trichia hispida*. Earthworms have introduced an assemblage of early Holocene shade-loving species into the marl (e.g. *Carychium tridentatum*, *Discus rotundatus* and *Aegopinella nitidula*), but the same assemblage was also found in the sub-soil hollow pictured in **7** (not part of **47**). The marl is overlain by a palaeosol dating from the Beaker period (Late Neolithic-Early Bronze Age) in which a peak in the curve of *Pomatias elegans* is found (**47**). This species is indicative of disturbance of shaded environments and, therefore, suggests that woodland clearance occurred between 2900-2300 BC. Thereafter the assemblages in the overlying colluvium indicate both arable and grassland conditions. These can be

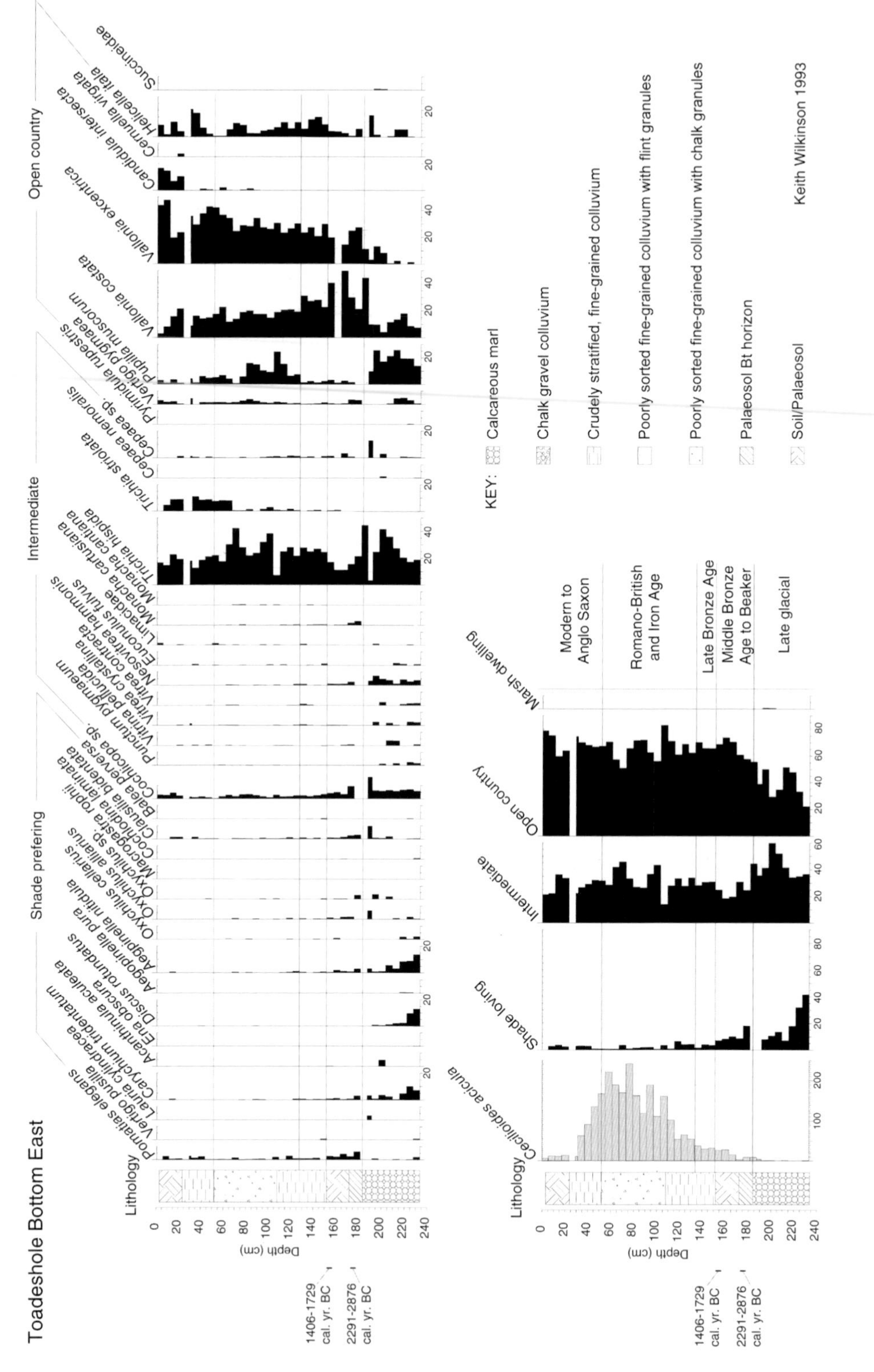

47 *A molluscan percentage histogram from Brighton, south-east England. The changes in representation of the different species reflect alterations to the environment and land use of the valley sides*

separated by looking at the variations in frequency of taxa such as *Trichia hispida* and members of the 'open country' group. The evidence for Toadeshole Bottom East suggests that clearance was initially for pastoral activity. Also of interest are the curves of *Trichia striolata* and *Candidula intersecta*. As is explained below the latter colonised the interior of Britain from the Roman period onwards and is therefore a good chronological marker. *Trichia striolata* is an occasional find in early Holocene deposits that formed in woodland. It then becomes rare in the palaeoenvironmental record following later prehistoric forest clearance, only to re-emerge in the Roman and later periods, by which time it is thought to have adjusted its ecology to find a niche in landscapes heavily modified by people.

It is no coincidence that all the examples that have been discussed relate to English sites. Analysis of non-marine Mollusca for the environmental information they can provide has remained largely a British phenomenon. One reason is that the technique was developed in Britain, but this is not the whole explanation. Much of southern Britain is characterised by limestone and chalk rocks from which calcareous sediments and soils derive, and therefore it is where well-preserved mollusc shells might be expected. However, the same situations are found elsewhere in Europe and North America. The real explanation is that Britain's status as an island since approximately 7200 BC, has meant that only limited colonisation by snail species occurred in the aftermath of the last cold stage from around 9500 BC. Therefore, Britain has an impoverished fauna of non-marine Mollusca of about 170 species (three of which are, at the present day, only found in Ireland), the ecology of each of which is relatively well known thanks to studies by naturalists since the mid-nineteenth century. On the other hand there are over five times the number of molluscan species in adjacent continental Europe, and their ecology is much less well understood. In Britain then, identification of mollusc shells and subsequent interpretation of what an assemblage might mean in terms of a past environment is a relatively easy undertaking, whereas in a continental situation it is not.

SAMPLING FOR NON-MARINE MOLLUSCA

Irrespective of whether mollusc analysis is carried out on waterlain or terrestrial deposits, the sampling, processing and analytical procedures remain the same. They are also largely unchanged from those developed by Brian Sparks and John Evans in the 1950s and 1960s respectively. Samples for mollusc analysis are usually taken through a sequence of deposits, typically a ditch or channel fill, or from a buried soil. The first stage in sampling is to clean the section face to remove any possible contaminating shells, just as the section face would be cleaned before sampling for pollen. Samples are then taken by removing blocks of sediment or soil in a constant thickness and placing it in labelled heavy-duty polythene bags or sample tubs for transport to the laboratory. Blocks are usually removed as a continuous column throughout a sequence or profile (e.g. see **7**). The thickness of the sample will depend on a number of factors, including whether the sample is from a soil or sediment (if a soil, which part of the soil profile), how rapidly the sediment is thought to have accumulated, the inferred impact of post-depositional processes, the amount of time and money available for the analysis and the questions concerning the past environment that are being asked (**48**). Samples taken from the A horizon of a buried palaeosol might be no more than 2cm thick as the potential chronological resolution is high. In other words, that soil surface is likely to have developed very soon before its burial and will provide information on the environment that immediately preceded the construction of an overlying monument, its coverage by a sand dune etc. On the other hand, sampling of a sequence of dry valley sediments may be carried out using samples of 10cm thickness. This is because such colluvial sequences tend to accumulate relatively rapidly, but only during sporadic high energy erosion

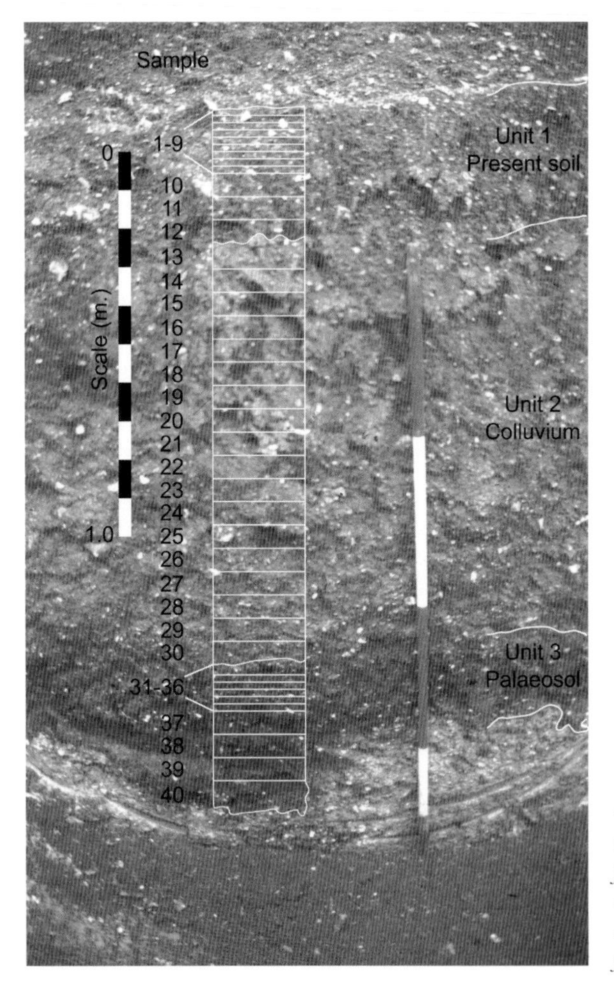

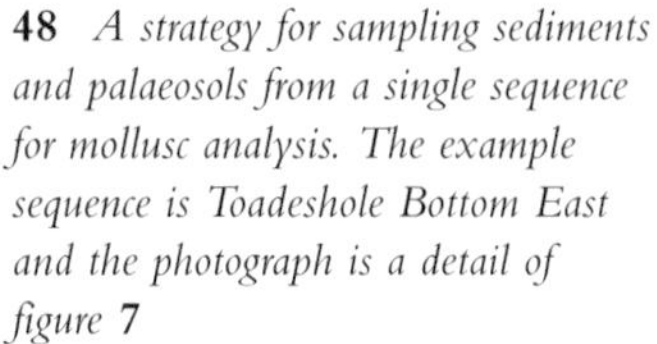
48 *A strategy for sampling sediments and palaeosols from a single sequence for mollusc analysis. The example sequence is Toadeshole Bottom East and the photograph is a detail of figure* **7**

events such as storms. Dry valley deposits also tend to be heavily modified by processes such as bioturbation, so the potential chronological resolution of a sample is low. Whatever sample thickness is adopted – and of course these can vary in different parts of the sequence/profile – no sample should be taken that crosses a boundary that separates one soil or sediment unit from another. To do so would be to mix shells that may come from two quite different environments, resulting in the reconstruction of a situation that never existed.

Thickness is of course only one property of a sample and the question remains as to how much material do we transfer to our sample bag? Evans' approach was to remove about 2kg of sediment/soil for each sample; the sample basis was weight. However, more recent studies recommend volumetric sampling, where a block of material measuring the chosen sample thickness, 25cm horizontally across the section and 25cm back from the section face, is removed (**48**). The advantage of this approach is that when it comes to quantifying the identified shells, this need not only be carried out in terms of relative species abundance, but also in relation to sample volume, which is an independent measure. We shall return to the representation of molluscan data later.

Once a sample arrives at a laboratory it can be processed relatively rapidly. Initially the sample is air dried and both its wet and dry weights measured. It is then placed in a bucket of water to which a small quantity of hydrogen peroxide (30 per cent H_2O_2 – commonly used in lower concentrations to bleach hair) has been added. This chemical breaks down the bonds – usually of an organic nature – that hold individual sediment/soil particles together and eases the subsequent sieving process. Once the sample has been left in the water/hydrogen peroxide mixture for a day it can be sieved to remove all fine particles. This is achieved by washing the sample – which is by now a slurry – through a 0.5mm mesh under a constant stream of water and air drying the material that is retained on the sieve (termed the residue). Frequently the sample will need to be soaked with hydrogen peroxide and sieved a number of times until all sediment/soil aggregates have been removed. The dried residue is then sorted under a low-power binocular microscope. Not all shell fragments are removed from the residue – to do so would, in the case of particularly rich samples, take an inordinately long time – but all those that can be identified and quantified, are. In the case of the gastropods this is the shell apex, and with some species the lip, while with the bivalves it is any part of the shell that includes a hinge fragment (**49**). Identification is carried out, at least while a new specialist is training, with the aid of a modern comparative reference collection and reference texts. Plates of the Limacidae and spherules of the Arionidae are also removed from the residue and are quantified at this family level. Attempts to identify the remains of slugs to higher taxonomic levels have proved unreliable using the low-powered microscopy favoured by **malacologists** (scientists who study molluscs).

MOLLUSC HISTOGRAMS

Once mollusc shells from a series of samples throughout a sequence or profile have been identified, the data can be plotted and an interpretation of environmental change made. As with the results from pollen analysis, molluscan data are commonly plotted as a percentage histogram (**47**). Just as in pollen analysis where plant genera and families are grouped by category, for example trees, herbs etc along the top of the diagram, molluscs are grouped by habitat. These habitat groups were originally developed by Sparks in the early 1960s and are still in use today (**Table 11**). In the case of land snails, marsh or woodland species are plotted on the left hand side of the diagram, those with catholic preferences (i.e. species which inhabit a variety of different environments) in the centre, and species that live in open conditions to the right. For a freshwater sequence, species that inhabit moving water environments are plotted on the left, followed by species living in ditch, catholic and 'slum' environments. Commonly – again in a manner similar to the representation of pollen data – summary columns relating to percentage variations of all members of each ecological group are represented on the right hand side of the histogram (**47**). It is this summary that is used in the first stage of interpretation. For example if the 'open country' category forms a high percentage then it is highly likely – providing a Holocene, north-west European context is being examined – that either a pastoral or arable situation prevailed when the land snail assemblage was forming. Conversely, a situation where the 'shade-loving' and

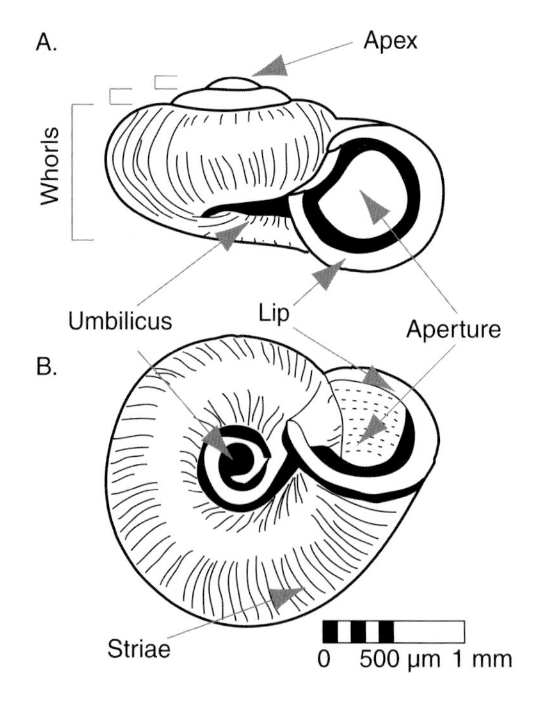

49 *Parts of the shell of gastropods commonly used for the identification and quantification process (the example used is* Vallonia excentrica, *a species characteristic of open conditions)*

Fresh water	
'Slum' group	Species tolerant of poor water conditions, ephemeral and stagnant pools with considerable changes in water temperature, e.g. *Lymnaea truncatula* and *Pisidium casertanum*
'catholic' group	Comprising Mollusca that will tolerate a wide range of habitats except the smallest and largest water bodies, for example *Lymnaea peregra* and *Pisidium millium*
ditch group	Species often found in ditches with clean moving water and abundant growth of aquatic plants, for example *Valvata cristata* and *Planorbis planorbis*
Moving water group	Composed of molluscs commonly found in larger bodies of water such as moving streams and larger ponds where the water is stirred by currents and winds. Species include the larger bivalves, *Valvata piscinalis* and *Lymnaea stagnalis*.
Terrestrial	
Marsh group	Species which live in damp, semi-aquatic environments, for example the Succineidae and *Vallonia pulchella*
Open country group	Species that inhabit dry land environments devoid of shade such as those found in ploughed farmland or in steppe. Species include *Vallonia costata*, *Vallonia excentrica* and *Pupilla muscorum*
'catholic' group	Consisting of species that tolerate a wide variety of terrestrial environments. These include *Trichia hispida* and *Cochlicopa* sp.
Shade loving group	Comprising species that live in shady environments such as woodland and scrub. Alternatively in artificial habitats that mimic these conditions, for example amongst rock rubble, in the bottoms of ditches etc. Species include *Discus rotundatus*, *Oxychilus* sp. and the Clausiliidae

Table 11 *Ecological groups used for interpreting land and fresh water Mollusca as defined by Brian Sparks*

'marsh' categories both have high percentages, means it is likely that the environment represented was either damp woodland, typically alder carr, or marshy scrub. High percentages in the 'moving water' category would indicate that the deposit from which they came formed in a river or stream environment, although we could not say for certain whether the sample site was in the main channel of such a feature, because of the significant movement of shells that occurs in this environment.

INTERPRETING NON-MARINE MOLLUSCAN RECORDS

After a broad reconstruction has been described on the basis of the summary data of ecological groups, the next stage in the interpretation is to provide a more detailed idea of the past environment: was open country arable land or pasture, was shade woodland, long grassland or just the result of a local depression in the landscape? Examining the variation in abundance of individual taxa rather than the summary data can fill in these details. Unfortunately, unlike insects, mollusc species are rarely specific to particular environment types. For example, taxa that live in open country environments commonly inhabit all niches of this type: arable fields, short grass pasture and even periglacial tundra. However, some species prefer one of these environments more than the others, which means that the relative frequency of different taxa can be used to infer which sub-category of open country environment was present. John Evans has used this observation to further refine interpretation of mollusc data using what he calls **taxocenes**. These are defined as groupings of mollusc taxa which are known to have lived alongside each other in the past (in other words the taxocene concept is only applicable to situations where shells have not been moved by sediment transport agencies, i.e. only to soils) and indicate a particular environment type. So far five dryland and eight wetland taxocenes have been recognised (**Table 12**). Mollusc data from palaeosols of newly investigated sites can be compared with these taxocenes and the environment interpreted accordingly. Despite the modern interpretative focus upon combinations of taxa as being indicative of a particular type of environment, so-called **indicator species** still have a role to play. Indicator species comprise the few stenotopic molluscs that are known, and which because of the tight environmental constraints in which they live enable relatively precise palaeoenvironmental reconstructions. Some examples of these include *Helicodonta obvoluta* and *Helicigona lapicida* which only inhabit woodlands that are not subject to extensive

Taxocene	Dominating species	Environment
	Dryland environments	
1	*Vallonia costata* & *Vallonia excentrica*	Stable vegetated surface, grassland
2	*Pupilla muscorum*	Instability (e.g. arable activity)
3	*Vallonia excentrica* (*Trichia* sp. & *Cochlicopa* sp. in lesser abundance)	Decalcification and impoverished vegetation,
4	*Vallonia costata* (with *Trichia* sp. or *Carychium* sp. common, and *Oxychilus* sp. and/or *Vitrea* sp.)	Instability in places that are sheltered and cool (e.g. ditches)
5	*Pomatias elegans*, *Pupilla muscorum* & *Helicella itala*	Mixed tillage (*Pomatias, Pupilla*) and grass fallow (*Helicella*)
	Wetland environments	
1	*Vallonia pulchella*, *Trichia hispida*, *Cochlicopa* sp. & Limacidae	Damp, low diversity grassland
2	As previous, but with *Lymnaea truncatula*, *Carychium minimum* & Succineidae (*Vallonia pulchella* and *Trichia hispida* dominate)	Wet grassland (wetter than Wetland taxocene 1)
3	*Zonitioides nitidus* & *Carychium minimum*	Wet woodland or long grasses, sedges or reeds
4	*Lymnaea truncatulata*	Mudflats
5	*Anisus leucostoma* & *Pisidium personatum*	Seasonal pools
6	As Wetland taxocene 1 plus abundant *Pupilla muscorum*	Damp grassland, possibly less damp than Wetland taxocene 1
7	*Lymnaea peregra*	Seasonal floodplains
8	*Valvata cristata*, *Pisidium casertanum* & other species from Sparks' slum group	Permanently wet, but subject to stagnation. Found at transition from aquatic to wetland habitats

Table 12 *Mollusc taxocenes as described by John Evans*

human alteration, *Ancylus fluviatilis*, which only lives on rocky substrates over 2m below the surface of relatively fast moving rivers and streams, and *Anisus leucostoma*, a species that prefers open, muddy conditions at the edge of aquatic features. A useful way of articulating changes in a mollusc percentage histogram is to divide it into 'mollusc assemblage zones' which are based on variations in habitat ecology seen in the assemblage and indicator species data. In the same way as those used in pollen diagrams, such zones can provide a framework in which changes in the environment can be made clear to the readers.

Although percentage histograms are by far the most common means of plotting molluscan data, there are problems associated with their use. Percentage histograms carry the inherent implication that all molluscan taxa portrayed on that diagram are interlinked – a product of the process of calculating percentages. However, ecologically this is rarely the case. For example, if numbers of shells of one taxa rise while the others remain at the same level through consecutive samples the effect of this in a percentage histogram is for the proportion of the latter taxa to fall in successively higher samples. Therefore, it would be tempting to interpret a diagram of this type to mean that a decline in the habitat of the second species was taking place despite the fact that in reality more favourable conditions for just one species had developed. Indeed, changes in environment of this type may not relate to the wider environment at all, but rather to a greater availability of food. Figure **50** clearly illustrates these problems of using percentage data. The percentage histogram (**50a**) suggests that shade increased rapidly at the expense of open conditions from around 100cm below ground surface (i.e. second sample from the base of the diagram), which might be interpreted as meaning the spread of woodland after site abandonment. However, in this case a second histogram was plotted on the basis of the absolute shell frequency (**50b**). This clearly shows that the decrease in open country taxa is a factor of the calculation of percentages. All open country taxa: *Pupilla muscorum*, *Vallonia costata*, *Vallonia excentrica*, and *Helicella itala*, are found in more or less the same quantity across this supposed ecological divide, and it is rather the sudden increase in numbers of shade-loving taxa that causes the dramatic variation in the percentage histogram. Knowing this, a rather different palaeoenvironmental scenario can be proposed. The large increase in the population of shade-loving taxa is likely to reflect the new ecological opportunity of a relatively newly dug, cool, dark, moist and vegetated ditch. The apparent time lag in colonisation by shade-loving taxa (the increase in shade lovers is only in the second sample) is because the initial ditch deposits were formed from eroded soil containing shells of open country taxa. The lesson of this example is that although percentage histograms are the easiest to produce and interpret, some kind of absolute frequency diagram should also be drawn where the percentage data reveal apparent dramatic ecological changes.

Two further potential constraints when interpreting molluscan data cannot be so easily resolved. The first of these relates to the spatial area that is represented

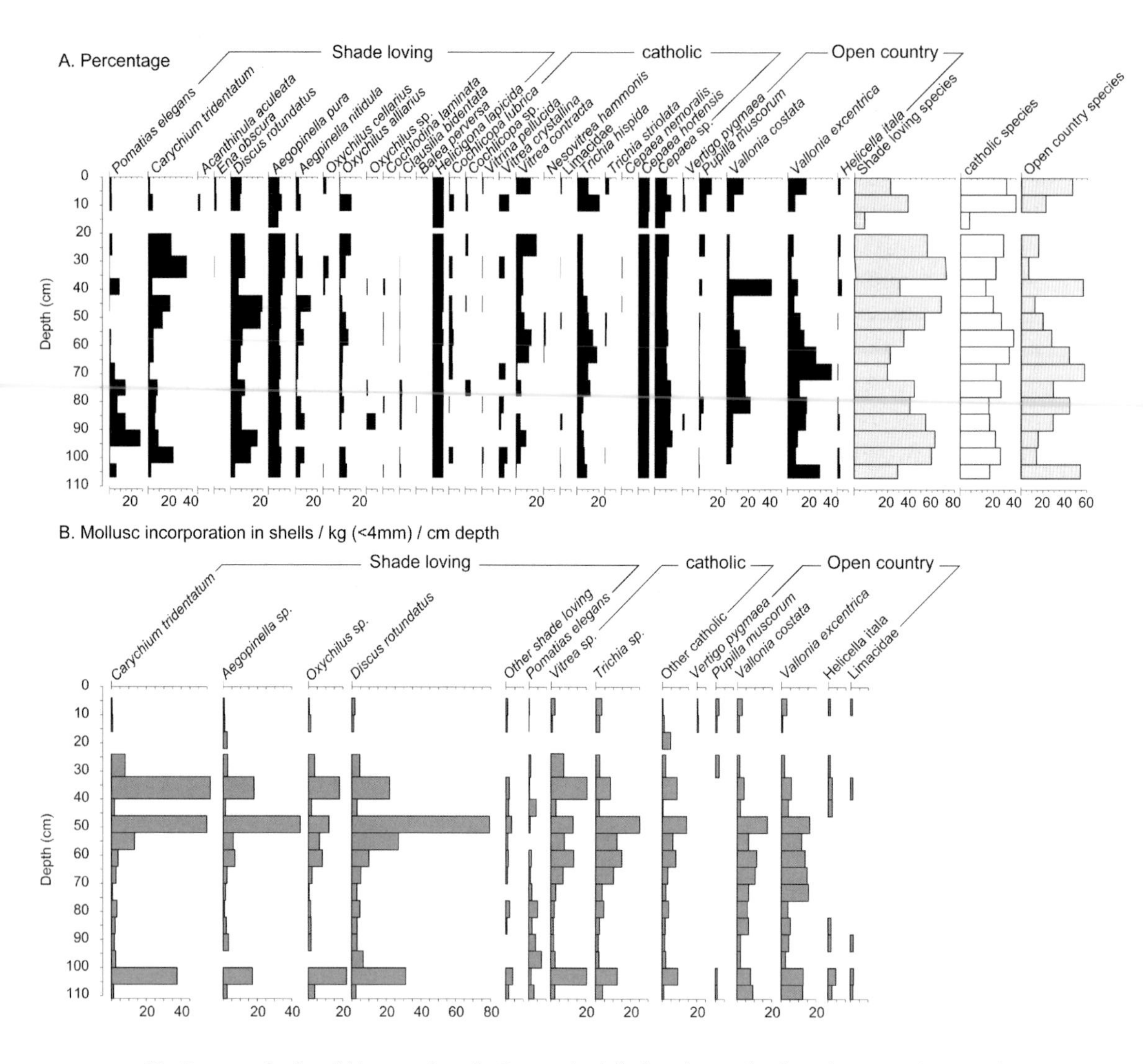

50 *Percentage land snail histogram from the Bronze Age hillside enclosure of Mile Oak near Brighton, southern England. B shows the same data, but plotted on an absolute frequency basis (number of shells of a particular taxa per kg sediment less than 4 mm – i.e. excludes size classes which are larger than the biggest snail shell)*

by the environment reconstructed using mollusc data. The second is associated with bioturbation (comprising floral- and faunalturbation), a topic that was partially covered in section 1 (**Table 5**). The calcareous geologies on which sites suitable for molluscan analysis most commonly lie also have some of the most biologically active soils. Floralturbation is only usually a significant problem in forested environments, where roots are large enough to cause a downward movement of shells, and where tree throw can cause a catastrophic mixing of the stratigraphy. The latter, and on occasion also the former, can be recognised in exposed sections, and either the sampling strategy modified accordingly, or allowances made when interpreting the data. Similarly, the faunalturbation caused by small mammals such as moles and certain rodents can also commonly be seen in exposed stratigraphy and be allowed for in the same way. Much more problematic when interpreting molluscan data is the effect of earthworms, which are highly active in calcareous soils. Earthworms in north-west Europe can transport material of up to 2mm in size downwards through a soil profile to a depth of 1.5m below the contemporary surface. Thus not only are later shells introduced into earlier deposits, but shells are also sorted by size, with those of larger taxa remaining at or near the surface. Earthworm mixing is a particular problem in interpreting mollusc data from palaeosols. Using evidence from modern soils, Stephen Carter has calculated that only the shells from the top 5cm of any palaeosols provide reliable evidence of the environment at the time of burial. Below that shells from different periods become irrevocably mixed and palaeoenvironmental reconstruction impossible. Luckily in Britain there is an independent check of the impact of bioturbation on sediment stratigraphy sampled for mollusc analysis and which is known to have formed prior to the Roman period. During the Roman period it is thought that the interior of Britain was colonised by a number of genera such as *Helix*, *Candidula* and *Cernuella*, which had ranges that previously extended to either the French coast, or to the coastal sand dunes of southern Britain. Therefore if shells of these species are found in prehistoric stratigraphy they can only have got there as the result of bioturbation (**47**). It therefore follows that the quantity of 'Roman introductions' that are found in each sample from a prehistoric layer are indicative of the amount of bioturbation that has taken place. Although data such as this cannot aid in any quantitative manner the reconstruction of a past environment, they can provide a constraint on over-precise interpretations.

In a previous section we saw how the size of a depositional basin determines the extent of the spatial area represented by a palaeoenvironmental reconstruction using palynology. The same is true of molluscan data, but with the added constraint that molluscs are slow-moving creatures and therefore do not travel more than a few metres during their relatively short life. Even following death, shell transport is over much shorter distance than pollen grains – unless movement is in a stream or river. Therefore the environments reconstructed from sub-fossil Mollusca are local ones, relating to conditions immediately surrounding a depositional basin, or extending perhaps no more than a radius of 10m from the sample site in the case of palaeosols. In other words the mollusc data from Toadeshole Bottom (**47**) relate to conditions within the valley, those from Mile Oak (**50**) to perhaps 10m from the ditch edge, and those from the Dover boat to varying parts of the lower Dour river catchment. The local nature of environmental reconstructions using non-marine Mollusca is a constraint upon interpretation, and for this and other reasons previously outlined, such data are not compatible with palynological reconstructions. However, molluscan-based reconstructions are often very useful in the interpretation of an archaeological site precisely because of the restricted spatial scale to which they relate. For example phases of site use and abandonment can be postulated at Mile Oak (**50**) based on the colonisation and clearance of long vegetation in the surrounding ditch – events that are seen only in variations in the mollusc assemblages.

Vertebrates

The use of vertebrate remains for studying palaeoeconomy is covered in detail in section 3. Here we will discuss only the use of fossil vertebrate remains for examining past environments. Nevertheless, as the same remains are used for both purposes, the reader is referred to section 3 for techniques of bone identification and quantification.

There are four main groups of vertebrates commonly found in archaeological situations: mammals, fish, birds and **herpetofauna** (the last comprising both amphibians and reptiles). Mammal bones, being the largest and most ubiquitous finds, dominate the archaeological literature. As with the invertebrate organisms that we have previously discussed, many vertebrate species are indicative of particular environment types. For example undomesticated horses (*Equus* sp.) are characteristic of grasslands, frogs (*Rana* sp.) of fresh water, and three-spined sticklebacks (*Gasterosteus aculeatus*) of streams. However, vertebrate species are rarely stenotopic and are much less specific in their environmental tolerance than invertebrates; they have much larger ranges. The African elephant for example (*Loxodonta africana*) lives in savannah environments, yet savannah is a broad descriptive term meaning a combination of grasslands of various types, stands of woodland and freshwater environments. Inferring the past presence of a savannah environment around an archaeological site from preserved elephant remains may be useful, but will tell us little of the density of woodland, the types of local grassland, or indeed the distance of the savannah from the site. The last of these is an important consideration in any use of vertebrate remains for studying environments for two reasons. Firstly, many vertebrate groups, particularly large mammals and many groups of birds, travel long distances and may range outside the environment of which they are supposed to be characteristic. Secondly, these same groups are used by human groups for food and raw materials, and are therefore likely to have been hunted and transported in the archaeological past. Thus, if we are to use large mammals, fish and birds in palaeoenvironmental studies we have to be certain that the remains that we study come from within their original environmental range (for which we need other classes of palaeoenvironmental data) and have not been transported over long distances by people. People of course are not the only hunters – and indeed were possibly not hunters at all for much of their evolutionary history – and therefore creatures such as lions, bears, wolves and foxes may move their prey, or rather parts of it, considerable distances to consume at leisure.

The discussion above leaves out certain types of vertebrate, namely small fish, small mammals, reptiles and amphibians. These groups are more likely than their larger relatives to be characteristic of a restricted environmental niche. It is for this reason that bones and teeth of organisms from these groups are more useful in palaeoenvironmental studies. Even so there are problems, many of which we have discussed in section 1 in relation to post-mortem transport. However, an aspect that we have not previously considered is that smaller vertebrates are the prey of larger animals. One particular hunter of note is the bird of prey, which frequently transports its catch over long distances and deposits the remains as faecal pellets in a completely different environment to that in which the hunt took place. For example, excavations of cave sediments have demonstrated that the small mammal fauna recovered is frequently the result of owl predation, and thus has very little to do with the cave environment *per se*. Nevertheless such small mammal assemblages represent a range of habitats surrounding the cave and can therefore indicate the variability of environment.

For all the above reasons, vertebrates are rarely used as proxy records of environments for Holocene period sites where other classes of biological remain are present. However, their study has proved a great deal more useful for the Palaeolithic period, where the robust nature of animal bones has often meant that they are one of the few classes of biological remain that survive in the archaeological record. Mammal bones are also routinely used as a means of relative dating for the Pleistocene period. This is because during the Pleistocene many species gradually (sometimes rapidly) evolved. Evolution has frequently resulted in changes in skeletal morphology and therefore fossils at the same evolutionary stage can be correlated between sites. A good example of this is dental changes in the water vole *Arvicola terrestris* in the Middle Pleistocene. From around 350,000 BP onwards this species gradually lost the roots of its molars, at the same time evolving a thicker enamel coating to these teeth. These distinctive changes are readily recognisable in the Pleistocene palaeontological record and are the basis for the definition of three subspecies

			Morphological trends	
Date (BP)	OIS	Taxon	Molar roots	Enamel thickness
150,000	6	*Arvicola terrestris* subspecies B	↓	↑
200,000	7	↑		
250,000	8	*Arvicola terrestris* subspecies A		
300,000	9	↑		
350,000	10	*Arvicola terrestris cantiana*		
400,000	11	↑		
450,000	12			
500,000	13	*Arvicola terrestris*		
550,000	14	*Mimomys savini*		

Table 13 *The vole clock: evolution of* Arvicola terrestris

(**Table 13**). Occasionally evolutionary change can occur so rapidly[2] that **speciation**, i.e. the birth of a new species, can occur. If it can be determined at what date the speciation event took place, we will have a good chronological marker. In this case all sites containing the new species will date to after the speciation 'event'. *Arvicola terrestris* is again a good example of this phenomenon as this species evolved from another water vole, *Mimomys savini* during the middle or late Cromerian period (**Table 13**, also **4**). As we shall see in the discussion below, this change is particularly fortuitous for archaeologists studying initial hominid colonisation of northern Europe. This is because the only proven hominid sites from the Cromerian period in northern Europe are found in association with *Arvicola terrestris*, suggesting that colonisation took place in or after the later Cromerian. These evolutionary changes in Pleistocene small mammals and their implications for reconstructing Pleistocene chronology have collectively been termed the **vole clock**.

Zooarchaeologists and palaeontologists have also been able to use known changes in distribution of vertebrate species during the Pleistocene as a sort of relative dating technique. As a result of the huge climatic changes that occurred over the long timescales of the Pleistocene, species ranges have expanded and contracted. For example the relatively warm conditions of the Eemian interglacial period (**4**) allowed hippopotamus (*Hippopotamus amphibius*) (but interestingly not people) to enter Britain, the only time in the Pleistocene that this animal reached this far north. Therefore a British site containing hippopotamus must be of Eemian date. Although we will not discuss the use of vertebrate data for dating any further (readers will find appropriate references in the further reading section), the examples we use in the following text are from Pleistocene sites.

BONE AND TEETH, AND THEIR SURVIVAL

During the life of a vertebrate animal, long bones such as the tibia, femur, humerus and radius consist of tubular structures made from a compound of calcium phosphate called **hydroxyapatite**, surrounding a proteinous core of **collagen**. All other bones in the animal's body are comprised only of hydroxyapatite. Hydroxyapatite is a mineral material and is extremely resistant to decay once an animal dies. However, the mineral is more susceptible to chemical change, a process that can often lead to fossilisation. Fossilisation occurs when all the pore spaces in the mineral part of the bone have been filled by salts of calcium or iron that have leached in from the surrounding sediments. Collagen too decays very slowly depending on the environment in which the bone lies, and may survive in the archaeological record for thousands of years. However, neither mineral bone nor collagen survive in acidic environments and in such situations decomposition occurs relatively rapidly.

If anything, teeth are even more durable in the archaeological record than bones (although unfortunately from the point of view of the zooarchaeologist, birds lack them). Teeth are made up of three substances – all of which are well known to us from visits to the dentist and toothpaste adverts. The core of a tooth is made from **dentine**, which is a relatively soft substance, but which, fortunately for the zooarchaeologist, is covered by a hard casing of **enamel**. The base of the tooth is covered by **cement**, which is the part most susceptible to damage and loss over time spans of hundreds to thousands of years. Therefore it is the enamel which protects the tooth from decay in the archaeological record just as it does in the living animal. Indeed even in situations where all other bone remains have been dissolved by acid-rich sediments, teeth may survive.

RECOVERING ANIMAL BONE

Bones are commonly recovered from archaeological sites, or off-site locations by hand collection during excavation. Bones are either collected according to archaeological context/stratigraphic unit or are plotted as individual finds by measurement according to site grid reference along with a spot height. However, a strategy that relies solely on either of these practices produces a bone collection that is biased towards larger mammals, and even in their case, towards the bigger body parts. To reduce this bias, large samples (i.e. of the order of 100kg or more) can be taken from each context of interest for sieving. This procedure is often combined with sampling for the recovery of small artefacts that would otherwise be missed by excavators. Sieve meshes of between 4 and 10mm are commonly used, and depending upon the geographic location of the site and the cohesiveness of the layers being excavated, water can be used to aid the process. Once fine material has passed through the sieve, the remaining residue is sorted by eye in the sieve mesh and all bone and small artefact fragments are removed for later identification. When sampling is carried out by this approach it is important to know the volume of sediment/soil that has been sieved, so that some idea of the concentration of bones can be obtained. The advantage of sieving for the recovery of small bones is that not only is it possible to retrieve a bone assemblage that is not biased towards larger remains, but that an assessment can be made of remains that have been lost in non-sieved areas. At many Palaeolithic and Mesolithic excavations, as well as cave sites of other periods, all excavated sediment is sieved using the smaller of the sieve sizes. In this way all animal bones – as well as micro-artefacts – are recovered for later study.

Bulk sieving of the type discussed above is useful, but many diagnostic bones and teeth of small mammals, birds, fish, reptiles and amphibians are less than 4mm in size. It is therefore important, even where 100 per cent of the excavated deposits are screened on site, that samples are taken for sieving through finer meshes. Samples can be taken either on a context by context basis, or using a sampling column excavated through the site stratigraphy as described for mollusc studies above. The decision on which approach to take will depend upon the questions that the small vertebrate remains are being used to answer. However, for both approaches large samples must be obtained (e.g. 40-60kg), unless bones are known to be present in particularly high densities. For sites of Neolithic date and later the most common approach is to combine sieving for small vertebrate remains with sampling for plant macro-fossils, and to process the samples using a flotation machine (see section 3). For earlier sites, samples are taken as described above and are washed through meshes of either 1mm or 500µm under a constant stream of water. For certain types of deposit, for example alluvial clays, the same sort of pre-soaking as described for mollusc analysis may be required to break down sedimentary bonds prior to processing.

Dried residues from flotation or wet sieving are sorted by eye for size fraction greater than 2mm and using a low-powered binocular microscope for finer material. The bones recovered are then identified by comparison with modern reference material and quantified as described in section 3.

AN EXAMPLE OF THE INTERPRETATION OF ANIMAL BONE FOR PALAEOENVIRONMENTAL PURPOSES: BOXGROVE, WEST SUSSEX

Eartham Quarry, near the village of Boxgrove, is about 8km from the town of Chichester. It is also Britain's best-known, and arguably most important Palaeolithic site. Following the discovery of handaxes in the early 1980s, excavations followed for much of the later 1980s and throughout the 1990s under the direction of Mark Roberts. The site is remarkable for the pristine preservation of discrete hominid activity areas, including for example handaxe production and animal butchery. However, the site is best known for the discovery of two teeth and a tibia of the extinct hominid *Homo heidelbergensis*. The dating of the site is based on the vertebrate fauna that was recovered. The presence of *Arvicola terrestris cantiana* and absence of *Mimomys savini* argues for a late Cromerian date or later. However, further species encountered at Boxgrove, for example the shrew *Sorex runtonensis*, the vole, *Pliomys episcopalism*, the cave bear, *Ursus deningeri*, the rhinoceros *Stephanorhinus hundsheimensis* and the giant deer, *Megaloceros dawkinsi*, all became extinct during the Elsterian cold stage (see figure **4**). Therefore, Boxgrove must date to the late Cromerian or early Elsterian, in other words around 500,000 years ago.

Hominid activity at Boxgrove took place on a coastal plain, which was located in the lea of chalk cliffs. Over the period of hominid use of the landscape relative sea levels fell, and the coastal plain became progressively a lagoon, and then an intertidal marsh. The cliffs were exploited by hominids for flint which was made into elaborately worked ovate handaxes, while the coastal plain was also home to herds of herbivores. Preservation of the archaeological remains and vertebrate faunas in their original resting place occurred because muds deposited by low energy tidal processes were periodically deposited on top of them. Palaeoenvironmental reconstruction at Boxgrove has largely focussed on vertebrate remains, together with molluscs and Foraminifera. Given the calcareous catchment and the lack of waterlogged deposits, preservation of other classes of biological remains was poor.

There are four broad depositional units on the Boxgrove site, which from bottom to top are: the Slindon sands, representing fully marine conditions, the Slindon silts, which accumulated in a lagoonal situation, then a palaeosol associated with a marshland environment, deposits associated with a lake, and finally chalk gravels and brickearth (Eartham formation) associated with periglacial conditions (**51**). Evidence of hominid activity occurs throughout this sequence, but is particularly focussed towards the top of the Slindon silts and in association with the palaeosol. From these deposits at least 89 vertebrate species had been recognised by 1999, including 11 species of fish, nine species of herpetofauna, 19 species of bird and 50 species of mammal. Many of these remains were recovered in the 65 metric tonnes of sampled sediment that was sieved through a 500μm mesh using an automatic sieving machine. The larger mammal bones were all individually recorded and removed by the excavators. The fragile nature of some of the larger bones meant that they had to be removed, together with a block of sediment, or even in foam jackets, to avoid damage.

The vertebrate fauna from the Slindon sands was rare and the bones that were found consisted mainly of fish. These comprised conger eel (*Conger conger*), either sardine or pilchard, and a species of flatfish. Sardines and pilchards are inshore shoaling fishes inhabiting shallow-water environments, while the conger eel occurs in rocky coastal waters, and, therefore, together the finds are indicative of coastal environment similar to those of the present day. In other words, the environment at this time consisted of shallow, sandy-bedded waters of moderate temperature. Lagoonal deposits of the Slindon silts also contain fish remains, but this time of diverse species including wrasse, thornback ray (*Raja clavata*), flounder (*Platichthys flesus*), three-spined stickleback (*Gasterosteus*

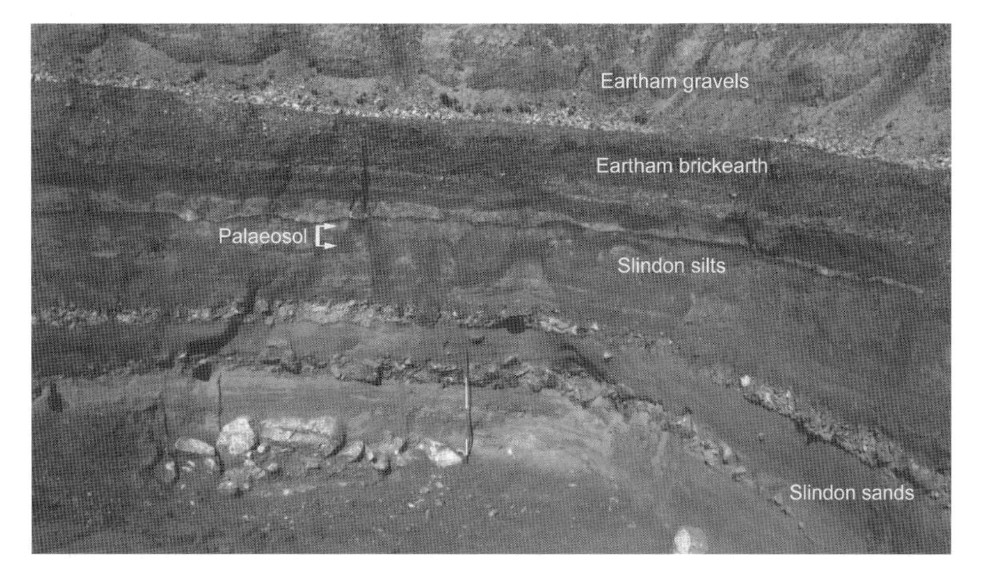

51 *The Boxgrove sequence photographed in 1998*

aculeatus) and either salmon or trout, which are all indicative of shallow inshore waters, together with cod (*Gadus morhua*) and blue-finned tunny (*Tunnus thynnus*), which are found offshore. However, the fish assemblage changes upwards through the silts to become increasingly dominated by species that can tolerate fully freshwater environments, such as stickleback, salmon and trout. In other words, the fish indicate that the lagoon had been cut off from the sea. These upper portions of the Slindon silts contain further indicators of fresh water in the form of newts (*Triturus* sp.), spadefoot (*Pelobates fuscus*), toads (*Bufo* sp.), and frogs (*Rana* sp.), which all require at least small pools. Several of the herpetofaunal species, such as the spadefoot and natterjack toad (*Bufo calamita*) suggest that conditions around the water were sandy, while others such as the slow worm (*Anguis fragilis*) and grass snake (*Natrix natrix*) indicate damp conditions with dense vegetation. Spadefoot and another amphibian, the moor frog (*Rana arvalis*) are not found in Britain today, but are common on the European mainland, perhaps suggesting that late Cromerian Boxgrove experienced warmer summers than it does at the present day.

Mammalian bones were mostly found in the top of the Slindon silts and in association with the palaeosols. Many mammal bones in the Slindon silts were associated with flint-knapping debris, and many had cut marks. The small mammals from these deposits are characteristic of grassland, but some, for example woodmouse (*Apodemus sylvaticus*) indicate denser vegetation in the immediate locality. Certain of the larger mammals, namely the wild cat (*Felis silvestris*) and roe deer (*Capreolus capreolus*) would appear to confirm this finding. Best preservation of mammalian bone is in association with the palaeosols. However, there are certain taphonomic problems here in that many large mammal bones have evidence of gnawing by carnivores. Trampling is also apparent, while weathering of many bones indicates that the surface was frequently exposed. Many small mammal bones show evidence of digestion by predators, suggesting that they were originally deposited in scats. Despite the mixing caused by these agencies, the mammalian fauna from the palaeosol analysed by Simon Parfitt paints a detailed picture of the environment. It would appear that vegetation had become more diverse since the lagoonal phase, although it was still dominated by grassland as represented by the pine vole (*Microtus subterraneus*), field vole (*Microtus agrestis*) and common vole (*Microtus arvalis*). Nevertheless, woodland seems to have been common as indicated by the discovery of bones of squirrel (*Sciurus* sp.), hazel dormouse (*Muscadinus avellanarius*), badger (*Meles* sp.), as well as two species of bats. Most of the woodland species, however, occasionally feed in grassland environments, and, therefore, it would seem likely that while the coastal plain was open, woodland occupied the chalk downland beyond the cliffs. The present day distributions of the small mammals found in the palaeosol have also been examined and suggest that the greatest similarity is with western Ukraine and eastern Poland. This information suggests that Boxgrove at 500,000 BP had a more continental climate than it does at the present day.

Birds of course are representative of wider environments than the other classes of vertebrate discussed. The bird remains found at Boxgrove are from the same species that characterise northern Europe at the present day, with the exception of great auk (*Pinguinus impennis*), a flightless seabird that became extinct soon after 1844 due to excessive hunting. The birds found at Boxgrove include waterfowl, seabirds, songbirds, a gamebird, a wader, the tawny owl (*Strix aluco*) and the swift (*Apus apus*).

OTHER TRACES OF VERTEBRATES

So far we have just discussed vertebrate bone, but there are other forms of evidence that indicate the former presence of mammals and birds. For example in either very low energy waterlogged conditions or desiccated environments, droppings may survive, which can be identified by comparison against modern-day examples. It is interesting in this respect that mammal scats are extremely well studied and whole books are published on their identification. Footprints may also survive in some conditions. We have already seen that hoof prints of red deer survived in Mesolithic muds adjacent to the site of Goldcliff in the Severn Estuary of Britain. These have been argued by Martin Bell to be the reason for the presence of human footprints in the same muds. However, more spectacular examples exist in the literature, of which some of the oldest are the prints of rhinoceros, elephant and antelope that are found in association with those of two hominids (probably *Australopithecus afarensis*) at Laetoli in Tanzania. These footprints were preserved in a bed of volcanic ash dated to about 3.8 million years ago, while the hollows created had in turn been filled by further ash dated to around 3.6 million years ago. Not only do these footprints provide the earliest definitive evidence of hominid bipedalism, but also some of the best evidence for the animals which occupied the same ecosystem as the hominid pair. The evidence from fossilised animal droppings and footprints is certainly of importance, but it cannot provide quantitative data. Instead, we can determine the presence of a particular animal in an ancient environment. From this point of view, although the finds may appear less spectacular, vertebrate bones and teeth collected on a systematic basis from a site are useful in the study of past environments.

Summary

In section 2 we have discussed approaches used to study past landscapes and the methods that can be used to understand them. We began by emphasising a key point first made in section 1: landscapes change over relatively short timescales. Few are stable on timescales of centuries or more. Therefore, it is dangerous to apply present-day landscape variables to a past situation. When investigating an archaeological situation we must try and understand how a past landscape both looked and functioned. The first step in this process is to determine the appearance of, and changes to, the physical landscape by examining preserved sediment sequences and soil profiles. For this reason we looked at how soils and sediments are described and what these morphological properties tell us of a past landscape. These ideas were then illustrated by examining studies carried out on three example landscapes: coastlines, rivers and inland ranges of hills.

Once the physical landscape and the changes it has undergone are understood, we can begin to populate it with plants. We looked at two methods of reconstructing past vegetation using preserved floral remains: pollen and plant macro-fossil analysis. Of these, the first is widely used for examining regional vegetation. This technique is largely responsible for the detailed knowledge we now have of the distribution of the world's biomes in the past. Plant macro-fossil analysis, when used for examining past environments (as opposed to economies – a subject that we shall turn to in section 3), has played a subordinate, but important role, particularly in helping to interpret pollen diagrams. We also saw how invertebrates can help in reconstructing past plant communities. Insect remains can often provide more specific information on plants growing in the past than remains of those plants themselves. However, beetle remains have other uses too, and we looked at how they have been employed in the reconstruction of Late Pleistocene temperatures. Perhaps more usefully from an archaeological point of view, we looked at their use in the examination of human environments in Roman and medieval cities. Another invertebrate group, the Mollusca, has also been used as a proxy for past environment. Although the level of detail of palaeoenvironmental information provided by land and fresh water snails is not as great as that of palynology or insect studies, the calcareous shells of these creatures do survive in the oxidising environments that characterise much of the temperate, arid and semi-arid zones, in which pollen and insect remains do not. Finally, we took a brief look at the use of vertebrate bones for palaeoenvironmental reconstruction. We saw that they have primarily been used in the study of Pleistocene environments, and that for later periods, when domesticated animals abound, it is the smaller mammals, reptiles, amphibians and fish that have provided the most useful information.

Having looked at the study of palaeoenvironments we now have an understanding of how the background against which people of the past operated is obtained. These same techniques also inform us of the interaction of people and landscape. However, it is now time to focus on the archaeological site and look at how we can study people's regular activities using biological data.

1. Note here that the word 'stream' when used by physical geographers has no size connotations, it simply means flowing water within a channel. Thus to a geographer the Nile or Amazon are just as much streams as a gently meandering brook flowing through the picturesque English village of a Turner painting!
2. The word 'rapidly' is used here in a geological sense. The change from *Mimomys savini* to *Arvicola terrestris cantiana* may have taken as long as 50,000 years.

SECTION 3

PALAEOECONOMY:

ANCIENT SUBSISTENCE AND PRODUCTION FROM BIOLOGICAL EVIDENCE

Introduction

In this section we examine the role played by environmental archaeology in understanding ancient economy. We begin by examining what economies are and how they were organised in the ancient world, before moving on to look at the proxies that environmental archaeologists use to study them. We finish by looking at the evidence that bioarchaeology provides for ancient farming and food processing techniques.

The nature of economies

The eighteenth-century economist Adam Smith expressed the view that the economy was the paramount factor that governed societies, a view that was shared by the great nineteenth-century sociologists Karl Marx, Emil Durkheim and Max Weber (of whom we will hear more in section 5). If we accept this view, and bearing in mind that one of the prime goals in archaeology is the deciphering of ancient society, then the study of past economies is of fundamental importance. By 'economy' here we are referring to the production and consumption of goods and services. In other words, the way that people organised the **production** of food, pottery, metalwork, houses etc. is one side of the economic equation. The other is how these things were distributed amongst the people and used (**consumption**) (**52**). By distribution the emphasis is not on how objects or produce were transported but rather on how they were shared out among the wider populace.

In order to explore these economic ideas we pose a series of questions that concentrate on those issues most pertinent to environmental archaeology – the breeding of animals, the growing of crops and the production of food, drink, textiles and other objects from them. The questions are based on an approach developed by Dee DeRoche for studying the British Iron Age. They aim to give rise to more detailed descriptions and understanding of past economies by environmental archaeologists. The questions lead the investigator from examining the biological components which relate to what was consumed and produced, through ecological and practical considerations as to the nature of the production sequence itself, to setting the resultant information within a social context. We will be looking at each one in the course of this section:

1. What was consumed? How was it consumed?

2. What was produced? What waste was produced? How was it produced? Can we assemble a production sequence?

3. What resources (including quantities) were used in production (land, structures, raw materials, tools, animal and human labour)? Where did these come from?

4. What time of year (or how often) were production and consumption activities conducted? Where did these activities take place? Who carried out these activities? What was the social context?

5. Who controlled production? Who owned the resources and organised the labour? For whom was the produce intended? Who owned the final product?

6. Who controlled consumption, exchange and distribution? How were these organised?

7. Does evidence for production and consumption vary between neigh bouring cultures or populations? Which aspects vary?

8. Does evidence for production and consumption change through time? Which aspects change? How widespread is such change? Through what social mechanisms does such change come about?

9. How did people learn about consumption and production? Was such knowledge restricted to certain persons?

10. How were production and consumption embedded into social ideologies?

Production and consumption

At the most basic level the division between production and consumption is the difference between *making* and *using* something. Production is not a single event but a sequence of events. A product can at any stage be exchanged and/or modified by further production. A cow, for example, may be consumed many times over. The mature cow can be regarded as a product. If she is bred,

52 *Comparing aspects of consumption and production in modern and ancient societies*

her calf is a product. If she is milked, her milk is a further product. If the milk is turned into yoghurt, cheese or butter these are yet more products, while if the cow is killed, the meat and skin may also be considered products. If the cow's hide is made into leather or the meat into beef sausages then these too are products. If the leather is made into a jacket and the sausage incorporated into sandwiches these are products. The fodder fed to the cow is a product of cultivation, consumed by the cow. Hence production and consumption should never be seen as two separate discrete events. Rather they are an integral mesh of activities and relationships.

A key element of understanding production and consumption sequences for any one society is the knowledge of the stages at which consumption (exchange or use) took place. Societies have conventions governing what are considered acceptable products, but they are not necessarily universal. For many traditional societies cereals are frequently exchanged as are grain and animals in a live state. Obviously this is very different to our own society where it is quite difficult to obtain live cows, sheep and pigs, or indeed whole grain for consumption.

Production sequences

The narrowest definition of **subsistence** refers to the activities of those individuals who farm for their food (**agriculturalists**) or hunt and gather (**foragers**) it from the wild. In its widest sense it refers to how anyone in any society gains the bare essentials of life, including food, clothes and shelter. Documentary evidence tells us that prior to the nineteenth century the majority of people in the world worked on the land or hunted on the plains and forests, producing or collecting the food they would eat, made their own clothes and built their own houses. The basic unit of production in these societies was the family who were largely self-sufficient. It is one of main tasks of archaeology to see if such a model can be applied to those periods for which we have less knowledge, i.e. prehistoric times.

We have divided subsistence activities into three stages to simplify the following discussion: primary production, procurement and processing.

Primary production consists of all those activities leading to the ripe crop being ready for harvest or the breeding and raising of animals to an age where they are ready to exploit (be that for meat, milk, fur etc.). This is of course a definition that is more applicable to agriculturalists, who concentrate their energy on primary production, than hunter-gatherers. The latter may, however, burn vegetation to promote hunting or to increase the yield of certain plants, but relatively little energy is expended in these tasks.

For the agriculturalist **procurement** activities involve the harvesting of crops and the slaughter, milking, bloodletting or shearing of animals. For the forager it is the gathering of wild foods and hunting of wild animals. In terms of time and labour expenditure the difference between foragers and farmers is the opposite to that outlined for primary production. Foragers spend a considerable part of their time in hunting or gathering. For farmers harvesting may take place only once to twice a year, often over a brief period of no more than a few weeks. The time required for the slaughter of an animal is also short compared to that taken to hunt.

Finally, we have **processing**. For the farmer this involves the conversion of harvested crops to clean grain, grain to flour or malt, and these components through various cooking procedures into food. Animals will be butchered removing the meat from the bone, perhaps skinned and dehorned. Milk may be processed into butter, cheese and yoghurt. For hunter-gatherers the processing of tubers, seeds, nuts and fruits may involve the mashing of fruit, the cracking and roasting of nuts, and grinding wild seeds. Some plant foods, such as the acorns enjoyed by many North American native groups, may require the leaching out of toxins.

Risk

Another aspect of subsistence is how communities manage risk, or in other words what farmers or hunter-gatherers do when things go wrong. For example when droughts cause crops to fail or alter the routes of migratory animals, or when severe disease decimates a herd.

Risk strategies refer to those schemes that people employ to minimise the impact of unpredictable events. Paul Halstead and John O'Shea identified four types of risk strategy for traditional societies; mobility, diversity, storage and exchange.

When resources run low or fail one option is to move to an area where such resources are more abundant. **Mobility** as a risk strategy is particularly useful in hunter-gatherer or semi-agricultural groups. If nuts or tubers are no longer available in the usual place the group uses its knowledge of the local environment to move to other areas where the resource might still be found. This is possible because flowering and fruiting of plants of the same species are often dependent on local soils, aspect, altitude and climate, meaning that even in a limited spatial area resources are not synchronously available. As plants are the food of the majority of game animals used by people exploit, mobility for hunters is simply a case of following the herds. This is done by the Iukagir of the Arctic today who follow reindeer, just as it was by people of Late Weichselian north-west Europe in pursuit of the same resource. This risk management strategy is also available to some farmers, for example, the slash and burn agriculturalists of the tropics, who are regularly victim to a loss of soil fertility due to leaching of former rainforest soils. This forces them to move every two to five years.

Another option when one resource fails is to switch to another; a strategy that is termed **diversity**. This is a strategy practised by foragers and agriculturalists alike. Having a more diverse diet disperses the risks involved in relying on a single resource. Natural environmental catastrophes such as severe winters, disease, overly dry summers, strong winds, or heavy floods, may destroy a single resource. However, while one resource may be badly hit, the catastrophe is unlikely to impact so severely on all possible resources. Diversity is then simply a case of following the old maxim of not putting 'all your eggs in one basket'. One way agriculturalists diversify is to grow two crops with differing ecological requirements simultaneously in the same field. Such crops are known as **maslins**. The idea with such an approach is that if conditions change then at least one crop will survive. Other alternatives are to utilise several fields, and sow a different crop in each, or alternatively to sow both spring and autumn crops. Agriculturalists might also utilise wild resources should their crops fail. These so-called 'famine foods' are a common feature of many traditional societies. Scandinavian farmers for example would resort to tree bark in times of crises. In other parts of Europe, chess, a weed of cereals, was also eaten. For the hunter-gatherers many resources may be only seasonally available, so that a diverse diet is necessary to survive each part of the year. Hunter-gatherers therefore employ the diversity risk strategy as a matter of course.

Another way of coping with variation in resource availability is to rely on a permanent **storage** of food against the risk of such events. Storage as a strategy is practised less by hunter-gatherers than agriculturalists, mainly as a result of the former's high mobility. Where it has been utilised by hunter-gatherer's it has been amongst relatively sedentary populations such as Japanese hunter-fisher communities who stored acorns. Meat is also occasionally stored by hunter-gatherers and joints wrapped in leaves under the ground are recorded amongst North American native peoples.

For agriculturalists who rely on storing food from harvest to harvest, using this risk strategy may be a case of simply adding an extra granary. When harvests are good the additional stores are filled, when harvests fail these reserves can be utilised. Halstead and O'Shea note that in many ancient societies centralised stores of grain were collected to cope with such events. They postulated the organisation of such stores led to the formation of social hierarchies in the Bronze Age Aegean.

It is not only plants that can be stored. Animals can be stored in processed or unprocessed form. Many pastoralists of eastern Africa for example will keep cattle as a form of reserved food, feeding them just enough to keep them alive.

Exchange takes many different forms. As a risk strategy it is the procurement of extra resources by direct payment or the promise to pay later. This may take the form of the simple swap of a sheep for a sack of grain or an exchange of labour or craft objects for food, as is practised by nomadic groups today. It may also comprise the redistribution of communal resources or gifts given by a chief in return for the homage of the receiver. As a survival strategy, these forms of exchange rely on some form of storage and it is, therefore, less practised as a risk strategy amongst hunter-gatherers.

Of all the strategies, exchange is by far the most important to the formation of social relationships. In theory farming can be a relatively self-sufficient way of life, especially where farmers can fall back on their own stores or wild food in times of crisis. However, as people begin to use exchange to aid their subsistence they become more dependent upon each other. The French sociologist Emile Durkheim expressed these ideas more clearly, coining the term **mechanical solidarity** to refer to societies where farmers are self-sufficient with little interdependence and **organic society** to refer to those societies in which there was great deal of interdependence.

Before moving on it is useful to note that Halstead and O'Shea state, that the risk strategies that are employed eventually become incorporated into more permanent subsistence patterns rather than being kept in reserve for times of need. A good example of this is the agro-pastoralists who began trading labour, crafts and animal products for plant-foods, but who eventually became nomadic pastoralists growing no crops at all.

Once storage and exchange become established in a society changes may occur to that society. It is therefore, worth taking a deeper look at exchange strategies so that we can understand how environmental archaeology can help in examining ancient societies. It should be noted that the exchange systems reviewed are not mutually exclusive and all may operate at different levels within a single society. The historical economist Karl Polanyi classified what he termed **modes of exchange** into three groups, while we have added a commonly used fourth. They are: market exchange, redistribution, reciprocity and inheritance. The main differences between each mode are the relationships between the participating individuals.

The important aspect of **market exchange** is less the existence of the market, but rather that such a mode of exchange may be undertaken by two individuals who are unknown to each other. This is the form of exchange most Western societies use today. We go to the supermarket and exchange money for those foods that we want to eat. Exchange using money is impersonal. Through exchanging money for goods, neither party necessarily expects to form social relationships. For purchase to work using this model many of us carry around in our heads a 'list' of prices of how much particular items should cost. However, coinage that we would recognise today is a relatively late development in most societies, only appearing in the Near East in the seventh century BC. Barter is a type of exchange system that does not use money. By using known quantities of goods people can exchange items such as shells, twists of wire, beads, metal ingots or tokens made of bone or clay that stand in for known fixed quantities of goods (referred to as 'primitive money'). However, not all items may be exchanged by barter. The Ghanaian Tiv of North-West Africa for example only exchanged by barter the goods and tools associated with subsistence, so while sheep and goats could be exchanged in this way, cattle as a symbol of wealth were almost never exchanged by such means.

Redistribution is the collection of goods (or money) from the wider society by a centralised administrative body. These goods, or goods purchased with the money, are then reallocated to the wider society and in this sense modern taxation systems are redistributive. Redistribution is strongly linked to control and power, and Polanyi has noted that those who controlled systems of redistribution would use the system to increase their own status. It is, therefore, a key part of the development of a hierarchical society and is consequently one of the most significant economic factors involved in the formation of the earliest civilisations. The population paid tribute in grain and cattle to the King, scribes then recorded the giving of such tribute, its storage in state and temple granaries, its processing and eventual redistribution to the populace. Such tribute allowed the state to support the elite families, the administrators and scribes who maintained the system, a full time military body (including guards and 'policemen'), who were responsible for enforcing the system, workers who processed the food for redistribution, as well as a large body of architects and other specialist personnel. Such tribute was also used to produce art, public buildings and other works that might help legitimise the system and the authority by which it was run. As with tax, tribute is frequently a proportion of an individual's or group's property or 'income' paid annually to a central source. It is not, however, the same as tax, as in the latter precise amounts to be paid are determined. Tribute, as the name suggests, can be more a payment of homage, a 'gift' of respect. The differences are blurred and certainly the ancient Athenians both prescribed how much tribute they expected to receive and were quick to use military might to exact such 'gifts' from members of their empire if they were not freely given!

The term **reciprocal exchange** was first coined by Marcel Mauss. In his most famous work *The Gift*, Mauss argued that gift exchange entailed the formation of social relationships. In giving a gift there are three social commitments, that of giving, that of receiving and the obligation to reciprocate. Reciprocal exchange may take place between individuals or groups of people. Gifts are made by one individual or group to another in the expectation that the exchange will be returned at some time in the future. Commonly the gift is given when a group have a surplus, so that the gift will not impact on that group. In some cases gifts are given to those in power to win their favour and support. So it is that the Roman historian Tacitus tells us of how tribesmen in Germany would bestow gifts of cattle and crops upon their chiefs. The chief accepts these as a token of the honour and position in society of the individual tribesmen. MacNiocaill records a similar situation for historic Ireland where farmers paid agrarian tribute to the local elite out of honour, although they were not bound to do so.

The fourth mode of exchange concerns that in which goods, resources or the rights to resources are passed from generation to generation through **inheritance**. The existence of inheritance, especially of land, has some important implications

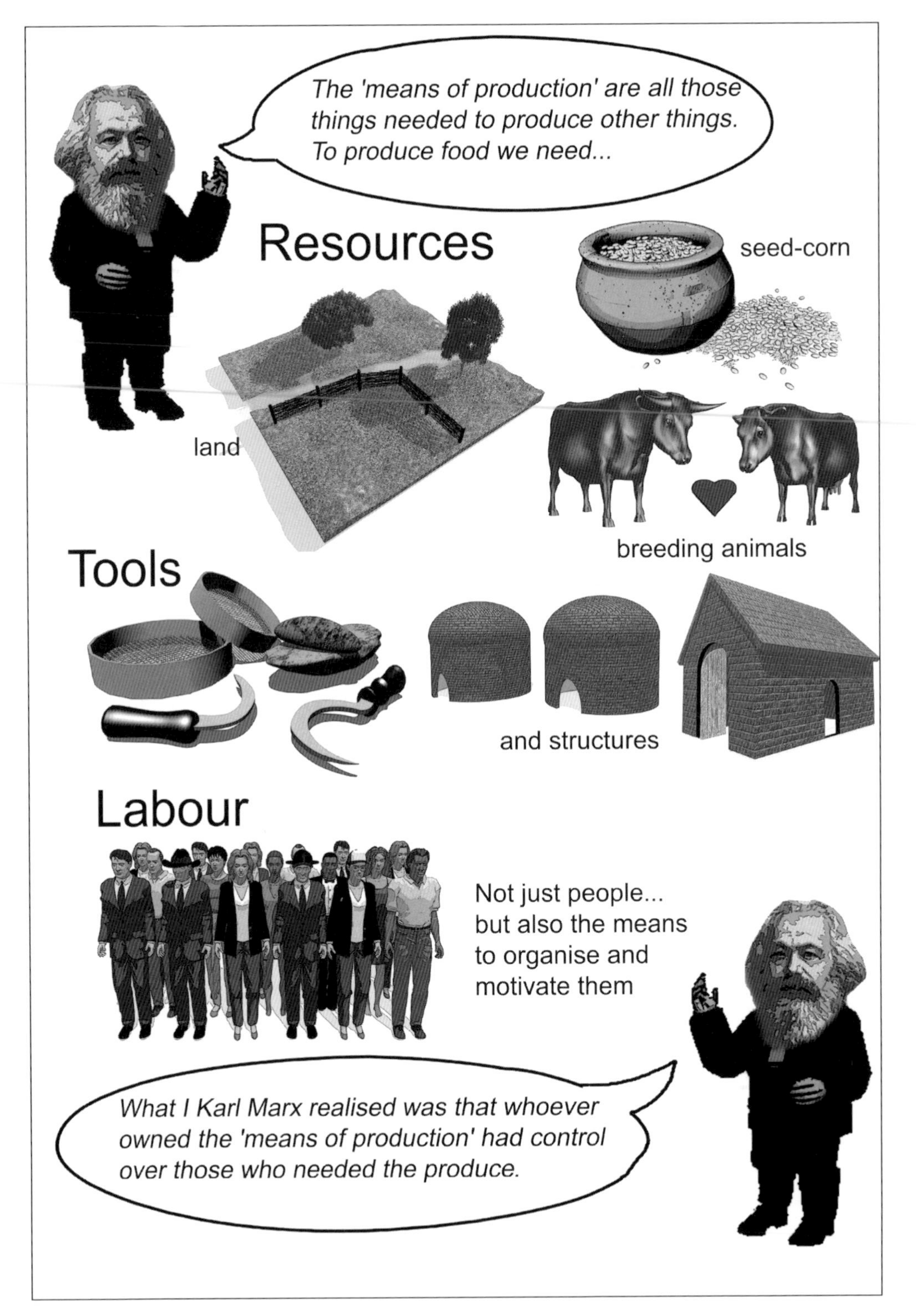
The 'means of production' are all those things needed to produce other things. To produce food we need...
Resources
seed-corn
land
breeding animals
Tools
and structures
Labour
Not just people... but also the means to organise and motivate them
What I Karl Marx realised was that whoever owned the 'means of production' had control over those who needed the produce.

for the control over the means of production (**53**). Where land is subdivided amongst a number of sons and/or daughters, wealth will also be subdivided. The alternative practised by the elite of feudal England into more recent times is to pass on the entire estate to the eldest son. This way the wealth of the estate is not subdivided and the power of the family remains undiminished.

Intensification of agriculture

We have looked at how the development of exchange systems was one of the main factors in the evolution of complex societies. However, in order to carry out exchange a surplus must be generated and for that to occur agriculture must be moved beyond subsistence level. In this part of the text, therefore, we discuss models for the intensification of agriculture; a subject that has been a constant source of debate over the last three decades of environmental archaeology.

Intensification is defined as increasing the yield from a fixed unit of land over a fixed period of time. When dealing with crops we are talking about the grain yield, with animals, increasing their numbers, or the net produce from these animals, in other words more meat, milk or skins. There is another type of intensification that often accompanies increased agriculture production, especially in those cultures which lack machinery. **Intensification of labour** is the increase in production achieved through expenditure of additional labour time and/or energy on a fixed unit of land. Increased labour may be used to protect and weed crops, resulting in a higher yield, or during the processing of a crop, spending extra time to make sure every last grain is obtained from the harvested product. Intensification is recognised archaeologically as the common reaction to situations in which populations increase, agricultural land is limited, where settlements become nucleated, or a combination of each. Where population increases, but where land is unlimited and settlement dispersed we might expect to see extensification instead. **Extensification** is the increase of yields through the expansion of agricultural land.

The terms intensification and extensification should be used with caution as it is often difficult from archaeological data – in particular biological proxies – to tell one from the other. Further, archaeologists also associate **intensive agricultural methods** with increased yields or production. Intensification and use of 'intense' methods are, however, not synonymous. Use of intensive methods does not always mean that production is increased. Where land is becoming both exhausted and in short supply intensive methods might be used to *maintain* yields. While the yield of the land has technically increased from what it would have been otherwise, there would be no overall increase in production.

A common result of intensification/extensification is the generation of surpluses. **Surplus** is defined as production beyond the basic requirements needed to feed the farming population. Surpluses are usually intentional, but may also occur accidentally when, for example, growing conditions are better than expected. Surplus production for obligatory taxes is likely to be calculated for by the farmer when planting. However, farmers are likely to sow for a greater yield than they need to anticipate bad harvests (another risk strategy). They may not even harvest such surpluses if the extra produce is not considered useful as payment, gift or to replenish surplus stocks.

The importance of surplus production depends on who controls it. Surplus production in small, largely self-sufficient individual households may allow for small-scale exchange, but is shown by many ethnographic studies to be relatively insignificant. For surpluses that are controlled by elites it is very different. Such surpluses allow for the payment of labour to farm more land, employment of craftsmen, purchase of prestige goods, hiring of men for military campaigns, building and repair of monuments and other structures, or the throwing of feasts. All of these may improve social standing.

Archaeologists of the mid- and late twentieth century have tended to assume that people in the past strove towards efficiency and intensifying production. Such attitudes are in part a result of approaching the past from modern perspectives. Modern Western society puts great value on making more (ideally for less effort), because these actions increase financial wealth, the principal ideological aim. Past peoples may not have had these attitudes or indeed incentives. In our society, if farmers produce more cereals or animals then they may sell them and increase their profit. However, consider farmers in a non-market economy who have enough land to farm and feed their own family. What benefit is it for them to intensify production?

Secondary products revolution

Gordon Childe's agricultural 'revolution' saw the development of agriculture as a response to population increases around various oases in the late Weichselian Near East as a result of increasingly arid climates. Childe argued that farming was an adaptation which allowed higher population densities to be supported in the vicinity of these features than through hunter-gatherer subsistence strategies. The successful adaptation was then exported from the Near East to Europe and Africa. However, early agriculture is thought to have been based entirely on grain and meat from domesticated animals. Other workers, however, also saw a second 'revolution' later in the history of agriculture. Andrew Sherratt argued that while animals were initially domesticated as a supply of meat, their use for milk, cheese, wool and/or for traction was a development of equal significance. Sherratt referred to this second development as the **secondary products revolution**. He believed this occurred at around 4500 to 4000 BC, much later than the first domestications in the Near East. Some of the evidence came from tools and slaughter patterns, but the prime mover was Mesopotamian art which depicted activities such as the production of milk and cheese. As with Childe's model for the original agricultural revolution, population growth was seen as the prime mover for the secondary products revolution. As such it can be seen as part of the intensification process. Animals were used in more than a single way and required a higher labour input to tend. Sherratt distinguished two stages of agricultural development that the secondary products revolution separated. The first saw the use of hoes and human power to transport the crop from the field. The second saw the use of animals for pulling ards and transporting goods.

The exact nature of the secondary products revolution has yet to be established, indeed whether 'revolution' is the right word is debatable. Evidence from Europe suggests that while the use of milk may have been a relatively early development, along with the use of animals to pull ards, both dated to some 5000 to 4000 BC, (**54**) that of wool might have not occurred until some time later.

Cultural implications of intensification

To change farming methods people need incentive, knowledge, instruction and a system that can support them should things go wrong. These aspects are not commonly considered in studies of agricultural change. This is at least in part because of the long time periods with which we deal in archaeology. There is a temptation to think that if something is better it will eventually win through. However, human history is littered with periods of stasis. For example, Acheulian handaxes were used for over 1.5 million years with little apparent change. Many will also remember a far more recent example where superior Betamax video technology was abandoned for the inferior VHS. Clearly here social pressures, albeit in the form of greater publicity and marketing for VHS, won out over increased 'efficiency' (**55**). So, as Alastair Whittle has argued, when we do see changes in subsistence activities we should ask 'What has persuaded these people to give up the certainty of their traditional practices in favour of a new but less certain way of life?' Until recent times then, change is the exception rather than the rule.

Biological evidence for ancient economies

We have divided this evidence up by the main specialist fields, two of which we looked at in sections 1 and 2. For each, we outline the mechanism by which material is preserved, and the sort of contexts (pits, hearths, postholes, buildings, ditches etc.) and environments in which it is preserved. Lastly, we discuss how the remains are recovered and analysed and which aspects of past economies might be investigated from each of them. The most commonly used proxies are charred plant macrofossils and vertebrate bone, and it is to these that we devote the most attention.

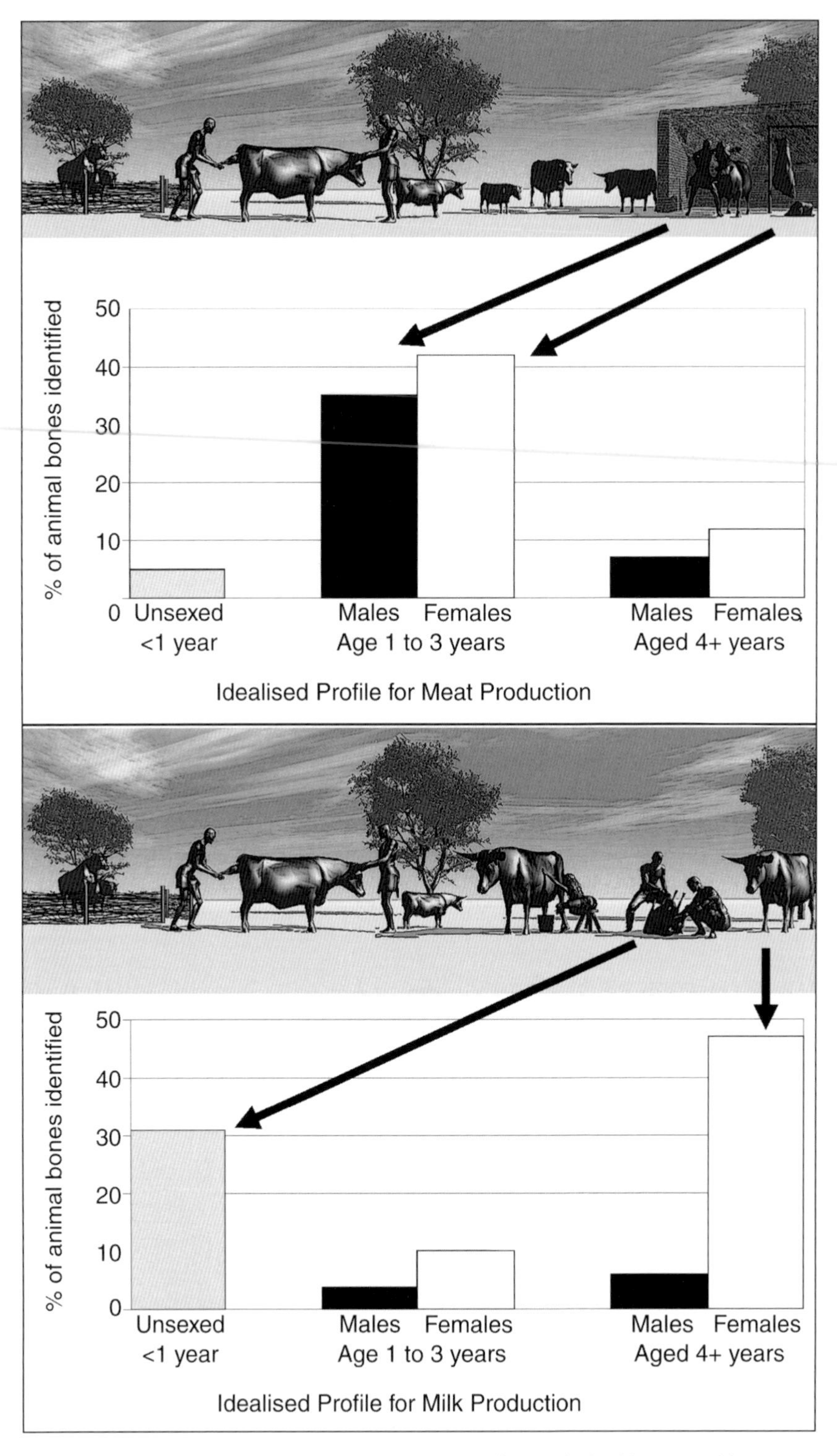

54 *How the transition to using animals for milk might affect an idealised bone assemblage*

55 *Despite numerous demonstrations and extensive publicity with regard to the improved yields it could bring, Jethro Tull's seed drill took over 100 years to become a common farming tool*

Archaeobotany

As we have seen in previous sections, plant remains do not readily survive on archaeological sites. For them to survive natural processes of decomposition must have been halted or arrested. Such a situation occurs where plant material is altered to a state or form in which it can no longer be broken down, where normal processes of decomposition cannot operate, or where the plant material in question is composed of material that is highly resistant to breakdown.

Carbonisation and charring

PRESERVATION

Charring is the most common of all archaeobotanical preservation types. It preserves plant material after it has been subject to burning, transforming the outer layer of the plant part (or in the case of carbonisation, all of it) from a carbon-based compound to almost pure carbon. Charring can preserve any type or part of the plant, but in reality fragile parts, such as leaves, flowers and lighter chaff are rarely found charred. Preservation by charring depends on the conditions of charring; the amount of oxygen in the fire, the temperature and the length of time the material is exposed. Good preservation occurs where temperatures are relatively low, where there is little oxygen and the fire smoulders for a long time. Charred material is very fragile and therefore, context is also very important for its survival in the archaeological record. The longer charred remains are exposed on the ground surface the greater is the chance that they will be destroyed by trampling, rain drops and rapid drying. When charred material is buried it will be vulnerable if it lies close to the active soil horizon for the reasons discussed in section 1. It is notable in this regard that in pits more fragile material is often under-represented in samples taken from the upper fills.

Potentially charred material can be recovered from any archaeological site where fires have burnt or to which fire waste has been brought. For this reason charred remains are noticeably more abundant on settlement sites than in field systems, ritual monuments and non-cremation burials. On settlements charred material can be recovered from almost any type of feature. The best-preserved material occurs where burnt material has been left *in situ* and then deeply buried, for example, in storage pits. Other good charring conditions occur on tell sites where ovens or house roofs have collapsed, smothering and slowly charring the material underneath. The collapse of mud brick walls or the levelling of such debris can then bury charred evidence protecting it from further damage.

Perhaps the most common type of charred plant remain is wood charcoal. As wood is frequently used to fuel fires this is no great surprise. Wood charcoal may easily fragment both in the fire and later in the soil, and so even though it is the most frequently recovered it may still be under-represented. Besides charcoal the most common charred remains are those resulting from the processing of food plants. In the Old World these comprise cereal grains, **chaff** (a collective term for straw and other cereal waste) and the weed seeds that were accidentally harvested. Sometimes burnt remains of nutshells, grains of pulse crops and parenchyma also survive (**56**).

56 *Fragments of charred hazelnut from a Neolithic site, near Salisbury in Wiltshire*

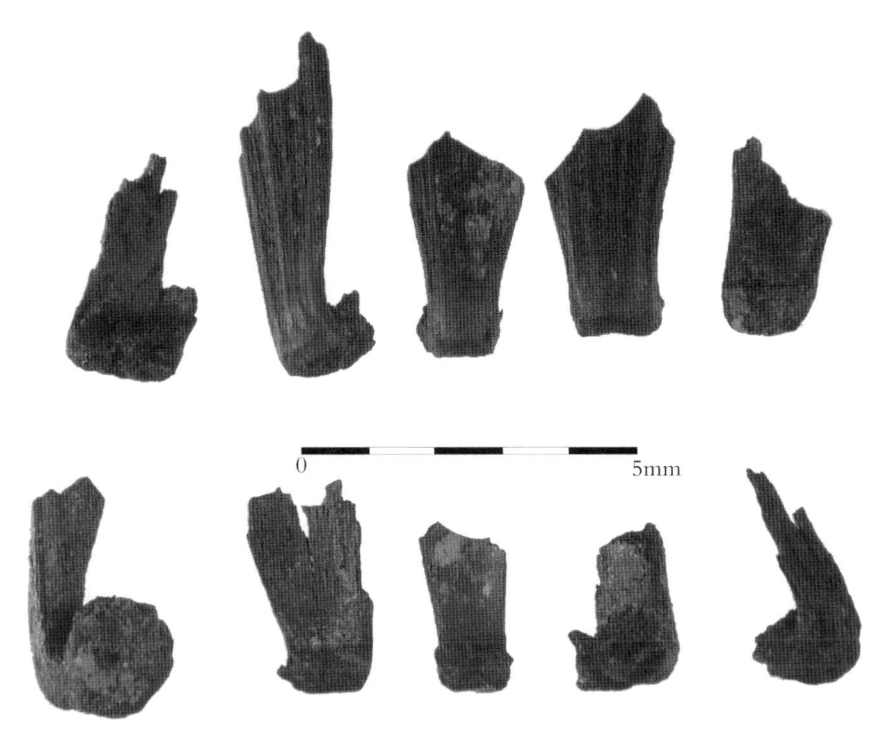

57 *Glume bases of charred spelt wheat* (Triticum spelta) *from Roman Ilchester*

Experiments have shown that certain plant parts survive charring, site formation and post-depositional processes much better than others. Those conducted by Sheila Boardman and Glynis Jones suggest that cereal grains have a noticeably better chance of survival than chaff. Seeds that are particularly dense or woody also tend to be preferentially preserved. While chaff is poorly preserved in general, **rachis** (i.e. the top of the stem on which the cereal ear develops) fragments of free-threshing cereals (barley and 'modern' wheats) and **glumes** (part of the spikelet holding each grain within the ear) of ancient wheats such as emmer (*Triticum dicoccum*), spelt (*Triticum spelta*) and einkorn (*Triticum monococcum*) also survive well (**57** – see also **62** for an explanation of cereal plant parts).

Richard Hubbard and Alan Clapham recognise three types, or classes of charred assemblages according to the nature of their deposition (**Table 14**). Identifying which type is present on a site greatly helps in interpreting charred plant remains once they have been extracted, identified and quantified.

Class C material forms the majority of the archaeological record and for this reason we will consider the taphonomy of such material in detail. A large proportion of the charred material that archaeobotanists study comes from hearths and fires. These features preserve plant parts only where they trickle down through the fire to become buried in the ashes. It is worth emphasising

Class A	Plant material that has burnt in the same location from which it was recovered. Assiros Toumba (Greece – see text below) is a good example but more commonly this category would encompass storage pits that catch fire and hearths or kilns, where the feature and its spent fuel survive in the archaeological record. In these cases the relationship between the context and the charred plant remains is very strong.
Class B	Plant material that is charred during a single burning event, but where the burnt material has been redeposited, either deliberately or accidentally. The act of redeposition means that the material may not specifically relate to the context in which it was found.
Class C	Material that comes from a number of temporally and spatially distinct burning events and activities. The material from these events is mixed together with settlement waste to become incorporated in archaeological features such as middens, rubbish pits etc. The remains have little if any relationship to the context from which they are recovered.

Table 14 *Charred plant remain classes according to Hubbard and Clapham (1992)*

then, that charring occurs only by fire and therefore the first and foremost activity charred plant assemblages from such features indicate is burning! From a spatial perspective the main human activity that charred material relates to is the use of hearths, bonfires and ovens. If the remains of the fire have been cleaned or otherwise moved, then the distribution of charred remains may reflect nothing more than the location of where such sweepings have been dumped. This may have been directly into open features such as pits or ditches or to a surface midden. It is worth pointing out that people of the past rarely lived amongst their rubbish, so both middens and rubbish disposal pits are commonly located outside the settlement periphery. Sweepings from the hearth may occasionally be found in field ditches if midden material containing charred remains was mixed with animal waste and taken to the fields.

RECOVERY AND ANALYSIS

Charred remains are recovered from bulk samples that consist of anything from 10 to 200 litres of sediment. Several organisations have recommended standard sample sizes for charred plant remains, but we suggest a more flexible approach where the size of the sample reflects the density of charred material and the size of the deposit from which it is taken. Where density is low larger samples should be taken, where deposits are large and rich a number of smaller samples through the deposit may be more appropriate. The absolute minimum sample size should be 10 litres where very high densities of charred macrofossils survive, but otherwise at least 30 litres of sediment should be taken. Layers and fills may both be sampled depending upon the questions the sample is intended to answer. The latter consideration is very important, as samples should not be taken where their purpose is not understood, or even known. Mixed or otherwise disturbed contexts should be avoided as the origin of the biological

King Alfred's College
Winchester

DEPTARTMENT OF ARCHAEOLOGY ENVIRONMENTAL RECORDING SHEET

Site Code: YAR 03	Trench: 3	Grid Ref.: 195/255	Sample No: 10	Context No: 354

SAMPLE PROPERTIES
1. Type of context sampled
2. Brief description of context properties
3. Container used for sampling
4. Type of sample; bulk or column
5. Sample volume
6. Approximate percentage of context represented in sample
7. Other comments

COMPLETE ALL CATEGORIES 1-7

1. Fill
2. 10 YR 3/2 Dark greyish brown silt containing frequent granular clasts of slate. Poorly sorted
3. 10 litre box (x3)
4. Bulk
5. 30 litres
6. c. 25%
7. Sample was cut from the vertical section face of the half sectioned pit and was accompanied by two other samples taken in the same way (nos. 8 and 9)

Plan No: 15	Section No: 10	Length: 0.75m	Wdth / Dia: N/A	Hgth / Dpth: 0.3m

Stratigraphic Matrix

353

This context is: 354

355

Reason for taking, and archaeological questions asked of sample: Do biological remains survive in context (354)? Bone was noted during sampling.

Pit [355] appears on the basis of its morphology and its artefactual inclusions to be of Roman date and dug for the purposes of rubbish disposal. Do the included plant remains confirm or reject this hypothesis?

What do the plant and animal remains recovered from the sample indicate of the economy of the site? Artefactual evidence from this phase includes a sickle. Does the pit indicate primary production of crops?
What do the biological remains suggest regarding the status of the sites inhabitants? Artefact finds from the pit include Samian pottery

Sketch of sample location and relationships:

78
299 - limit of bioturbation
8
352
9
353
389
355
354
10
c. 0.5m
North facing section of cut 355

Provisional date of sampled context: Roman

Other details

Sampled by: J. Smith

Sampling date: 6/8/03

Checked by: A.N. Other

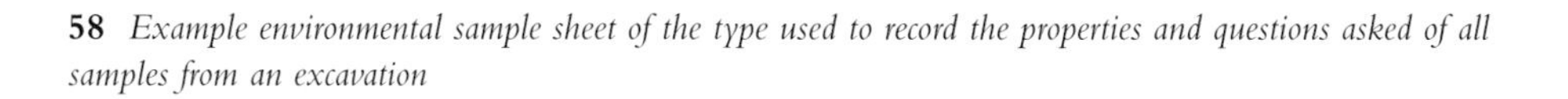

58 *Example environmental sample sheet of the type used to record the properties and questions asked of all samples from an excavation*

remains in them is unknown. For each sample a record sheet should be filled out by the person taking the sample (**58**). This will guide both the person processing the sample and the archaeobotanist at later stages of examination.

Thought should be given to spatial sampling to look at variation across a site. For example, the excavation of a settlement mound at Assiros Toumba, Greece revealed a house that had been burnt to the ground in the early Bronze Age. During this destruction, an upper floor and a roof both collapsed onto the fire creating ideal charring conditions. The botanical study of the resultant remains, carried out by Glynis Jones, is one of the best known archaeobotanical studies for the archaeological information it produced. Vessels and sacks containing crops were stored in an upper room and recovered from where they fell in the lower room. The sampling of the deposits had to take into account the possible spatial distribution of the material, so that the location of each stored crop, and even its approximate position in the room above, could be reconstructed.

Charred remains are extracted from the sedimentary matrix by a procedure called flotation (**59**). Samples are placed on a 1mm or 500μm mesh suspended within a container (usually an oil drum-sized tank) of water. The sediment rests on the mesh while the charred remains, being less dense than water, float to the surface. Water is then introduced from beneath the sample through fine jets. This has the effect of increasing the water level in the tank, and at the same time encouraging denser charred particles to float. As the water level rises further it

59 *A flotation machine in use. The machine uses an electric pump to recycle water from the second settling tank*

60 *A flot recovered from midden deposits beside a Roman villa in Somerset, England. The flot contains wood charcoal, snail shells and small mammal bones*

passes over a lip in the tank, carrying with it the charred remains, which are then caught in a single or series of sieves placed beneath the lip. The finest of these sieves has a mesh size of 250-300µm. The charred material (often accompanied by modern plant roots) caught in the sieve is termed the flot, while the mineral material contained within the mesh inside the container is called the **residue**. As well as stones the residue may contain larger animal bones and artefacts.

The flot is air dried and is then examined under a low-power stereo-microscope. Wood charcoal may be picked out for further identification using high-power microscopy, while all the grains, seeds, bits of chaff and anything else are removed for identification and quantification. The flot may also contain snail shells, fish scales and small animal bones (**60**). The flot record of the former is biased towards whole shells (which retain air, and therefore float) and species whose shell is not dense. Their collective ecology is difficult to interpret and, therefore, studying snail shells from flots should be avoided. Any small animal bones recovered on the other hand are of greater use and can be examined as discussed in both section 2 and later in this section. The residue is also dried and is then sorted to remove identifiable biological remains, including bones, shells and plant remains, together with artefacts. They are then identified and quantified as described elsewhere in this section. Despite the widespread use of the flotation machine in archaeology, recent experiments conducted by Dominique de Moulins have revealed that in some models less than 50 per cent of the charred plant material in a sample is retrieved in a flot. It is, therefore, extremely important that the same attention is given to sorting the residue as it is to sorting the flot.

In particularly rich deposits it may not be necessary to use flotation and sometimes dried samples may be placed directly under the microscope. For such richer deposits we may be interested in the exact spatial and stratigraphy relationships of the charred material. In these cases blocks of sediment that can be micro-excavated in the laboratory are sometimes preferable to bulk samples. For example, samples from a site at Ham Hill in Somerset, investigated by Joy Ede revealed grain arranged in perfect ears, even though the glumes had not been preserved. Micro-excavation of storage pits at Danebury, Hampshire by Martin Jones and Wandlebury, Cambridgeshire by Rachael Ballantyne, revealed that glumes and grain still lay where they were burnt on the base of the pit in the form of whole spikelets, but separated from the ears (**61**).

Charred plant material is most commonly identified using gross morphology and diagnostic internal and external features. These may include the texture of seeds, and the presence and location of points where the seed attached to the plant. The identified grains, seeds, chaff, nut fragments, tubers, are then quantified for each sample that is taken. The resultant data are tabulated on a sample-

61 *Whole spikelets from Wandlebury Hillfort, Cambridgeshire. The outer two are of Emmer wheat* (Triticum dicoccum) *and the central example is spelt wheat* (Triticum spelta)

by-sample basis with individual counts for each identified component (**Table 15**). Occasionally presence or absence, or approximate abundances of identified species may be used where a rapid scan of the flot is being carried out.

The presence of individual species may be informative, but it is also useful to compare sample data spatially and temporally both across a site, and between sites by examining the proportions of key components. For Old World crops this is commonly the types of cereal grains and chaff and the categories of weed seeds that are found (**62**). Comparison may be made using simple abundances, percentages or by using more complex statistical techniques. In doing this it is important that samples are examined by class. Samples of Class A and B can be readily compared in this way, but more of the variation in Class C samples is likely to be the result of taphonomic processes. For example, samples from shallow features such as gullies or postholes are biased towards grain when compared to those from the bases of pits, where more fragile material is well preserved.

As any single plant may produce several thousand seeds the composition of any single sample has the potential to be highly variable. However, despite this potential problem charred plant assemblages often show a remarkable degree of consistency for any single site and period, even for Class C material. That this is the case is probably a reflection of a much higher degree of mixing from different activity recognised by archaeologists and archaeobotanists alike.

The identification and analysis of roots and tubers (e.g. yams, potatoes, turnips) is carried out using very different techniques than those described for seeds. Fragments of the root or tuber must first be recognised within the sample by virtue of having a non-vascular structure. The term **vascular** refers to the bundles of stems that characterise wood charcoal and gives rise to the striated appearance of such remains. As we saw in section 2, non-vascular material is referred to by the term parenchyma, literally meaning that it is soft plant tissue. Identification of such material involves the use of high-powered scanning electron microscopy (SEM). The structure of the charred material is then compared to that of modern specimens of known taxonomy.

ECONOMIC INFORMATION FROM CHARRED PLANT REMAINS

Given their relationship to fire, charred plant remains at the most basic level only provide information on what was used as fuel. It also identifies those aspects of food processing that produce large amounts of durable waste that is unsuitable for other purposes (such as animal fodder). On most Old World sites this waste is from cereal agriculture, although crops such as legumes are also relatively common finds in samples from archaeological sites. Vegetable crops are rare in the archaeobotanical record, and along with fruits and nuts will only occur where the ends of roots, fruit stones or nutshells are thrown into the fire. That such remains represent the processing of food for human consumption is probably a reasonable assumption, although still a point of conjecture. How

Context	**1237**	**1521**	**109**
Feature	Possible Granary	Well	Ditch
Volume (litres)	12	12	6
Species			
Rumex sp.	-	1	-
Rumex cf. *crispus*	1	-	-
Vicia/Lathyrus sp.	6	16	1
Vicia sativa	-	2 cf.	-
Vicia sp.	-	3 cf.	-
Lathyrus sp.	1	-	-
Medicago lupulina	2	-	-
Medicago/Trifolium sp.	1	1	-
Lithospermum arvense	5	11	-
Tripleurospermum inodorum	-	1	-
Lolium/Festuca sp.	1	-	-
Poaceae indet.	4	3	-
Poa sp.	1	-	-
Avena sp.	-	-	3
Seed indet (<2.5 mm)	2	-	-
Cereals			
Hordeum sp. (hulled)	-	-	4
Hordeum sp. (undiff.)	-	2	-
Triticum undiff. (grains)	15	1	9
T. dicoccum/spelta (spikelet	2	1	-
T. dicoccum/spelta (grains)	5	1	-
T. dicoccum/spelta (glume)	31	2	1
T. spelta (glume bases)	1	3	-
T. aestivocompactum	-	-	9
Cereals undiff. (grains)	24	14	14
Cereal undiff. (rhachis)	-	1	-
Cereals undiff. (culm nodes)	2	-	-
Cereals undiff. (culm	-	1	-
Root/Tuber (indet)	1	-	-

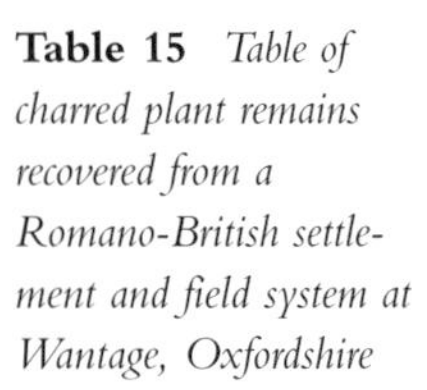

Table 15 *Table of charred plant remains recovered from a Romano-British settlement and field system at Wantage, Oxfordshire*

such 'foods' were cooked and prepared for human consumption, whether as bread, cakes, doughballs, gruel, or beer and so on, is almost never known for certain, although occasional charred examples of these products do occur. Querns are a more common find and where found they indicate that cereals were processed to produce flour.

This may appear to be an overly pessimistic view of the contribution made by charred plant remains. Indeed, rather than providing definitive evidence of diet and how food was eaten, charred plant remains can provide much more useful information regarding cultivation and processing techniques. Much of the information for the former comes from weed seeds; whereas the latter may be deduced from both weed seeds and cereal chaff. Unfortunately, at each stage, processing removes more and more weed seeds, so the cleaner the grain when charred, the less we can say of cultivation practices. Weed species may be characteristic of environmental factors such as soil types, manuring or culti-

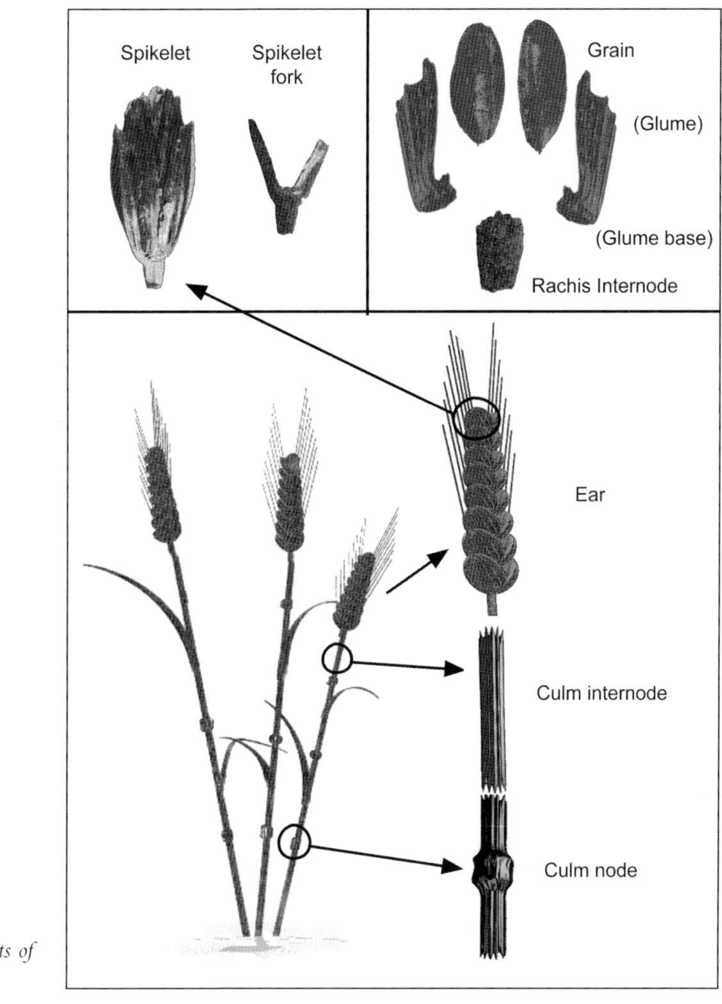

62 *The component parts of hulled wheats*

vation practices such as tillage, sowing times and weeding. Charred assemblages can further reveal how crops were procured (harvested), processed and stored. Such information might also reveal something of the social organisation of these activities and where they were conducted. Given what we have said about there often being only one hearth (and so only one major source of charred material) within each household, we should not expect to see any great spatially distinct evidence for production activities besides those relating to the hearth and the deposition of its waste. We will return to look in more detail to what charred plant remains tell us about ancient production towards the end of this section.

Mineralisation

Mineralisation occurs when minerals carried in solution are deposited around plant cell surfaces or within inner voids, effectively encasing the plant structure.

Mineral replacement may also occur, when inorganic compounds take the place of decaying organic structures. The most common form of mineralisation on archaeological sites is as a result of calcium phosphate precipitated from cess. Therefore, mineralised remains are commonly found in cesspits, latrines, sewerage systems and other less than salubrious conditions (**63**). However, mineralisation by other materials such as the metals bronze and iron, by glauconite (a type of mica) and by gypsum have also been recorded. In spite of the range of minerals that can preserve plant material in this way, mineralisation is not a common form of preservation as it requires a combination of a mineral supply, the presence of water, anaerobic (oxygen-free) and relatively warm conditions. Plant remains are differentially preserved by mineralisation, and indeed some remains are not preserved at all. While mineralisation can in theory preserve most types of plant material, the remains that are found in reality are limited by three factors. Firstly, to those edible species whose seeds and pips are often ingested and so associated with cess. Secondly, seeds with hard or woody seed coats – especially those containing silica or calcium – have a relatively high chance of preservation compared to those with no or little seed coat. Lastly, mineralised seeds, like charred seeds, are selectively destroyed depending upon how robust they are. Consequently, the most common types of seeds preserved by mineralisation are fruits with hard or woody seeds that are readily swallowed, such as those of figs, grapes and blackberries/raspberries. Although other material, such as seeds of the Apiaceae (including fennel and coriander) and the Rosaceae (plum stones and apple pips), also survive by mineralisation, cereal grains are rarely preserved by this mechanism.

63 *Mineralised seeds and fish bone from sixteenth-century London. Plum stone* (Prunus domestica), *fig* (Ficus caria) *and cornflower* (Centaurea cyanus)

An important aspect of mineralisation is that it preserves species that would otherwise be poorly represented in the archaeological record. The fact that some mineralised remains have passed through the human gut, especially those within coprolites, also provides direct information on plants that were actually eaten. Mineralised seeds also have a more distant relationship to waste from crop processing and therefore they reveal less information on how foods were produced. However, given the unusual nature of many of the foods preserved, they may reveal information that relates to the social status of the individuals inhabiting the site.

Mineralised material is best recovered by wet-sieving. However, as such material occurs often alongside charred material it is often extracted with a flotation machine. Therefore, if it is suspected that mineral plant remains are present a 250μm mesh should be used inside the flotation tank to collect the residue as mineralised plant remains do not float.

The analysis of such material is undertaken in a similar manner to that of charred material, with a greater emphasis on recording presence rather than comparing the proportions of different taxa through time. High-power microscopy techniques are on occasion employed to identify plant material that has been digested and/or destroyed, by comparing the structure of the tissues where they are preserved to those from modern specimens.

Plant impressions

Plant impressions occur where plant material that either formed part of an object, or lay against it, has been destroyed but its shape has been retained as a pseudomorph. Such impressions are most frequent in clay used for pottery, daub or mud brick. Ghost impressions of large wooden objects may also survive where objects have been buried under heavy loads of sediment, such as those seen in chariot burials of northern England or the case of the bull harp's sounding box from the excavations of the Royal Cemetery at Ur in Mesopotamia.

Many specialists find it useful to make a casting of the plant impression, although this is not always necessary. For pottery, castings are often made with a rubber latex solution that can easily be pulled away from the impression. For impressions in sediments, plaster-of-Paris may be preferred, with the sediment being chipped or dug away from the cast. As we only have the impression, identification to a high taxonomic level must rely on gross morphology alone. The finer grained the matrix in which the impression is found, the better the resolution of impression. Therefore, because potter's clay is mostly less than 2μm, identification of impressions from ceramics may be to a species level. Such high levels of identification are rarely possible for impressions in mud brick and daub.

It is almost impossible to study evidence from impressions in any quantitative way and the knowledge of a particular plant's mere presence is all that can be achieved. However, on some sites, the impressions may be the only form of botanical data that is recovered and in such circumstances these data take on a greater significance. For the majority of sites, however, the macrofossil record is a more reliable proxy for plant exploitation.

Waterlogging

We have dealt with waterlogged plant remains in detail in section 2. For this reason we only consider their use in examining past economies here. Waterlogged plant remains are recovered from archaeological features that contained water when they were dug, such as wells, cess-pits, sewers, moats and ditches, although in some cases quick inundation may also preserve material and therefore originally dry contexts may contain waterlogged plant remains too. Water then, must have been present almost continuously since deposition for the plant remains to survive.

Waterlogging has the potential to preserve a much greater range of material than where plant material is preserved by charring or mineralisation. Like charring, woody material is preferentially preserved, but unlike charring it is not always dense plant material that survives. It is significant that waterlogging rarely leads to the preservation of cereal grains and other grass seeds. This is partly because such remains are composed of softer tissues, but the grains are also selectively eaten by organisms living within or close to the water. On the other hand cereal chaff frequently survives, as do more woody seeds, nuts and fruit stones (particularly in urban environments), as well as twigs, bark and woody thorns and buds. Worked wood may also survive in such conditions. Needless to say, away from urban environments plants that grow in or close to water are often well represented, meaning that we often have an array of pondweeds, sedges and rushes in our samples.

Despite almost blanket preservation, waterlogged remains offer relatively limited economic data if not recovered from settlement sites (**64**). Within settlements waterlogged plant remains are often of taxa that are rarely recovered in the charred record, such as fruits, berries, pips, as well as worked wood relating to structures, bowls or other objects. Waterlogged plant remains from the Swiss lake villages for example revealed not only evidence for the consumption of cereals, wild fruits and use of timber, but also a whole door, indicating the exact method of construction. Waterlogged plant remains have also been found in the guts of bodies found in north-west European peat bogs such as Lindow Man from Cheshire and Tollund Man from Jutland, Denmark. Such plant remains reveal the last meals taken by the individual before death, which, as we review in section 4, seems to have been a weed infested gruel.

Desiccation

Plant remains preserved by desiccation are common finds in those countries bordering the Sahara desert, and the deserts of western North and South America. Desiccation occurs where moisture levels are too low for the

64 *Cones of alder* (Alnus glutinosa), *and a seed of water-lily* (Nuphar lymphae). *Bronze Age* c.*1600 BC, remains recovered from riverine deposits in south-west England*

organisms responsible for organic decomposition to survive. In arid environments it is not only rainfall and ground moisture that are absent or insignificant, but also the soil itself. Animals that consume plant material are not completely absent in such arid environments, as insects such as termites, ants, weevils, beetles and mammals such as gerbils may all be present. However, many of these creatures only consume fresh plant material, so once buried, plant remains will have a much-increased chance of preservation. Particularly well-preserved desiccated plant remains have been found as stomach contents of mummified burials in the deserts of Chile, in coprolites in caves in Utah and Nevada, and as general scatters in Pharonic period settlements in Egypt. Desiccation does not result in total preservation. As with waterlogging, plant material composed of soft edible tissues, such as cereal grains, legumes and grass seeds are selectively eaten by insects. So in a reverse of the situation discussed for charred remains, grains may not be preserved, while chaff may be present in huge quantities. Desiccation can also preserve remains not otherwise found in the archaeobotanical record including whole fruits, flowers, leaves and vegetables, for example onion skins.

The biggest threats to desiccated material are changes that introduce water. In Egypt, the construction of the Aswan dam, and the increased use of irrigation away from the Nile floodplain has been particularly problematic.

For mummified burials, small samples from the stomach or clothing may be taken for the specific task of identification of food remains and materials used for equipment and clothing. As few plant species grow in regions where desiccated plant remains are found the material that is recovered from settlement contexts is predominantly economic in nature. In these circumstances samples are taken from secure datable contexts that are thought to have been quickly infilled or buried. As with charred material, where deposits are rich in plant remains, it is preferable to take small spot samples, especially where they may enable the spatial distribution of activities to be determined. Such samples may be put straight under the microscope for identification without processing. Those containing some sediment may first be sieved. Where deposits are sampled that have low concentrations of desiccated material, a mixture of dry sieving and flotation strategies may be preferable. Firstly, coarse fractions are dry-sieved to 2-4mm to recover plant parts that are still articulated, for example leaf fragments, and other material that may be broken up by water. The remainder of the deposit is then put through the flotation machine and desiccatted plant remains will float off in the same way as charred remains.

One note of caution about the examination of plant remains from sites subject to desiccation needs to be sounded. As all plant remains can be preserved they may be easily reworked into later deposits. For sites where we are dealing with short occupation phases this may not be an issue, but if we are looking at material from a site used for hundreds or thousands of years this type of contamination can be a serious problem. Fortunately, in most of these cases such modern contamination can be easily seen as the material has a much 'newer' appearance.

Desiccated plant remains are identified using a mixture of low- and high-power microscopy as previously discussed for other categories of plant remain. A detailed case study of the use of desiccated plant remains to address archaeological questions relating to the Workmen's Village at Amarna, Egypt is outlined in section 6.

Freezing

We are very familiar with freezing as a means of preservation of plant remains as the vast majority of us will possess a freezer in which we place vegetables and other plant foods for long term storage. However, permanent freezing is also a feature of some parts of the northern latitudes and some of the world's highest mountain ranges, and, in these, well-preserved plant remains are relatively common finds. They are, however, much less common on archaeological sites, a fact which is perhaps of little surprise given that such environments are marginal for humans. In frozen environments low temperatures render most decomposing organisms (especially those which are not mammals) inactive.

At cold environment archaeological sites plant material most commonly occurs in burials or other sites of death. We have the bodies and grave goods of the Pazyryk nomads in southern Siberia and Ötzi the iceman, a unique occurance of a man who died in the Italian/Austrian Alps near the Similaun glacier. As with other examples of burials, the information from frozen bodies is often personal and concerns the last meal of the deceased, the use of plant materials for clothing and artefacts. Ötzi is also unique in this respect as he was not buried, but rather died whilst undertaking normal daily activities. This means he can be taken as representative of how Neolithic people would have dressed and the objects they would have used. When we look at the artefacts recovered from Ötzi, it is salutary to think just how few of them are recovered in more conventional archaeological settings. These included not just clothes, but tools, containers, pouches, fire tinder and a backpack.

There is no set procedure for the sampling of frozen botanical remains and a case by case approach should be adopted. Plant remains can be recovered from such samples by 'de-frosting' and are then identified as outlined for waterlogged plant remains in section 2.

Phytoliths

The word phytolith is a literal derivation of the ancient Greek for *plant-stone*. Phytoliths are silica 'skeletons' that form in the cells of plant stems, leaves and other tissues as a result of uptake of this mineral from groundwater. Phytoliths are an important part of many plants, but in particular of the grasses, giving them stability, as well as sharp edges (as in the sedges) or 'stinging' hairs (stingy nettles) which can ward off browsers (**65**).

When plants die and begin to decompose, phytoliths remain, as silica is an extremely durable element. Phytoliths will survive in almost any environment which is not subject to mechanical decomposition processes, and which is not strongly calcareous. Also, phytoliths only form where silica is carried in groundwater in high concentrations, which fortunately for the archaeobotanists is in most geological situations.

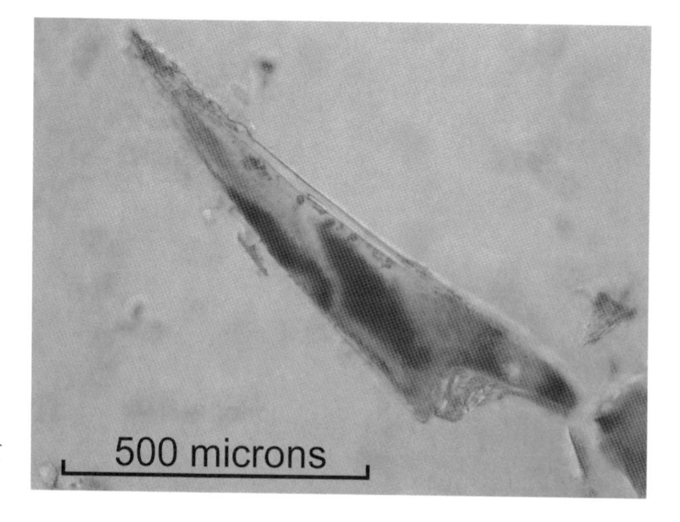

65 *Hair phytolith from the Indus city of Harrapa (c.2500 BC)* Photograph courtesy of Marco Mandella

In undisturbed deposits phytoliths can survive intact and in these situations they may retain the structure of the whole plant. For example, at the Neolithic site of Çatalhöyük in Turkey, baskets made of reeds with bodies of babies laid in them were quickly buried in clay. The outlines of the baskets were then preserved in the form of articulated phytoliths, even though the grass stems had decayed.

Phytoliths potentially survive in any type of deposit, but samples are best taken from deposits that have been well-sealed to minimise potential contamination. Being microfossils large quantities of phytoliths may be present in a given unit of sediment. Therefore, samples of about 50g are commonly taken, frequently as sub-samples from monoliths as has already been described for pollen analysis in section 2. As when sampling for pollen it is advisable to take a number of phytoliths samples across or through any deposit of interest. Similarly, if artefacts are being sampled, control samples should be taken from around the artefact to demonstrate that the phytolith assemblages are associated with the artefact rather than being ubiquitous through the sediment from which the artefact was recovered. It is also useful to take control samples from the topsoil to check that phytoliths derived from modern vegetation are not mixed with the archaeological deposits.

Separating phytoliths from their organic and sedimentary matrix is not a straightforward process and many different procedures are used. Firstly, the sample is treated in various ways to break down clays and remove organic material. This may involve soaking in solutions of hydrogen peroxide, sodium hydroxide or sodium hexametaphosphate ('calgon'), which is then followed by microsieving. Phytoliths are separated from other mineral remains by the addition of a heavy liquid and spinning in a centrifuge. The phytoliths can then be decanted and mounted on a microscope slide for further study.

The identification of plant species or plant parts (e.g. leaves, stems etc.) from phytoliths relies on distinctive types being representative of a single plant species or part, or alternatively articulated phytoliths that preserve an identifiable cell structure. Otherwise identification may only be at low taxonomic level and not to a specific plant part.

Once identified, phytoliths are quantified by counting diagnostic types. By comparing phytolith types between samples researchers can begin to assess whether particular phytolith suites are associated with certain plant groups or whether certain parts are associated with particular context types or artefacts. For instance, a storage pit might be expected to contain phytolith suites characteristic of awns, glumes, palaeas and lemmas, those phytoliths associated with cattle dung might include types characteristic of leaves and culms (**62**).

INTERPRETING PHYTOLITH ASSEMBLAGES

One of the main uses of phytoliths in archaeobotany has been in the search for early domesticated plant species in the New World. For example, Deborah

Pearsall and Dolores Piperno have examined domesticated maize. They could not identify a *single* distinguishing phytolith for this plant, but found that certain types of cross shaped phytolith occurred more readily in domesticated maize leaves than those of wild grasses, and that one of these types was also larger in maize. Such differences allowed domesticated maize to be identified from archaeological phytolith assemblages. Using this formula they concluded that maize was present within the Las Vegas-type cultures on the coast of Ecuador at 8,000 to 7,000 BP, a date much earlier than had been previously thought. However, given that the earliest known maize cobs in this region are some 3,300 years later their conclusions are not supported by all archaeobotanists.

Another use of phytoliths results from the association of particular phytolith types with certain catagories of artefact or ecofact. For instance, if we are unsure whether querns were used for grinding tubers or nuts rather than processing grasses it may be possible to obtain an answer by examining cracks on the quern's surface. Alternatively, we may be unsure if humans of the late Mesolithic/early Neolithic were subsisting from domesticated cereals or from other wild plant foods. Again an examination of phytoliths from the relevant sites can solve such a problem. Finally phytolith analysis may help reveal activity areas on a site or contexts associated with activities such as the storage or processing of cereals.

Archaeozoology

Vertebrate bones

We have seen in section 2 where vertebrate bones are most likely to survive and the methods that are used to recover them. In this section we will examine how bones are used to study ancient economies. Something we did not cover in section 2 was how bones from archaeological sites may be context specific. This is to say that they can be deliberately placed by people. In such cases bones of animals, like those of humans, must be carefully excavated and recorded before they are removed. Even with partial skeletons the position of bones may provide valuable information as to whether they were likely to have been articulated before burial.

It may prove useful for the interpretation of archaeological bone to divide assemblages into similar groups to those used in examining charred plant remains based on the nature of their deposition (**Table 16**).

As with plants the first data to be recorded are the species that are present in the bone assemblage (**66**). Identification of animal bones, as with other classes of bioremains, relies on the use of comparative reference material from securely identified animals. Not only should the species or genera of the animal be identified, but also the specific bone as it relates to the skeleton. Whether

Class A	Whole animals or parts of animals are deliberately placed in the features or deposits where they are found. In these situations we are referring to bones that have entered the deposit either in an articulated form or otherwise before they have been defleshed. The bones will show no signs of weathering, gnawing, fragmentation or butchery. There is a direct relationship between the archaeozoological material and the context
Class B	Deposits that result from either a single activity or continuously accumulated through a single repeated activity. In these cases bones may have been butchered before they entered the deposit. Examples include caches of the same bone part, for example horn cores or scapulas or alternatively fully butchered remains of a single animal. Such finds are likely to relate to the context or the general area of the context.
Class C	Assemblages relating to a number of different activities spread out over unknown time periods and spatial areas. Such assemblages are those that accumulate as bones rather than as joints or whole carcasses. The bones are likely to have been discarded on the surface, redeposited from *Class A* or *B* deposits, or within middens that have been redeposited. There is little direct relationship between the assemblage and the context in which it is found. The bones may be highly weathered and may show evidence of fragmentation and gnawing (particularly by dogs).

Table 16 *Vertebrate deposition classes*

it is from the left or right side is also a matter of considerable importance as we shall see later. It is also often important to determine what sex that animal was. Next the zooarchaeologist will need to record 'artefacts' or 'signatures' on the bone, such as butchery marks, signs of gnawing, weathering and the degree of fragmentation, as well as any pathologies that have left their mark. Pathological information might relate to injuries that damaged the bone but healed while the animal was alive, or they might relate to periods of disease or dietary stress (**67**). Other morphological traits within the skeleton, such as over-developed muscle attachments or long slender bones may be indicative of the animal having been used for traction.

In order to compare all these traits the zooarchaeologist needs a way of usefully quantifying information. As with archaeobotany, quantification is a source of much controversy within zooarchaeology, and, therefore, as with plant remains it is necessary to present data in a meaningful way so that contexts, sites and periods can be readily compared and characterised. Several approaches to quantifying animal bones have been adopted by zooarchaeologists. The first is the **N**umber of **I**dentified **Sp**ecimens (**NISP**) which, as the name suggests, calculates the number of identifiable bones (or fragments of bone) from each species. Such counts can be performed by context, phase, site or indeed any other archaeological subdivision. While some authors see such an approach as meaningful many focus on the various problems associated with it. These are many. Firstly species that have more bones in their bodies may be better represented than those which have fewer. In many ungulates, such as

66 *An assemblage of sheep bones from Oram's Arbour Iron Age enclosure, Winchester. Scale 10cm long*

67 *Periodontal disease in a cattle mandible from Oram's Arbour. Infection has caused bone reaction and receding alveolar margins. Scale 10cm long*

horses, cattle and sheep, evolutionary changes have fused bones or bones have been 'lost' to produce the characteristic single or double hoof. Such animals have fewer bones than are seen in other animals, for example dogs. Another problem is that the preferential preservation and recovery of larger bones may lead to a bias in such estimations towards larger animals which have a greater number of larger bones. This problem might be exacerbated if, for example, medium-sized animals such as sheep and goats were eaten while very young, and larger animals such as cattle were kept for much longer periods as dairy animals or indicators of wealth. Such a situation is not uncommon for many populations throughout the world. A further source of error is associated with the ease of identification. Clearly species with more diagnostic bones are more likely to have their bones identified to species and are thereby over-represented in the zooarchaeological record. The example of eels given by Terry O'Connor is a clear case in point. Eels have highly distinctive vertebrae with a double aperture for a double rather than single spinal cord, making them far easier to identify than other fish. If the bones are badly fragmented, diagnostic bones of eel are more likely to be identified even from the smallest fragments, therefore increasing the apparent importance of eel.

Some of these problems are overcome if we are just assessing the relative numerical importance of each species between sites or periods where taphonomic processes are believed to be uniform. For instance, while we may be less sure that cattle were more important than pigs, an increase in the number of cattle bones to pig bones within a large number of samples/features still shows a potentially important change taking place. However, one problem is not resolved in such studies and is the main reason for the rejection of the NISP by a large number of authors. The NISP does not take into account bias where whole or partial skeletons of single individuals are encountered. Clearly, a whole sheep skeleton, if excavated and sieved could produce a sum of bones more than enough to dominate over other counts where articulated skeletons have not been recovered. Over whole sites that are rich in bones this problem may resolve itself, but when NISPs are calculated for small assemblages on a site-by-site or period-by-period basis the room for bias is much larger. For this reason other methods of quantification are preferred by most zooarchaeologists.

One method that is specifically designed to overcome problems with NISP, calculates the **M**inimum **N**umber of **I**ndividuals (**MNI**) of each species of animal present in an archaeological unit. This is possible as each bone only occurs once in the skeleton and most have diagnostic elements that allow them to be identified to their exact position, including the side on which they occur. It is frequently much easier to identify the bone and its position than the species of animal from which it derives. To calculate the MNI, the zooarchaeologist looks for the bone type that appears most frequently for each species. For instance, we may have eight complete cattle skulls from a site, but if we have 10 left tibias (the rear lower leg) then we know that at least ten cattle are represented and that two skulls are 'missing'. Many bones are paired into right and left versions (hence the need for telling which side of the body each bone comes from). For instance an assemblage that had a right cattle tibia and a left tibia would in strict terms have an MNI of one. However, it may be obvious to the examiner that the left tibia is so much larger than the right tibia that they cannot have possibly have come from the same animal. If this is the case this would give us an MNI of two.

As has already been stated MNI counts may be calculated on a sample-by-sample basis, for phases or for the site as a whole. Attempting the latter is problematic given the potential for animal remains to have been deposited at different times.

MNI counts, like those obtained by NISP are fraught with problems. While they are often preferred by zooarchaeologists to NISP, the reasons, as Terry O'Connor has argued, are often less than convincing. With smaller assemblages there is still a problem in MNI of over-representing certain species. For example, we could have 50 different fragmented bones of sheep in ten pits that might easily span some 200 years of time. However, because they are highly fragmented only three bones can be seen to be of the same type, giving an MNI of three for the whole site. However an unusual deposit within a single pit might contain seven leg bones of chickens (five right and two left), giving an MNI of five. Therefore even though sheep bones are much more common on the site, MNI estimates would suggest that chickens are of greater importance.

A more worrying aspect of MNI concerns the consistency of data obtained by different workers. It is a well-known fact that the more a body of data is manipulated the less reliable it becomes. NISP has the advantage over MNI in that the variation between different analysts is likely to be minimal. With MNI some variation between researchers may be a result of differences in effort spent cross-comparing right and left bones.

It is worth considering the impact that the classes of assemblage presented in **Table 16** has on the data. Clearly on sites consisting mainly of *Class A* deposits MNI are likely to produce more reliable results than NISP, while on those consisting mainly of *Class C* deposits, NISP counts are probably more reliable. For sites producing *Class B* material, MNIs for single species may be proportionally higher than counts produced using NISP if the same bone is represented many times in a single deposit (as is often found with horn cores for example).

Occasionally the zooarchaeologist may use yet other methods of quantification, particularly when estimating the importance of an animal to the diet. One of these is **bone weight**. The weight of bone in relation to the weight of the body as a whole is a relatively constant ratio for each animal species. This means that if the weight of bones for each species is measured, an approximation can be made of the meat that would be yielded. For more accurate estimates conversion tables can be used, where analysis of the exact ratio of meat to the weight of each bone type has been worked out for the species in question. Yet another method is to use estimates of meat yield from animals of each species and multiply that by the MNI. However, there is a basic assumption in the use of meat yield estimates, namely that animals represented in the zooarchaeological record were consumed. Certain animals such as cows, goats and sheep may have been used for their milk, while other animals may have been kept as pets. Only if butchery marks are found on the bones can we be certain that an animal was eaten.

Sexing animals can also be important in understanding the use of animals in past economies. It relies on the use of diagnostic characteristics. In general mammalian males are larger than females, although obviously animals may vary in size according to their specific breed. Other obvious differences for some species are the development of horns or antlers, with males often possessing larger and/or more ornate examples. Some mammals may have different distinguishing features; for example boars often have extended canine teeth or tusks that may help distinguish them from sows.

Another method of sexing animal bones relies on measuring the length to breadth ratio of certain bones. One particularly good set of indicators is the lower limb bones or hand and feet bones (the metatarsals and metacarpals) of ungulates. They are often abundant enough on a site to make reliable comparisons. In female ungulates these bones are usually both smaller and proportionally narrower than those in males. The distribution of measurements of this type overlap when plotted, but where a large data set can be compared, either across space or time, separation is often obvious. Bones belonging to castrated males, however, may upset such measurements, as these are often longer and so proportionally thinner than the normal long bones of males.

Ageing animals is also important in many studies. One of the best means of estimation is by using the teeth. There are several markers. These are firstly the age at which tooth eruption occurs and secondly the age at which transition from deciduous or milk teeth to permanent adult dentition begins. The age at which this transition occurs for each tooth is approximately fixed depending upon species, although some variation has been noted. Once the adult teeth are established then a second measure can be used. During the life of the animal the molars gradually wear down. If the amount of wear is consistent between different populations then it is possible to compare wear patterns from modern sheep of a known age with mandibles from archaeological sites. Eşref Deniz and Sebastian Payne undertook such study of sheep and goats in Turkey. The major problems highlighted by their study were that differences in diet or soil type could produce different amounts of wear. The solution would seem to be to calibrate for tooth wear against the eruption stages, where such information is available for each site examined.

Another method of ageing vertebrates that tells us whether an adult or juvenile animal is represented is the measure of **epiphysical fusion**. Epiphysical fusion refers to the degree of 'joining' between the terminal articulating ends of the bones (the **epiphysis**) and the main shaft or body of the bone (the **diaphysis**) or for the degree of suture closure in the cranial plates in the skull. For most animals this fusion takes place in different bones at various ages within the first three to four years of their lives. While each bone in the body fuses at different ages according to species, there is some consistency, meaning that as long as zooarchaeologists know which species and bone they are examining, an estimate can be made of the approximate age of that individual at death. The biggest problem with the use of epiphysical fusion for ageing is that each bone can only give a minimum or maximum age for the animal it came from. For example, in sheep, the distal (part closest to the feet) tibia fuses at between 12 to 15 months. If we find an archaeological specimen of an unfused tibia, the best we can say is that the animal it came from was younger than 15 months at death. If the tibia was fused we can say the animal was older than 12 months. To age animals more precisely we are reliant on finding articulated bones from the same individual and obtaining minimum and maximum ages from at least two epiphysical fusions. A further difficulty lies in the effect of poor diets which prolong the period of fusion, although in such instances the bone itself may be malformed indicating such a case. Another problematic situation concerns castrated animals where epiphysical fusion often takes longer to occur than is seen in non-castrated males.

ECONOMIC INFORMATION OBTAINED FROM ANIMAL BONES

The most obvious source of economic information provided by animal bones is that the very presence of a species indicates their use by the inhabitants of the site. The trick for the zooarchaeologist is to try to understand what this use was. Were animals used for their meat, skins, horns, milk, or for traction? By examining signatures on bones the zooarchaeologist may be able to establish evidence for butchery, cooking or the use of the animal for traction. The spatial distribution of certain bone types in large numbers may indicate specialist – and/or the separation of – activity areas. In turn, the distribution of different body parts may be indicative of exchange networks. The age and sex profile of the animal populations may be indicative of the breeding and slaughter cycles, and how animals were kept. This in turn may suggest the raising of the animals for meat, milk or wool (**54**). As most wild and many domesticated animals breed at specific times of the year, the age profile may further be used to suggest any seasonal component to the site. For fish bone, where the nature of bone growth leaves distinctive rings, such remains may be particularly useful for estimating the season in which those fish were exploited. We will return to these matters later in this section.

Marine molluscs

Mollusc shell, being composed predominantly of calcium carbonate, has the same requirements and limitations for preservation as that seen for animal bone. It goes without saying that shellfish are rare in inland situations and where found in these contexts they must have been brought inland by people. Marine mollusc exploitation by prehistoric hunter-gatherers is often highly conspicuous in the archaeological record and sometimes the only records we have of actual 'settlement' activity.

SAMPLE RECOVERY AND ANALYSIS

Shells recovered from non-shell midden contexts are collected in much the same ways as bones, either by hand during excavation or in sieved samples. Columns of samples similar to those previously outlined for land snail analysis may be taken through shell middens. Each sample is soaked in water and then wet sieved through a 1mm mesh. All shells are removed from the residue following drying for further study. Quantification is on the basis of the same criteria as explained for terrestrial molluscs in section 2.

Beyond noting the presence of certain species, several analyses may be undertaken of shell fish. Changes in the exploitation and/or the availability of shellfish may be seen through simple changes in their frequency through shell middens. More detailed analyses may include the thin sectioning of the shells to examine seasonality. As with fish bones the shells of shellfish grow incrementally, depositing new bands on the shell edge or surface. The laying down of new shell may be affected by tides, sea temperatures and/or the availability of food, meaning that in certain seasons growth may be more rapid, while in others it slows or ceases altogether. For many species of temperate climates, shell growth declines or stops during winter. This means that by examining the rings of a shell the season of collection may be ascertained.

Another more complex technique of examining seasonality involves calculating ratios of oxygen isotopes ($^{18}O/^{16}O$) within each of the shells bands. We have already discussed the use of oxygen isotopes from Foraminifera as a proxy for water temperature, and the same principles hold true for marine molluscs. By measuring oxygen isotopes from growth rings it is possible to tell whether these were laid down in winter or summer. For example, examining shells of limpets from Nelson Bay in South Africa, near Cape Town, Nick Shackleton was able to determine that the bay was exploited by hunter-gatherers only during the winter months some 9,000 to 5,000 years ago.

ECONOMIC INFORMATION FROM MARINE MOLLUSCS

Marine molluscs have provided a dietary supplement for many different groups – and not only those inhabiting coastlines. The word 'supplement' is used advisedly as work by Betty Meehan on shell middens produced by native Australians has suggested that even the largest midden represents a relatively small proportion of the animal protein used by these people. However, marine molluscs have been found in other circumstances where they provide rather different evidence concerning past economies. For example, at Bishopstone,

Sussex, an Iron Age field system and settlement on the coast of southern England, remains of marine molluscs were found in ancient ploughsoils. Martin Bell, the excavator of the site, suggested that the small marine mollusc shells had been attached to seaweed that was brought to the site as fertiliser. Occasional shells of *Rissoa* sp., were recovered by one of the authors from the Mesolithic shell midden at Culverwell on the Isle of Portland. The size of these shells (less than 2mm) is too small for them to have been deliberately collected, and here too, the most likely explanation is that seaweed was brought to the site, perhaps as food.

Before we leave molluscs we must not forget that not all edible molluscs come from the sea. Many large species of land snails have been used for food in the past. The most famous is *Helix pomatia* the Roman edible snail, although many of the species of *Helix*, including common garden snails *Helix aspersa* have been also been consumed – often in error!

Biomolecular studies

Biomolecular studies examine the molecular structure of plant and animal remains recovered from archaeological sites with the purpose of identifying what the organism was and understanding something of its life history. Such studies may be applied to human and animal bones to try and understand what was eaten and indeed where that food came from. Biomolecular studies may also be carried out on materials of unknown origin, for example residues from pots, in order to determine what that material originally was, and what it was used for. This area of archaeology is rapidly expanding out of environmental archaeology into a sub-discipline in its own right. Here we briefly review the main techniques in the belief that this may be the last chance to include them in an environmental archaeology textbook – the biomolecular archaeology manual can only be a few years away!

Analytical techniques

Depending upon the nature of the study that is being made, a variety of **spectrometry** techniques including mass spectrometry, infrared absorption spectrophotometry, fluorescence spectroscopy, nuclear and proton spectrometry are employed. In addition, various chromatographic methods are also used, in particular gas chromatography.

Mass spectrometry is perhaps the most commonly used technique. It works by converting the molecules within a substance into gaseous form and ionising the gas in a beam of electrons. The newly charged particles (ions) are then separated by applying an electric field. A sensor measures the amount each ion is deflected, and from this its atomic mass and charge can be calculated. These data allow the chemical element to be determined, while by plotting the frequency of each particle type (based on atomic mass), the relative abundance of each element can be calculated.

Chromatography relies on converting solids into a liquid or gaseous form. In liquid chromatography chemicals are used that absorb the various constituents (chemical elements and compounds) of the sample at different rates, so that as they spread through the medium they separate into bands. Gas chromatography involves heating the sampled substance until it becomes a gas and then passing this through a narrow, coiled tube. This tube contains materials that cause variations to the flow of the different component parts of the gas. As each constituent part arrives at the end of the tube it is then detected using a variety of techniques such as mass spectrometry, (gas chromatography-mass spectrometry – GCMS).

Residue analysis

Residue analysis is normally carried out on deposits left in ceramic vessels. It is also possible to examine the pottery fabric itself as frequently (especially with coarse wares) the pot may have soaked up residues of the liquids that it once contained. The analytical method by which the sample is investigated will depend both on which of these circumstances applies and also on what the investigator thinks may be present or wishes to test for. The starches of various plants species may be identified using a variety of chemical tests, while starch granules may sometimes be identified using high-powered microscopy, including scanning electron microscopy. Starch granules are often of different shapes and sizes as well as having individual morphologes depending upon the plant species from which they derive. The usual approach is to compare granule shape of archaeological starches with those of modern reference examples. Such techniques have allowed Delwen Samuel to identify cereal starches within ancient Egyptian pots. Experiments with modern material showed that malting produced starch granules with distinctive pitted and channelled surfaces. As similar granules were found in archaeological vessels, she was able to suggest they had been used for the consumption of beer. In some of the archaeological examples, gelatinisation of the starch granules could also be seen, indicating that the cereals had been heated while moist. Such analysis can not only serve to link ceramics with their contents but can even be used to decipher the sequence used to make ancient beer.

Research into starches has also been a useful tool for tracing plant foods such as tubers which rarely leave any other archaeological traces (parenchyma is relatively rare because it preserves only by charring). Starches are abundant in tuberous food and so provide an ideal subject matter for such studies. Using starch granule morphology, Donald Ugent, together with Sheila and Thomas Pozorski, was able to identify desiccated tubers, dated between 4,000 and 3,200 years ago from the desert coast of North America as an early domesticated potato.

A large body of work has also been carried out on the identification of various lipids (fats), acids and amino acids (proteins) from ceramic vessels. This type of analysis has enabled investigators to identify milk, butter, animal fats (including in some cases the species of animals), and a variety of vegetable oils including olive oil, mustard oil and grapes, as having been the contents of pots. For example, a small fragment of the yellowish residue present in a Neolithic jar found in the Hajji Firuz Tepe village (northern Iran) and dated to 5400–5000 BC was analysed by means of infra-red absorption spectroscopy. The spectra produced a plot indicating that the residue contained calcium tartrate. After comparison with ancient and modern reference samples, and following confirmation of the result by means of chromatographic and ultraviolet (UV) spectroscopic methods, the sample was announced as the earliest chemical evidence for wine. Tartaric acid occurs naturally in significant amounts (>1 per cent) only in grapes (*Vitis vinifera*), while the calcareous environment of the site had converted the acid into the compound calcium tartrate. The shape of the jar confirmed its use as a liquid holder. Oil or resin from *Pistacia atlantica*, better known as the terebinth tree, was also identified in the residue using the same techniques. This would appear to make good sense as terebinth resin was used in the ancient world to inhibit bacterial growth in wines and improve both taste and 'nose'.

Isotope analysis

A few chemical elements exist in only one form – that which is given in the periodic table. However, many elements exist in a number of different forms (**isotopes**), separated on the basis of relative atomic mass (RAM). Nevertheless, for any single element, isotopes have identical atomic numbers. This means that isotopes are distinguished on the basis of the number of neutrons in the atom (RAM is the number of protons plus neutrons, while the atomic number is the number of protons). Where isotopes do not undergo radioactive decay they are termed **stable isotopes**, and where they do (as for example with the ^{14}C [carbon-14] isotope used as the basis for radiocarbon dating) they are called **radioisotopes**.

All animals are effectively what they eat, and vertebrate bones for instance are formed from the various elements and compounds contained within food. The constituent chemical elements in each type of food will be represented by isotopes present in various proportions depending upon their source. The ratio of isotopes for each element is then reflected in the bones of the consuming animal. By comparing the ratios of various stable isotopes in bones with known isotopic ratios in modern food sources we can begin to piece together what that person (or in theory, animal) had been eating when that bone formed. As bone material is continually replaced during the life of an individual the isotopic ratio will only represent diet in the last ten years of their life. The main isotopes used in such dietary analyses are carbon (^{13}C and ^{12}C) and nitrogen (^{15}N and ^{14}N), both of which are extremely common elements in the natural world. The principles of carbon absorption are relatively simple. In the atmosphere ^{13}C is about 100 times less abundant than ^{12}C. The ratio of ^{13}C to ^{12}C (termed $\partial^{13}C$) is altered by plants during photosynthesis, with plants absorbing generally more than ^{12}C than ^{13}C. So $\partial^{13}C$ in plants is greater than in the atmosphere. The amount by which photosynthesis affects this ratio, however, varies according to whether plants fix carbon (the process by which carbon dioxide and water is converted to a carbohydrate and oxygen) into a 3-carbon molecule (the Calvin cycle associated with C_3 plants) or 4-carbon molecule (the Hatch-Slack cycle, C_4 plants). C_3 plants incorporate more ^{12}C into their tissues and have a much lower $\partial^{13}C$ than that found in C_4 plants that incorporate less ^{12}C (**68**).

Most terrestrial plants are of the C_3 variety. However, several savannah grasses including millets and maize are of the C_4 type. Marine plants (including plankton) also have a much higher ratio of $\partial^{13}C$ being closer to C_4 plants. These are passed to the animals that eat the plants, so humans who eat a great deal of maize or millet will have higher $\partial^{13}C$ than those eating wheat and barley or other food resources. Isotope ratios derived from plants are passed through the food chain, and could, therefore, in turn reflect the diet of the animals that are eaten. Thus it is possible to identify whether humans were eating marine foods as they will also have much higher $\partial^{13}C$ than those who do not. Analyses of European populations carried out by Mike Richards and

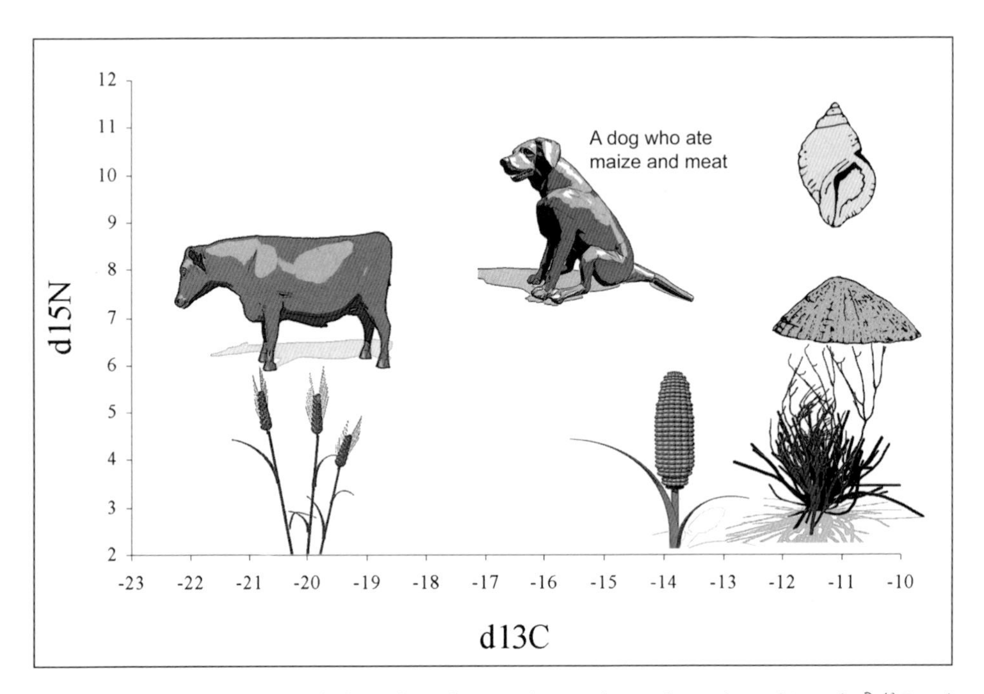

68 *The approximate position and relationships of certain plants and animals in relationship to the $\partial\ ^{13}C$ and $\partial\ ^{15}N$ isotopes in them*

colleagues have shown the high importance of marine resources in Mesolithic diets compared to those of subsequent agriculturalists.

Nitrogen isotopes have been similarly examined, but these, rather than providing information on the contribution of marine and non-marine food, tell us instead, of the relative contribution of plant and animal food. There is a trend for increasingly higher $\partial^{15}N$ ratio, the more animal products (including milk, cheese etc) that are eaten. Therefore $\partial^{15}N$ ratios also tell us about trophic structure. Isotope measurements are usually made on collagen, as in only this part of the bare can contamination from carbon and nitrogen that is in the soil be determined.

Economic information from residues and isotopes

Residue and isotope analyses have an extremely important role to play in detecting elements of consumption and production that are otherwise invisible in the archaeological record. Liquids for example are notoriously difficult to detect archaeologically using traditional macro- or micro-fossil based techniques. We are often left to hazard a guess whether crops such as olives or linseed were used for oil, or grapes and cereal for wine and beer respectively. Residue analysis also allows us to see how certain foodstuffs were consumed. Without such analyses we often have only the artefactual evidence itself, which is often equivocal; for example do strainers really indicate the use of milk? Lastly of course these types of analysis also provide valuable information on the use of ceramics. For all the pottery we find on archaeological sites we are often at a loss as to what vessels once held and how they were used.

Genetics

The existence of genes which carry information from one generation to the next has long been known. All genes contain strands of deoxyribonucleic acid, better known as **DNA**, which are effectively the code that programmes an organism. The arrangement of DNA sequences is *similar* in each species, but *unique* to each individual of that species. DNA thereby not only allows the identification of the particular organism, but can also enable individuals of the same species and even subspecies to be told apart. Study of ancient DNA, therefore, has a great potential that is only now beginning to be realised. The earliest stages of DNA study in archaeology saw a great deal of attention focused on mapping the movement of early forms of our own species, *Homo sapiens* around the globe, but more recently it has been applied to the processes of domestication.

Genetic studies in archaeology can be undertaken by one of two routes. Firstly, it is possible to study the modern genetic properties of a group of related organisms and from this to piece together their evolutionary history, and estimate, for example, when their last common ancestor lived. This is the route for studying our own ancestry. The second route is to extract genetic material from the archaeological remains of past organisms. At present it is the first line of study that has been most productive so far in the study of the origins of agriculture. Modern DNA of cereals has revealed both information pertaining to their wild progenitors, the number of times each species of animal or crop has been domesticated, and the possible areas of the world in which domestication took place. Maize has also been studied using modern DNA. The origin of maize has been a controversial issue since the 1930s and two schools of thought exist as to its origin. One held that maize had been domesticated from a group of wild grasses known as teosinte; the other held that maize was derived from a now extinct 'wild maize'. In the 1980s John Doebley carried out a pioneering genetic study that was to shed new light ton the situation. He collected six varieties of teosinte, and several subspecies of two of the annual teosinte species. He then compared the genetic information from these with modern domesticated maize. He found that only one subspecies, *Zea mays* subsp. *parviglumisi*, was genetically closer to maize than it was to the other varieties and subspecies of teosinte. This subspecies is only known from Mexico, and here three distinctive wild strands have been identified. Doebley then compared the DNA of *Zea mays* subsp. *parviglumisi* in these three regions to domesticated maize, and narrowed the area of domestication down to the Central Balas Valley just to the south-west of Mexico City.

A similar study was conducted for einkorn wheat, *Triticum monococcum* subsp. *monococcum*. Here the wild ancestor was known, *Triticum monococcum* subsp. *boeticum*. However, the distribution of wild stands of wild einkorn stretch from the Anatolian peninsula (Eastern Turkey), across northern Syria and into Iraq. An international team from Norway, Germany and Italy led by Manfred Heun, looked at 68 modern 'populations' of domesticated einkorn from Europe and the Near East. They found that they were all genetically similar. The next stage was to compare these to 261 wild einkorn populations stands stretching across their known range. It was found that populations in the Karacadag mountains of south-east Turkey were not only genetically distinct from other wild stands, but were also similar to the domesticated populations. Therefore, it is almost certain that this wild einkorn is the modern descendant of the populations that produced the first domesticated einkorn.

Emmer wheat has also been subjected to rigorous genetic analysis, but a rather different story emerged. Robin Allaby compared populations of modern domesticated emmer and wild emmer. His analysis showed that rather than just a single domestication event, modern emmer populations are the result of at least two distinct separate domestications.

Applying biological evidence to economic questions

Having reviewed the biological evidence we will look at how this has been used along with other archaeological evidence to answer some of the economic questions posed at the beginning of the section.

An examination of production

While almost any aspect of production could be studied using biological evidence we will concentrate here on those relating to the most important: food. Almost all foodstuffs derive from plants or animals, salt being the obvious exception. Firstly, we need to consider the sources of variation in production sequences. The first of these are the species of plants and animals involved. The exploitation of plants clearly follows very different sequences to that of animals. However, for any single species its physiology and ecology limit the number of possible ways in which it may be successfully exploited. Many plant resources are slave to annual climate cycles so that seeds, fruits and nuts are only produced at certain times of the year. Similarly, many domesticated animals will only become fertile, breed and produce offspring in certain seasons. While domestication has altered some natural cycles, the amount of change induced has often been limited. The physiology of plants and animals means that processing can only be achieved in a limited number of ways. For example, from an examination of the practices of several traditional societies, Gordon Hillman has noted that most grain crops, including wheat, barley, sorghum, rice, maize, millets and even pulses follow very similar patterns of exploitation. For each the desired result is clean grain. However, the harvested crop consists of a mixture of grain, chaff or pods, stems, leaves and weeds. In all these cases the grain must be separated from these other elements by a mixture of sieving, winnowing and pounding/threshing activities. For each grain crop the number of ways in which clean grain can be obtained is limited to only a few possible sequences. Many animals too can only be processed in a limited number of ways to remove the meat and extract the marrow and other organs.

The physiology and ecology of plant and animal species, therefore, give the environmental archaeologist a firm footing from which to begin investigating patterns of economic exploitation. In discussing the subject further we will concentrate on Old World societies, for which the four main domesticated animals are cattle, sheep, pig and goat and the domesticated cereals, wheat and barley. We will also focus mainly on archaeological studies in Britain, although the techniques and concepts discussed are applicable outside this region. First we will look at reconstruction of production sequences, then turn to the bioarchaeological evidence that we have for individual stages that these comprise.

Arable production

Soon after flotation, as a technique for recovering charred plant macro-remains, appeared in the 1960s it was realised that assemblages from the whole Old World varied little either spatially or temporally. Cereal grains, the remains of chaff and seeds of wild species, (many of which are found today as common weeds of crops) dominated. In the study of artefact production it has been noted that the order of events, for example, in the production of a flint tool, a pot or metal object, often follows a well-defined sequence. Each stage modifies the object being produced and simultaneously produces waste that is the evidence for that stage. Such sequences and their studies are referred to by the term **chaîne opératoire**. As already stated the number of methods by which crops can be produced is limited, which means that if we know what produce and waste is produced in each part of the sequence (**Table 17**), we can reconstruct the process. We can also determine from which part of the process our archaeological samples derive. In the following text we will look at each in turn.

Studies of weeds growing in cereal fields have shown that the species found depends upon geology, soil type, drainage, climate and the specific cultivation practices employed by the farmer (**cover picture**). We therefore have a proxy for determining where past crops were grown, provided that weed seeds survive in the archaeobotanical record, that they can be identified to a high taxonomic level and that we have sufficient information regarding their ecology. The use of weed ecology to reconstruct past arable environments also relies on the principles of uniformitarianism which we reviewed in section 1. The assumption in weed ecology studies is that as plants had the same physiological characteristics in the past as the present, they grew in similar environments. For example, certain species of thistles and buttercups are to be found in poorly managed wet pastures, and grassy waysides and riverbanks that are not mowed or ploughed but occasionally grazed by cattle. If we find seeds of these species in archaeobotanical assemblages accompanying cereals, we may then assume that fields were located in the environments associated with the modern weed plant. However, another aspect of the study of ancient weeds is the assumption that plant associations have also been constant over time, allowing very specific past environments to be inferred from archaeobotancial assemblages.

Two approaches have been used in the study of ancient weeds to reconstruct crop-growing environments. **Phytosociology** defines plant communities by recording the presence and dominance of individual species

Primary Production	*Procurement*	*Processing*
Locating the field	Harvesting	
Previous land use		Storage
Clearing and the field		Grinding, Pounding, Brewing
Manuring		Cooking, baking
Tilling the field		
Sowing the crop		

Table 17 *Stages of crop processing as defined by Gordon Hillman*

69 *Reconstruction of Francis Pryor's interpretation of Early Bronze Age field systems as excavated at Fengate, Cambridgeshire*

within them. Each plant community is given a name, which defines the species present and their ecological characteristics. Two groups have received particular attention. The **Chenopodietea** are defined by the presence of species of the Chenopodiaceae family. This group is seen as being indicative of crops grown in well-tilled and manured fields, sown in spring and in rows. The **Secalietea** group is characterised by wilds oats and seeds of brome grass. It is characteristic of crops that are grown on poorly manured, infertile soils, sown in winter. In contrast to phytosociology **Autecology** studies the interaction of plant species with environmental factors. The aim is to understand how physiological characteristics enable that species to survive, compete with other plants for space both above and below ground, and successfully reproduce. By growing the weed plants in laboratory conditions, the plant's reactions to various controlled factors such as water availability, soil nitrogen, temperatures and varying degrees of daylight can be studied. When the weeds are found in archaeobotanical assemblages that also include crop plants, these laboratory data can be applied to the archaeological field. The development of the two approaches since the 1920s has been closely mirrored by their use in archaeology. Phytosociology has been favoured by mainland European archaeobotanists, such as Willem van Zeist and Karl Behre, while autecology has been favoured by archaeobotanists working in Britain, such as Martin Jones, Marijke van der Veen and Gordon Hillman.

LOCATING THE FIELD

The evidence for the location of past fields may come from weed seeds that are distinctive of certain soils. Weeds that are specifically indicative of the soil structure, its chemistry and pH, fertility and water conditions are particularly useful for these studies. Species such as stinking mayweed, *Anthemis cotula*, for example are today exclusively associated with clay soils, while corn chamomile, *Chrysanthemum segetum* is associated with sandy soils. The former species has been used by Martin Jones to suggest an increasing cultivation of clay soils in southern England beginning around the time of the AD 43 Roman invasion. By the medieval period this species is a common component of almost all archaeobotanical assemblages. Corn chamomile is often found on British medieval sites, along with cereal grains of rye, oats and barley. The latter assemblage type has been interpreted as indicating that marginal land was being farmed to support the higher populations that are known from the historical literature. Seeds of sedges and rushes are indicative of the cultivation of wetter soils. Rush seeds for example were found along with cereal remains from the wall slots of roundhouses on top of the acropolis at the Etruscan city of Cerveteri near Rome. These houses pre-date the Etruscan city and are of the Villanovan (Iron Age) period (900-700 BC). The houses are situated on a plateau between the Rivers Marmo and Mola, surrounded on many sides by steep cliffs. It would appear that ancient Villanovan inhabitants grew their crops in the wetter areas of the river valleys, bringing them up to the settlement from where they were stored and processed. The chaff from the crops, along with the seeds of weeds that grew with them, was discarded and thrown into hearths. It was this which later became incorporated in voids where the house wall once rested.

No ancient fields have been yet located on the Marmo or Mola floodplains. However, evidence of ancient fields located in river floodplains is commonplace in Britain as a result of both aerial and ground survey. Even when identified the debate is whether these fields are used for the grazing of animals or for cereal agriculture. Only the discovery of ard marks in the subsoil can prove this one way or the other. Ard marks representative of past agriculture are not just found under alluvium, they may be preserved under colluvium or lynchets, (mounds of soil themselves built up through repeated movement of soils by ards or ploughs). While dating these marks may be problematic, those preserved under monuments will have at least a known minimum age. The earliest known ard marks in Britain were made some 5,000 to 6,000 years ago, and are preserved under an Early Neolithic long barrow at South Street, Wiltshire (**70**). Although they cover most of the excavated area under the mound (some 100m^2) this must still be considerably less than the original field size. Similar ard marks are also commonplace under burial mounds in Bronze Age Denmark.

Another, albeit less firm, form of evidence for the location of arable fields comes from manuring scatters. In the past manure often consisted of a mixture

70 *Neolithic ard marks preserved under South Street long barrow in Wiltshire*

of dung and straw, combined with general domestic and household waste, for example hearth sweepings, food debris, and sometimes even human faeces, combined with broken artefacts. Little of the organic material survives, but inorganic artefactual material is often preserved where not subject to mechanical abrasion. Sherds from such scatters then provide a useful means of distinguishing both the extent of past fields and their approximate dates using field walking. Where field walking has been conducted around Romano-British settlements in Britain it has often been found that manuring spreads are located in the immediate vicinity of the site, suggesting that settlement and field were closely tied.

PREVIOUS LAND USE AND FIELD CLEARANCE

The previous vegetation history will to some extent determine the weed species composition of an arable field. Many soils contain what is known as a **seed-bank** comprised of seeds lying dormant and ready to grow as soon as conditions become suitable. For **annual** plant species (those that live for just part of the year) the seed-bank is an important means by which re-colonisation of the field can occur. In arable fields ploughing disturbs the seed bank and thereby enables germination. Species vary in the degree to which they rely on the seed bank. Some, such as poppies, produce thousands of seeds that lie dormant within the soil and only when exposed to air and daylight as a result of disturbance do they germinate. This is why we have poppies today to remember the victims of the First World War. Shelling on the Western Front brought poppy seeds to the surface that had lain dormant from when the land had been under cultivation prior to the war. Many other weed species that were once a common sight in wheat fields before the seventeenth and eighteenth centuries do not seem to rely on the seed-bank at all.

Just as certain annual arable weeds may increase their prominence in a field the more times it is cultivated (such as poppies) so others species may go into decline. Studies of arable fields previously under pasture have shown that over a twenty-year period the number of **perennial** species (plants that live for more than a year) gradually declined with each successive year of cultivation. Many perennial plants do not produce seeds every year and so if cultivation (by ploughing or weeding) should remove the plant they have limited means of re-colonising the field. We can, therefore, make a generalisation that the longer cultivation continues, the more that perennial species will decline and annual species will increase. Of course, as Karl Behre has stated, where fields are put to ley or pasture, then perennial weeds will increase in number. This provides the archaeobotanist with a tool for looking at ancient farming practice by using the ratio of perennials to annuals. Fields may also be left **fallow** where crops are not grown on them. A common misconception in the archaeobotanical literature is to argue that such practices increase the number of seeds of perennial species. Technically fallow fields must still be ploughed and harrowed (a method of cutting and chopping-up weeds). As the soils in such fields can be disturbed throughout the year, such practices would actually decrease the number of perennials species rather than increasing them.

MANURING

The reaction of various weed species to manure has been recorded in detail and forms a particularly important aspect of weed studies. The manures in modern agriculture are very different to those that would have been available to ancient farmers, the contents of which we have already described. A prime constituent of all manures, however, is nitrogen. Available soil nitrogen (nitrogen in a form that can be easily taken up by plants) in temperate environment varies seasonally. As soils warm in spring, soil micro-organisms become active in breaking down organic material (nitrification and ammonification) into nitrates that can be taken up by the plant. During autumn, as soils cool, these organisms become less active and as nitrogen becomes leached by rainfall it is transported deeper into the soil profile. Release of nutrients in farmyard manure follows a similar cycle. Marijke van der Veen has suggested that large quantities of seeds from species of the Chenopodietea indicate that intensive manuring was carried out. A decline in the same was interpreted by Martin Jones as indicating overcropping or a decline in soil fertility. Many species of plant, especially those of the Chenopodietea, use the nitrogen cycle to promote the germination of their seeds from the seed-bank in spring when nitrogen levels increase, hence the association of this group with spring-sown cereals. The problem is that because the Chenopodietea are often associated with fertile soils they are also used to indicate high fertility kept up through manuring. So in practice where we find the Chenopodietea dominating archaeobotanical samples we do not know whether this means that highly fertile soils, manured highly fertile soils, or soils that would not have been fertile had they not been manured, were exploited.

The previous discussion demonstrates that the best form of evidence for manuring still comes from spreads of pottery sherds around archaeological sites. Such evidence is often lacking for prehistoric sites where sherds, even if once present, are less durable and so do not survive as well as their historic counterparts (in northern Europe at least). However, at South Street the mound preserved not only ard marks but also stray pottery sherds that Peter Fowler has suggested might indicate that manuring was part of the Early Neolithic agricultural cycle. The possibility of detecting manuring through the analysis of buried soils, using micromorphology and/or chemical techniques such as phosphate analysis might provide our best hope of detecting manuring during prehistoric periods.

TILLING AND WEEDING THE FIELD

Tillage is carried out to aerate the soil, which increases the rate of nitrification and ammonification, and thereby provides a good growing medium for the crops roots. Some forms of tillage can also destroy weed species, although, depending on the method of tillage employed, weeding by hand is often a more effective means of removing weeds. In most modern agricultural fields a plough is used to till the soil, and a harrow (an implement pulled by a tractor with large numbers of metal discs placed at right angles to the ground surface) to cut and chop up weeds. Ploughs have coulters: iron blades that cut the soil, followed by what are known as mouldboards - large 'angled' blades that cut the soil and turn over the sod. Such ploughs are very effective at turning over all the soil and destroying perennial weeds. Perennial species are literally up-ended, their roots exposed to the air, and their leaves and stems buried under the sod. Such ploughs bury the seeds of annual species while exposing seeds previously buried. These then germinate with the crop so that the annual species may produce more seed and thereby, reinforce the pattern of relative frequency of annuals to perennials that we have previously discussed.

Harrows are used where crops are sown in rows. The discs on the harrow cut down weeds growing between the rows of germinating cereals, leaving the cereal plant untouched. Such methods will destroy most annual weeds, but will be especially devastating to perennial ones or those annuals that do not form long-term seed-banks.

Ards differ from ploughs in that strictly speaking they have neither coulters nor mouldboards. Instead, they have shares, pointed blades of wood, sometimes tipped with metal or sometimes even stone, which cut vertically through the soil. The effect of these implements on the weed flora is very different to that of a plough. Perennial species are either missed or pushed aside, while seeds of annuals are less likely to get buried. Karl Behre has postulated that the ratio of perennial to annual weeds might also reflect the type of tillage that was carried out (**71**). Ards were used from the Neolithic until at least the early Roman period. For example, an ard share was found at Ashville, Oxfordshire, which dates to the Roman period, while ard marks have been recorded by George Lambrick at Fen Drayton some 2 to 3km away. That the ard marks at Neolithic South Street, at Bronze Age Gwithian and at Roman Fen Drayton all have a criss-cross pattern shows a continuation of the same basic form and method of cultivation for over 5,000 years. Criss-crossed ard marks are also seen from mainland Europe, for example the middle Bronze Age site at Twisk in North Holland, demonstrating the wide geographic area over which this method of cultivation was carried out.

The middle to later Roman period saw the introduction of the plough. At Warren Villas in the English Midlands, Mark Robinson recorded asymmetrical ploughmarks dating to the second or third centuries AD buried under overbank alluvium. Coulters from such ploughs have been found from several later Roman sites. Mouldboard ploughing is usually carried out in linear strips rather than criss-cross and according to Martin Jones the same might be true of Roman asymmetrical ploughs. Roman ploughmarks from adjacent to several British villa sites and Hadrian's Wall are all 'one-way'.

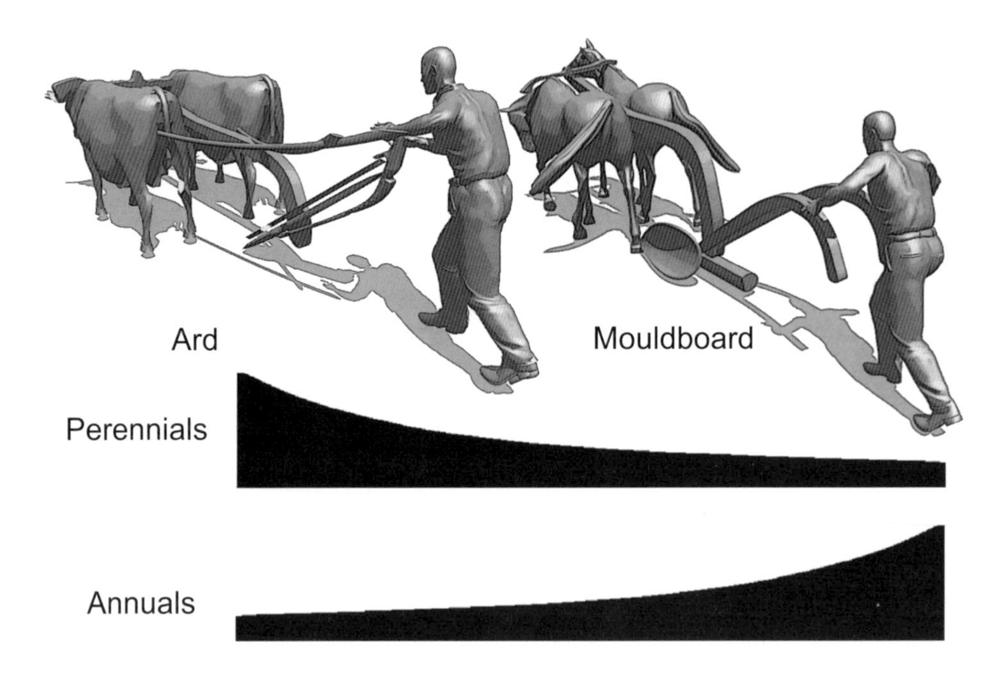

71 *Variation in annual and perennial weeds using mouldboard plough and ard cultivation*

Further tillage methods have been recorded in the archaeological and ethnographic literature such as the use of mattocks or spades. The former are well attested from Egypt, while spade marks have been noted at Bronze Age Gwithian in Cornwall. Wooden spades have been recovered from later prehistoric Denmark, although they were probably used for peat cutting rather than cultivation.

It has been noted that some species of perennial weeds, notably false oat grass (*Arrhenathermum elatius* subsp. *bulbosus*) and creeping buttercup (*Ranunculus repens*) reproduce from fragments of root stem and so in fact benefit by more effective tillage. This brings us to an important point concerning perennial species. It is not the just the presence of a species in a field that we are interested in, rather the effect of cultivation practices upon their ability to produce seeds and hence occur in the archaeobotanical record. Perennial species could proliferate under certain cultivation practices, but if these adversely affect their ability to produce seed they may still remain invisible to us. For example, it is possible that false oat grass rarely produces seeds in arable fields, so that while its tubers are often found archaeologically, its seeds (which technically could be identified) seldom are.

The effect of weeding on the ratio of certain annual and perennial species is a matter of speculation. It is likely that weeding will preferentially remove those species with longer growing seasons. Weeding is then more likely to remove perennial species, annuals without a persistent seed-bank, and those species that only germinate in autumn. Some perennials, however, have deep roots and are also capable of reproducing from them and so are unlikely to be removed by all but the most rigorous of hoeing. The efficiency of weeding also depends on whether the crop is sown in rows or not. Crops planted in rows may be weeded while the crop is still growing, while those planted by the broadcast method are harder to weed once the crop is established.

Marijke van der Veen has used the ratio of annual to perennial weeds to differentiate between two groups of Iron Age sites in northern England. One group (Group 'A') consisted of small isolated defended hillforts, the other (Group 'B') of non-defended lowland settlements in close proximity to the River Tees. Plant assemblages from Group A were dominated by annual weed species; Group B by perennial species. The interpretation was that fields close to the Group A sites were more intensively cultivated with fields being dug and hoed regularly, while those of Group B represented more extensive systems, with fields being ploughed [arded] once or twice before sowing. The Group A sites produced evidence for a system of farming known as cord-rig in which ridges are made up by spade or ard. Peter Topping has suggested that such ridges had a beneficial effect on the growing of cereals, implying (perhaps inadvertently) that cereals in such system were planted in rows. It is possible then that intensive arable systems through one means or another led to the predominance of annuals at these sites.

SOWING THE CROP

Cereal crops may be planted either in rows or by the **broadcast** method, in which grain is evenly scattered across the field by hand. As we have already seen, the importance of the method chosen concerns the amount of weeding that can be conducted, but it also affects the amount of space between plants. By being ploughed in rows, cereal crops can be more effectively weeded and have a greater space in which to develop. This is especially important for autumn-sown crops where the frost causes them to produce more tillers. Tillers are the number of stems produced from a single seed. Where the crop does not tiller a single stem is produced with one ear, meaning that tillering is desirable for increasing yield. Another method of increasing tillering is by grazing animals on crops, something that is recorded by ancient Mesopotamian writers, although it is a practice that may be less successful in Britain. It is probable that such practices would have the effect of increasing perennial weeds and decreasing annuals.

A point worth making is that while autumn sowing has the advantages of producing higher yields, in northern latitudes such crops are more vulnerable to heavy frosts and autumn floods and rains.

Historical and ethnographic evidence suggests that the broadcast method of sowing was the most common in the past. The effect on the weed flora caused by the widespread introduction of the seed-drill during the eighteenth and nineteenth centuries (**55**) provides us with a tool for investigating the ancient method of sowing. Edward Salisbury tells us how certain weed species, once a common sight in medieval cornfields, became increasingly rare in the eighteenth and nineteenth centuries. These weeds included purple-flowered corn-cockle (*Agrostemma githago*), the pheasant's eye (*Adonis annua*), blue cornflower (*Centaurea cyannus*), darnel (*Lolium tetulentum*) and brome grass (*Bromus* sp.). The disappearance of these weeds has often wrongly been attributed to improved ploughing. Rather it is the cleaning of the grain and the sowing in rows that diminished them so. These species survived in the cereal fields by virtue of mimicking the crop. They would grow to the same height and their seeds would ripen at the same time so that they were harvested with the crop. The seeds were loose and the same size as the grain so that they were difficult to separate from it. Consequently they were scattered across the field with the seed-grain using the broadcast method. Like cereal plants these species are annuals and their seeds, like cereals, have no dormancy mechanisms. Given the right amount of warmth, sunlight and water, their seeds would germinate with the crop, blooming before the corn ripened and producing seeds in time for the harvest. Such adaptations, however, were to prove their undoing. Seeds of plants that were left in the field would germinate before the next sowing and be destroyed by the plough or weeding (**72**), while those germinating between the rows of cereal plants were easily removed by hand and/or harrow. With no seed-bank to fall back on, every plant removed effectively reduced the population in

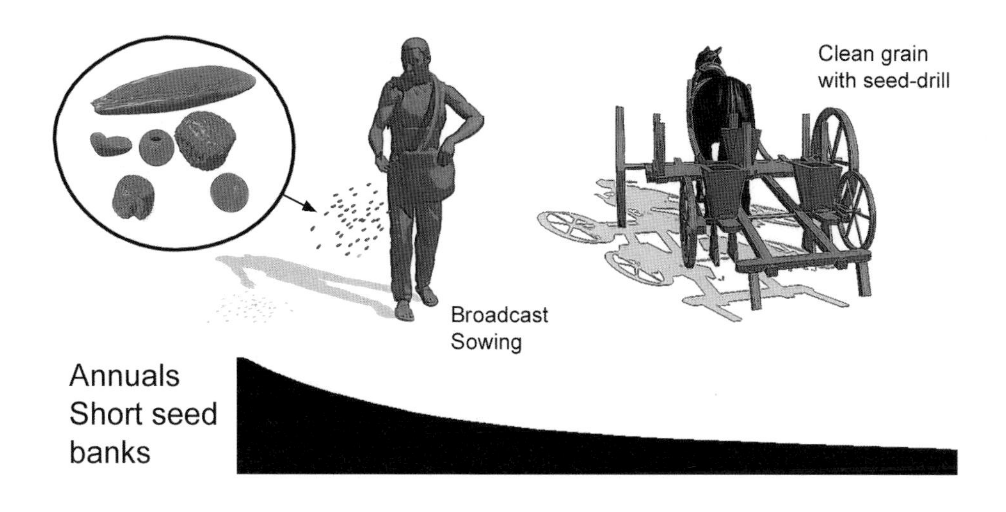

72 *Annual weed species that form short term seed-banks and which are diminished by modern planting methods*

the field. That many of these weed seeds are such common components of archaeological charred assemblages probably testifies to the widespread use of broadcast sowing prior to the nineteenth century.

We can investigate what time of year a crop was sown by examining the relative proportions of the Chenopodietea to the Secalietea. Many species of the Secalietea such as bindweed (*Fallopia* convolvus), oats (*Avena* sp.) and brome grass (*Bromus* sp.) germinate from autumn to late winter/early spring. In the case of bindweed, chilling is required for germination in early spring. Seeds therefore tend to germinate in spring, but like other species that require chilling, for example cleavers (*Galium aparine*), many germinated seedlings are destroyed before sowing of cereals occurs in later spring. The twining nature of these plants and their large seed mean that it can climb its way up autumn-sown cereal plants. Experiments at Butser Iron Age Experimental Farm showed cleavers to be entirely absent from spring-sown crops. The reason suggested by Martin Jones was that the seeds of cleavers only germinated before the field was tilled and so the seedlings were destroyed before the crop was sown in spring. Another of Jones' observations is that perennial species do better alongside autumn-sown crops. Perennials often need a longer growing season, which they do not get in spring sowing as if they are disturbed in spring then it is only a matter of months until they are disturbed again by the harvest. This means they have much less chance of flowering and setting seed.

Many seeds of species of the Chenopodietea only germinate in spring, or rather only plants germinating in spring survive to flower and produce seed. Autumn-sown crops will lack light in the spring, which discourages germination. Therefore a high incidence of the Chenopodietea group in an archaeobotanical assemblage might indicate spring sowing.

Careful readers will have spotted a problem. The relative quantities of both annual and perennial weeds, and Chenopodietea and Secalietea can be indicative of several different aspects of ancient cultivation. To explore this further let us return to Marijke van der Veen's work in northern England. As we saw, her Group A sites were dominated by annual species of nitrogen-rich soils (mainly those of the Chenopodietea) and Group B by perennial species of nitrogen-poor soils (predominantly seeds of heath grass, *Danthonia decumbens*, formerly *Siegina decumbans*). Group A was seen as indicative of intensively cultivated soils with a high amount of manuring. Group B was seen as extensively cultivated soils that were also cultivated intensively without adequate replacement of nutrients. Given our discussion we might ask why variation between the two Groups cannot be attributed to differences in the time of sowing. Van der Veen, however, dismissed this idea on the basis of questioning farmers in the Netherlands. They saw that the time of sowing had less effect on the weed flora than the type of ploughing, manuring and weeding. Nevertheless as we have seen modern farming practices are very different to ancient ones, and may not therefore provide a reliable analogue.

HARVESTING

Harvesting is one of the most important events in the agricultural year. When a crop is ready for harvesting it must be cut within a relatively short period, a few days at most. The farmer thereafter has a few weeks to get the grain processed to the state in which it is to be stored while it is still dry. The harvesting and processing of cereals for storage represents one of the most labour-demanding times of the year. The need to mobilise large numbers of people to perform such tasks is paramount to the success or failure of any community.

The time of harvesting may vary according to the time of sowing and local climatic conditions. In Britain the harvesting of autumn-sown cereals takes place in early summer while that of spring-sown cereals takes place in late summer. Barley is often harvested while it is still slightly unripe to prevent the ear shattering and releasing its grain. While this practice might effect the representation of weed species, many species are capable of setting seed between June and September, while others, by virtue of germinating with the crop will also ripen with the crop. The absence of seeds of weed species from the harvest will otherwise depend on the method of harvest. Gordon Hillman has outlined the ways in which harvesting techniques can be investigated on the basis of weed seeds found in archaeobotanical assemblages. He outlines three methods of harvesting to which we have added a fourth.

Harvesting the ears by plucking
Harvesting the ears using *mesorias*
Harvesting the entire cereal plant by uprooting
Harvesting by cutting the culm with a sickle or scythe

Harvesting by plucking the ears is likely to produce no weed seeds at all and similarly no culm nodes. Therefore, archaeobotanical assemblages from such harvesting should be free of culm nodes (straw) and weed seeds.

Leonor Peña Chocarro has recorded a unique method of harvesting einkorn used by farmers in the Sierra Subetica in Andalucía, Spain. The ears of the wheat were harvested by gripping them between two rods called *mesorias*. The rods were then clamped and pulled, removing the ears and catching them together with weeds in baskets. Such a method would produce an assemblage of grain and rachises, but no culm nodes or cereal roots (basal culm nodes), together with seeds of the weeds that grew at the same height as the cereal ears.

Harvesting by uprooting produces grain and rachises along with culm nodes, culms and basal culm nodes. Such assemblages are relatively free of weed seeds and only seeds of twining species are present (**73**) including bindweed and cleavers.

Harvesting by sickle will remove cereals, rachises, some culms and weed seeds according to the height of harvesting. Harvesting high on the culm will only bring in taller-growing weeds, such as goosefoots and those species that mimic the crop, for example – brome grass and cornflower. Harvesting low on the culm will bring in not only these weeds but also seeds of low-growing species such as clover (*Trifolium* sp.).

The evidence from Britain is that from the late Bronze Age into at least Roman times many archaeobotanical assemblages contain seeds of low-growing species such as clover, seeds of free-standing and twining weeds, and many straw nodes and stems. This suggests harvesting was carried out with sickles low down on the stem. Some sites also have evidence for sickles or billhooks made of iron, while Bronze Age sickles are also known from Britain. In Italy the early Bronze Age lakeside settlement at Fiave produced evidence for sickles made of wood with flint blades set in resin. There is little artefactual evidence for harvesting using sickles from the British Neolithic, but at the Stumble in Essex Peter Murphy found both culm nodes and free-standing weeds, which suggested that British Neolithic crops may also have been harvested with flint sickles. It is notable that many British Iron Age sites produce evidence for the basal culm indicative of uprooting. The frequency of free-standing weeds in Iron Age assemblages suggests, however, that they were harvested using a sickle. Two explanations for the Iron Age differences offer themselves. The first is that not everyone had a sickle and some harvesters simply uprooted the crop. The second is that as sickles become blunt there is a chance of uprooting the cereal plant in an attempt to cut through the stem.

Crop processing

When compared to ecological studies on the cultivation of cereals, crop processing has been very well studied by archaeobotanists and ethnographers alike.

73 *Three methods of harvesting and their effect on the seed composition in archaeobotanical assemblages*

PROCESSING FOR CLEAN GRAIN

The aim of growing the crop is to obtain clean grain that may be further processed into food. After harvest the farmer is left with a great deal of undesirable (and certainly unpalatable) weeds, chaff and straw. Ethnographic work undertaken by Gordon Hillman and Glynis Jones in Turkey and Greece respectively, has demonstrated that there are relatively few ways (in the absence of modern machinery) in which the processing sequence can be carried out. Consequently it is relatively safe to assume that peoples in the past carried out more or less the same activities in the same sorts of ways as people in traditional societies today. Each stage produces both a well defined 'product' and waste. Archaeobotanists have argued that these can readily be identified in the biological record, meaning that in theory it is possible to determine for each assemblage, how far it had been processed, or, if waste, at which point it had been produced.

The crop processing sequence varies slightly depending on the cereal crop type. Here we distinguish between what are known as **free-threshing cereals**, our modern bread and durum wheat, and barley, and the ancient **hulled** or **glume wheats**; emmer, spelt and einkorn. For free-threshing cereals the grains are easily broken away from the rachis. In the case of free-threshing wheats and naked barley the palea and lemma that enclose the grain also break away from the grain. In the case of hulled barley the palea and lemma remain tightly attached to the grain. For glume wheats the grain does not become free and remains in what are known as **spikelets**. Each spikelet consists of a rachis fragment and two glumes that tightly enclose (in the case of emmer and spelt) two grains complete with their paleas and lemmas (**62**). In the case of einkorn, as its Latin name *Triticum monococcum* suggests, the glumes enclose one grain, although on occasion there are two.

At its most basic, processing consists of activities that break things apart and activities that separate things out. The sequence is reconstructed in figure **74**.

The preliminary 'breaking' is called **threshing**. The aim is to break the ears from the straw. In the case of free-threshing cereals it also breaks the grains away from the rachis. For glume wheats the ear is split into separate spikelets. The way in which threshing is undertaken is perhaps the most variable part of the process. It may be achieved using sledges fixed with flints underneath, which are dragged by humans or animals across the harvested crop. Animals may also be used to trample the crop, the crop may be beaten with wooden poles or flails, or individual sheaves may be held and beaten against a hard or toothed surface. The result in all cases is a pile of debris. This is then raked over. Many of the coarser weeds are removed along with much of the straw and some unbroken ears by hand picking, the straw being used as animal fodder and bedding.

The pile is then winnowed. This is conducted by throwing it into the air in the presence of a light breeze. The breeze carries away the finer

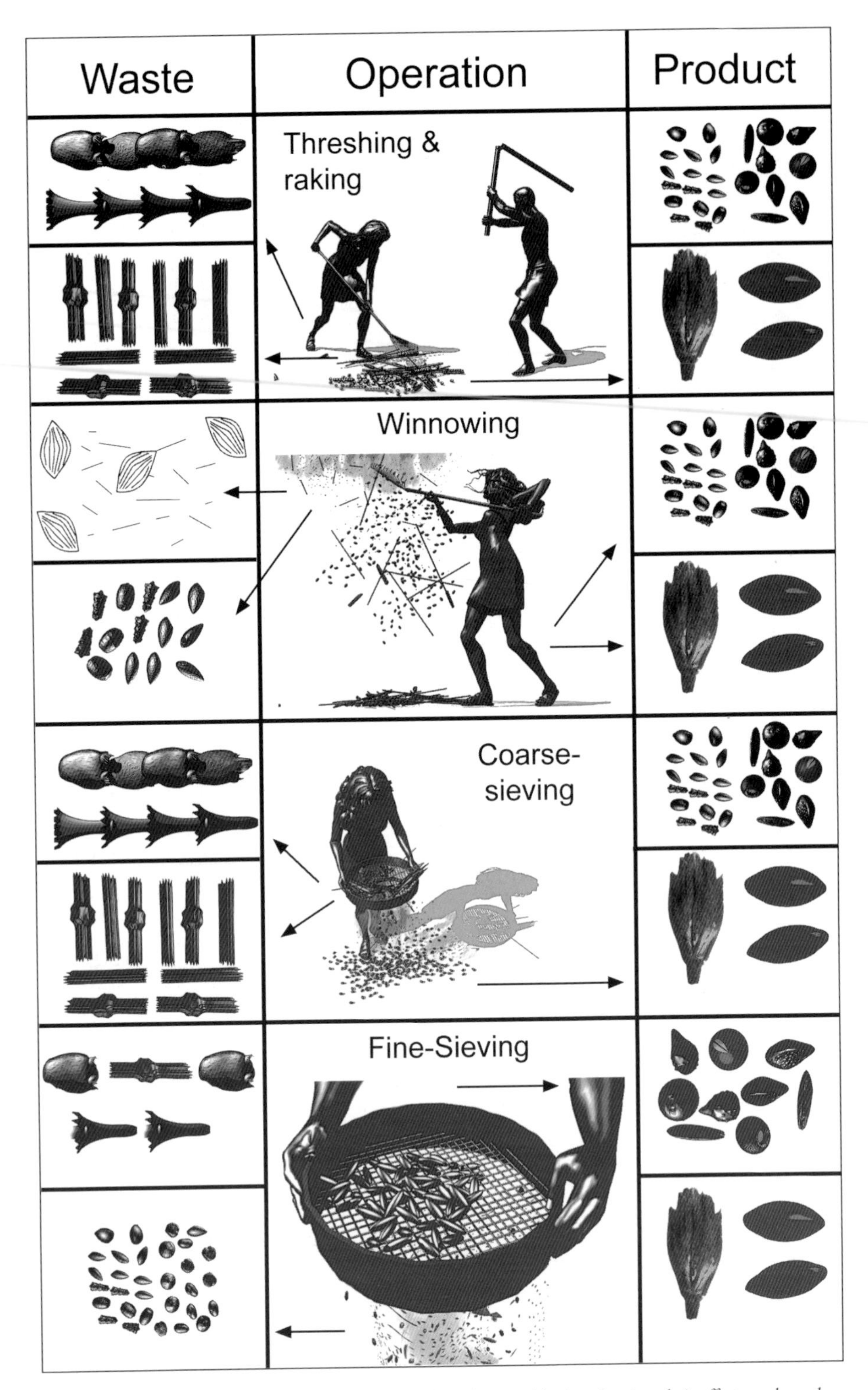

74 Above and opposite: *Crop processing stages for wheat and barley showing their effect on charred plant macrofossil assemblages. The activities themselves are described in this section*

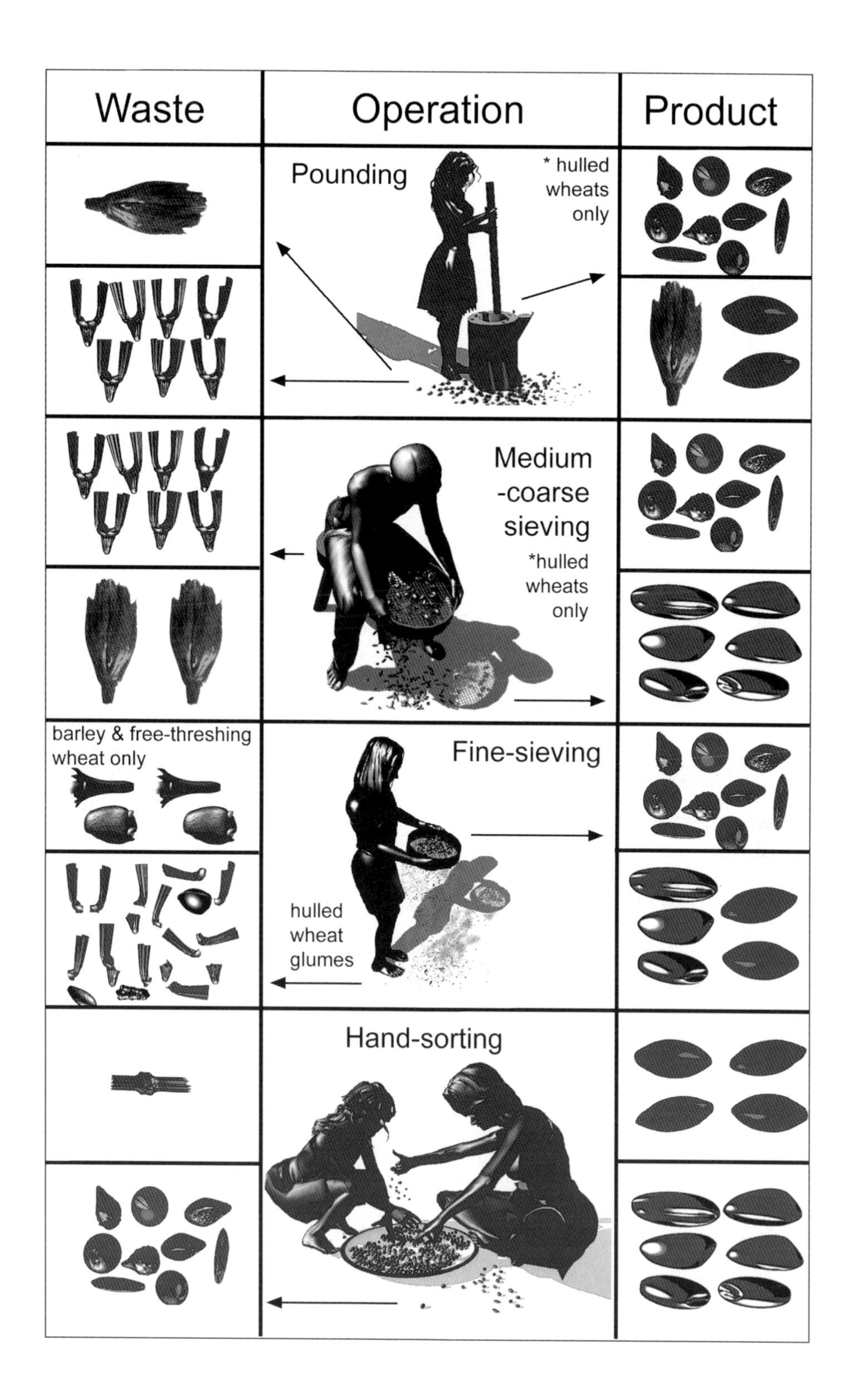

Waste
Operation
Product
Pounding
* hulled wheats only
Medium -coarse sieving
*hulled wheats only
barley & free-threshing wheat only
Fine-sieving
hulled wheat glumes
Hand-sorting

components, paleas, lemmas, awns, some single rachises and the light weed seeds. The remaining components fall to the ground or into a winnowing basket/sieve. This stage is especially important for free-threshing wheats and naked barley as it removes much of the light chaff. In the case of hulled cereals the palea and lemma have not yet been released.

The remains that are left are sieved through a coarse mesh. The sieve retains unbroken ears, weed seed-heads, broken bits of straw and, for free-threshing cereals, unbroken rachis fragments. The grain (or spikelets) and remaining weed seeds all fall through along with any single free-threshing rachis segments and a few straw culms. The coarse debris retained on the sieve is discarded.

Fine sieving is undertaken using a 'wheat-sieve' with a mesh of 2-3mm. This mesh retains grain in the case of free-threshing cereals and hulled barley, or, spikelets in the case of hulled wheats (in which case by virtue of the larger size of the spikelets a larger mesh size *c.*4-5mm can be used). Gordon Hillman suggests that this stage is optional for hulled wheats (while some authors have even omitted it) and it may instead be carried out to remove the glumes. This would take place when the glumes are removed as part of a single sequence for the storage of fully cleaned grain. Both archaeological evidence and ethnographic evidence have shown that fine sieving is commonly carried out where the crops are stored as spikelets. In either case the material passing through the sieve is discarded. Following this stage we have **semi-clean grain** or, in the case of glume wheats, **semi-clean spikelets**.

The next sequence applies only to glume wheats and hulled barley. In order to get clean grain the grain must be released from the spikelet or in the case of hulled barley, the palea and lemma. This is done by pounding or pummelling the spikelets or hulled barley grain in a mortar. For hulled barley, this may be followed by further winnowing and sieving to remove the paleas and lemmas. Alternatively the grains may be placed in water with the light chaff floating to the top. For glume wheats the grain is sieved through two sieves and then winnowed. The first sieve has a mesh that is larger than the grain but smaller than the spikelet (around 4-5mm), this retains unbroken spikelets that can then be pounded again, but also some broken spikelets complete with palea and lemma in which the grain has 'popped' out. At the same time the sieve may be used to winnow the light chaff by throwing the spikelets up into the air before it is shaken. Alternatively a winnowing basket may be used. These stages may be cycled until most of the grain is removed. The grain is still accompanied by bits of broken glume bases, some small weed seeds and fragments of rachis. To remove these the grain is sieved with the 'wheat-sieve'. The grain is retained, while remaining small weed seeds, glume bases and bits of chaff pass through the sieve.

After this stage, hulled (glume) and free threshing cereals are at the same stage once again. The semi-clean grain will still have some bits of chaff or weed seeds of a similar size to the grains within it, along with the occasional culm node or grain sized stone. These remaining items are picked out by hand, although some of the weed seeds may not be worth the effort of removing. The final stage is the preparation of the grain into food. Much of what is done will depend on how the grain is to be used. It may be malted for use in beer, it may be boiled and crushed into groats, it may be roasted and eaten as it is, or of course it may be ground upon a quern or between mill stones to flour.

These procedures sound efficient, but analyses carried out on archaeological bread, dough and bog-bodie's stomachs often show the operations did not result in clean grain and bits of chaff, unground grain and weed seeds still remain in the flour.

It was once argued that sites involved in the growing (producers) of cereals could be distinguished from those receiving (consumers) cereals. Two models were developed to define the differences. Gordon Hillman argued that people in the past exchanged grain in the form it was stored in, in other words as semi-clean grain or spikelets. Consequently, all the waste from the early stages of cereal processing would only be present on those settlements *growing* grain. However, he also suggested that, as cereals were often processed in the field, the evidence for these processing stages might be absent from settlements. Martin Jones argued much the same, but noted from ethnographic studies in Africa that grain was spread just about everywhere on farming settlements and concluded that it would inevitably end up in the fire, producing charred assemblages rich in grain. However, he also postulated that those *receiving* grain would not have the same cereal debris in their settlements. Instead the grain would be carefully stored and processed so that the resultant assemblages would be rich in chaff and weed seeds. On examining assemblages from southern England, Jones found exactly this pattern between sites that from other sources of evidence were thought to be involved in cultivation and those that were not. However, subsequent work by Marijke van der Veen and one of us (CJS), found problems with these hypotheses. Many settlements emerged as consumers and very few as producers. As prior to the eighteenth/nineteenth centuries in England the vast majority of the population were involved in farming, this seemed very odd indeed! Also some of the settlements categorised as consuming sites according to Jones' criteria had produced independent evidence for their involvement in growing cereals, including the discovery of sickles, and manuring scatters around them.

Most charred plant assemblages appear to come from routine processing of grain taken from storage. As Hillman and Jones have observed, any processing conducted before storage is likely to be carried out in the field. Even if not, the waste from these activities unless used as fuel is unlikely to enter settlement fires. The most probable source of charred material, therefore, will be those activities that are carried out on a regular basis, in other words the taking of cereals from storage, their processing and the throwing of the waste on the fire. Variation in charred plant assemblages is, therefore, more likely to relate to differing storage practices than consumption/production. As we have discussed, archaeobotany commonly tries to assign charred assemblages to a single processing stage. However, if assemblages result from an amalgum of all those stages needed to obtain clean grain for storage, then the charred plant remains will only tell us how much processing had occurred prior to that stage. So if we have only glumes, grain and large weed seeds we may conclude that storage was as semi-clean spikelets. If on the other hand we have large numbers of small weed seeds, straw culms and stems we may conclude that cereals were stored either as sheaves or as partially threshed ears.

An examination of large numbers of Iron Age assemblages containing hulled wheats from Britain revealed that most were richer in glumes than grains. Given that grains tend to be more readily preserved, this must argue that storage was in spikelet form. Indeed, Gordon Hillman has speculated that for the prehistoric period in Britain at least, this was the case, while sites such as Danebury demonstrate conclusively that later prehistoric crop storage was as semi-clean spikelets.

THE SOCIAL, SPATIAL AND TEMPORAL CONTEXT OF CROP PROCESSING

Using charred assemblages to assess how crops were stored we can begin to examine social, spatial and temporal distribution of processing activities. The cleaner the crops that are stored, the more processing that was conducted in the short period after harvesting. Less processing undertaken immediately after harvest means that more processing must be carried out on a daily basis. The gap between harvest and storage is often a very short space of time, a few weeks at most. Therefore, the period during early to late summer sees a tremendous pressure on labour within any agrarian community. As we have already seen storage for agriculturalists presents a way of spreading labour expenditure over longer periods. By storing crops in a relatively unprocessed form, farmers who do not have access to large amounts of labour may spread the processing workload across the year. For those who farm as part of larger community where labour can be pooled and moved from crop to crop as it ripens in order to harvest it, processing can be carried out *en masse*, to a point where the grain is relatively clean. They can then divide it amongst themselves and store it ready to process as and when needed. So relatively unprocessed, then, grain may be indicative of society organised by family, and highly processed grain, suggestive of organisation at the level of the settlement.

Animal production

Unlike crops the breeding and husbandry of animals is much more dependent on the eventual intended use of the animal. Again we shall divide the production sequence into three parts, that concerning the raising of the animal, that concerning its slaughter or the procurement of milk, blood and wool, and that involving the processing of the carcass, its butchery and use of its meat, marrow-bone and bone.

THE IMPORTANCE OF GOOD BREEDING

Most animals do not breed or give birth throughout the entire year. Rather the times at which females become fertile are controlled by external factors. For example, in the case of the wild mountain sheep, *Ovis canadensis*, breeding takes place in November, with a gestation period of around 180 days. One to two lambs are usually born from March to April. Most of us who have ever taken walks in the countryside in spring will be aware that the process of domestication has not had much effect on the breeding cycles of sheep. Ewes are still visited by the ram in November to lamb in March and April. In this case it is probably safe to assume that ancient farmers would have made the provision for breeding similar to those of their modern cousins.

With cattle, however, we no longer have the wild ancestor (the aurochs) to know when mating and calving took place. Although the breeding cycle of some cattle has been changed, many still follow what is likely to be the more traditional breeding cycle. This comprises of mating in June, with a 300-330 day gestation period required to calve – and like sheep – one or two offspring are produced in March and April. Pigs, however, are somewhat different. The gestation period for wild pigs (*Sus scrofa*) is around 112-115 days, with litters of between three and 12 piglets being produced between February and April. However, domestication has had a considerable impact, and some pig populations can reproduce the whole year round.

Several studies have addressed the issue of where animals were kept in the archaeological past using phosphate studies to investigate stabling within houses and by examining ditches and postholes within the outlines of past fields. At Cefn Graeanog in north Wales, phosphate analysis was conducted on several huts, with samples taken in close spacing across the excavated floors. The analysis conducted by John Conway showed that a feature interpreted as a drain running down the middle of the house contained high concentrations of phosphorous. The feature was interpreted as a drain running between two animal stalls.

On Dartmoor, divisions called 'reaves' have been found which are probably to be long strip fields divided by walls of stone rubble, now overgrown with grass. These are thought by Andrew Flemming to have been used during the Bronze Age for pasture. The moors also contain smaller square fields located in areas more suited for arable production. Similarly on the Cambridgeshire fen margins field systems were laid out in the later Bronze Age which later became buried beneath overbank alluvium. A droveway between two such fields was excavated revealing hoof prints produced by animal trampling. Phosphate analysis also suggested the presence of many animals. The excavator, Francis Pryor, noted features within the excavations that were similar to features in modern fields, such as drafting-gates, collecting pens, funnels or crushes and a race. Such features are used for funnelling sheep and then separating them into pens according to whether they are to be bred or slaughtered.

Using such information we can begin to address questions concerning where animals were raised and kept, while by assuming that natural breeding cycles have remained uniform, we can suggest when certain activities took place. The social context of such activities is more difficult to ascertain, although Francis Pryor estimates that at Fengate near Peterborough, we are not talking of hundreds, but of thousands of animals. This suggests highly communal systems, and perhaps communal ownership or at least management of herds. At Cefn Graeanog the presence of simple stalls within the household is suggestive of a very different context in which a few animals, presumably owned by Cefn Graeanog's occupants, were stalled overnight and/or during the winter.

PROCUREMENT

By using the techniques discussed earlier the archaeozoologist can assess the ages at which animals died, and even the approximate time of year in some cases. In one such analysis of a Neolithic causewayed enclosure at Abingdon, Oxfordshire. Tony Legge proposed that pigs had been killed all year round. If, rather than being used for their meat, animals are kept for their milk, they will have to be bred to guarantee that milk is constantly produced. This inevitably produces large numbers of young males who in turn consume milk supplies that could otherwise be used by people, while because only a single or few males are needed to maintain the herd, they have little economic use themselves (**54**). Animals intended for wool do not need to be bred, although it is much easier to manage and produce wool from females than boisterous males. Animals kept for meat are less likely to be kept for prolonged periods as with time the meat becomes tougher and less palatable. It is also the case that keeping animals past maturity means that resources are consumed without extra meat being produced. In modern systems the animals are kept until the fat to meat ratio satisfies modern consumer guidelines, in other words 'too old too much fat, too young not enough meat'. For this reason good modern beef breeds go for slaughter at about one and half to three years of age.

The idealised population structures can be compared to zooarchaeological data sets to see whether the herds most closely resembled those used for meat, wool or milk (**54**). Using such an approach Peter Bogucki has argued that the slaughter pattern and the presence of ceramic strainers (usually interpreted as cheese sieves) in early Neolithic Europe indicates that herds were maintained for dairying by 5400 BC. At Schalkenburg, a Neolithic site in Germany, 49 per cent of all animals made it to maturity and 20 per cent past the age of four years. From these figures it was suggested that animals were being used for milk or wool.

The main weakness to this type of approach lies in its acceptance of a universal ideal of what the best population is when breeding animals for milk, meat or wool. For this we are reliant on modern data, but as we have already stated for plants, this may not be a reliable analogue. For example, not all breeds take the same length of time to mature, while a poor diet will prolong this period. The White Fulani cattle of northern Nigeria are not usually slaughtered until the age of five. The Fulani are often reluctant to kill female animals until they are past breeding age, which is between two and ten years, although they are milked during this time. The Miranda cattle of central and north-east Portugal are not bred for milk at all and are instead primarily used for traction. However, because they are used in breeding, cows are often not sent to the abattoir until they are seven to ten years old. The Ganda of Uganda also keep cattle, but not for milk as that is considered taboo. However, animals are only slaughtered for meat after they have become barren. For bulls this may be as late as five years, for cows as old as eight to ten years. Perhaps the weakest aspect of the use of modern analogues regarding age stratification in herds, however, is that they ignore the fact that many peoples such as the Masai exploit cattle for blood in addition to milk. Cattle are an important sign of status and they are therefore only killed for feasting to increase an individual's standing within the society. For these reasons it is often better to rely on other evidence for the identification of milk in the archaeological record such as the analysis of residues from pottery. As an aside it would be useful to see residue studies consider the exploitation of blood rather than just concentrating on foodstuffs most commonly used by modern Western cultures.

It is worth considering the use of present practice as an analogue more carefully. Suppose a culture developed a liking for young calf meat, and the settlements bred calves in large numbers for this purpose. Archaeologically the population would look like a modern dairy assemblage. This was exactly the sort of problem Terry O'Connor faced when examining sheep skeletons from Roman York. Large numbers of juveniles were present but no method of reliably sexing them was available to determine whether they were mostly male. After considering dairying as a possibility, the still relatively young age of the few mature sheep found at the site led O'Connor to suggest that the local Romans had developed a liking for young lambs, perhaps in part for the soft leather their skins would produce. It was also noted, however, that the lambs had all been slaughtered at the same time of the year, in May or June. It is possible, therefore, that the bone assemblage might result from animal sacrifices, a topic that we shall examine in section 4. O'Connor states that such a kill pattern does not seem to accord with any universal notion of effective economic practice that we understand today.

The processing of animals

There are two ways of looking at butchery. One is to see it as exploitation, i.e. getting the most from the animal carcass. The other is the view taken here, that butchery in any economy past or present is fundamentally about sharing. After an animal is killed it must be butchered to share out the carcass, between both living people and sometimes spiritual beings (see section 4). For cereals, the methods used in processing, the waste from earlier processing stages, and the way in which the crop was divided may all be invisible. This is not the case with animal husbandry. Individual skeletal elements may all be identified, while sometimes the method used to separate that element from the rest of the body, or remove meat and marrow, may leave its marks on the bones. Sometimes these will be decisive chops, cutting the bone straight through. Sometimes the bone is broken or smashed open for the marrow (**75**). Sometimes niches or cuts may indicate where limbs were detached or meat cut away. The study of butchery marks upon animals may, therefore, reveal a great deal about slaughter and processing. Such studies may indicate whether the animal was processed for meat, horns, skins, glue or grease. The butchery of animal carcasses can leave distinctive marks relating to the tools and the methods that were used. At its most simple level the zooarchaeologist can distinguish between 'cuts' made by knives and flint blades (**76**), and chops made by heavier cleavers or flint axes. The other, albeit rarer, contender are marks made by saws. Study of the marks on a bone using a low-power binocular microscope or magnifying lens can clearly show the nature of the implement used and sometimes the direction of the cut. Flint implements tend to leave multiple 'V'-shaped notches, whereas metal tools leave distinctive and often deeper 'U'-shaped profiles.

Butchery may be learnt and copied or just guessed at. Its practitioners may have developed great expertise and show skill in making their cuts. In such cases the cuts may separate meat from bone swiftly and efficiently, leaving little mark. The bone evidence may equally show the inexpert hand and a number of cuts may have been needed to achieve the butcher's aim. Of course, much butchery will depend on the intentions of the individual and the nature of the social practice concerning meat. It will depend on how much, how fast and how accurate processing of the meat was required to be. In societies where many carcasses are to be quickly processed for transport, exchange or storage, and good implements exist, bones may simply be hacked straight through and the joints sent on their way. Mark Maltby has suggested that differences seen in Roman Britain between town and countryside reflect such priorities and technology, with urban population using heavy choppers to quickly dismember and distribute carcasses. Where such heavy implements are not available carcasses may be quickly separated by good knife work. Depending on the ultimate destination, certain cuts may be required. For example, meat which is to be minced does not need to be butchered with the same precision as what are considered to be the best cuts.

The carcass must be drained of blood to avoid contamination. It must also be hung, as areas in contact with surfaces will rot more quickly. Burning on, or of, bones implies that they were most probably 'roasted' with the meat on. Large quantities of unburnt bone would indicate that the meat was either boiled or cut from the bone before cooking.

Other than edible products, animals can also be exploited for their fur, wool, skin or horn (**77**). Despite what has been said of the evidence from Schalkenburg, there is no positive evidence for the use or production of wool within Neolithic Europe at present. Certainly the evidence for the iceman's clothing gave no indication that wool was used even though he was popularly believed to have been a shepherd. The increase in the numbers of zoological remains of sheep/goat during the Bronze Age is taken as an indication of the introduction, or at least increased use, of wool. Anthony Harding has drawn attention to the fact that wool and other animal products do not survive water logging to the same extent as vegetable textiles, such as flax, meaning that wool may be under-represented in the archaeological record. From this base he argues that the reduction in textile finds from Bronze Age contexts in Alpine lake villages indicates a switch from the use of plant fibres to wool for the production of clothing.

Skins like all organic parts tend to rot away with age. To be of use they must be preserved. Like meat they may be preserved or cured for a while by salting and drying, but to last for any length of time they need other special treatments. The word tanning is derived from the use of what are known as **tannins**. Tannins are contained in many plant parts, including seeds, nuts and barks. To prepare the skin it is usually salted or dried after being removed from the animal. In drying the skin is stretched and rubbed with brains and fats, which have a softening effect. The skins are then washed and re-soaked in water after which they may be placed in a solution of lime, which facilitates the removal of any remaining hair.

Evidence for the husbandry of animals purely for their skin is almost as slight in the zooarchaeological record as it is for keeping animals for wool. The best bone evidence for such practices is cut mark evidence targeted at joints and around the head. However, this will not indicate that the entire herd was exploited in this way, but merely the animal in question. Indeed, as with wool, the best evidence of animal husbandry for obtaining skins is the artefactual record.

Summary

In this section we have looked both at how past (mainly agricultural) societies carried out their subsistence activities and at how such activities can be recognised in the bioarchaeological record.

Having established the background and examined some of the main issues with which bioarchaeologists have become embroiled we turned to the main subject of the section, namely the use of biological data to understand past economies. We examined the main categories of botanical data that are used for this purpose, but focused particularly on plant macro-remains – and indeed, mainly on the examination of seeds. Preservation of these proxy records is by

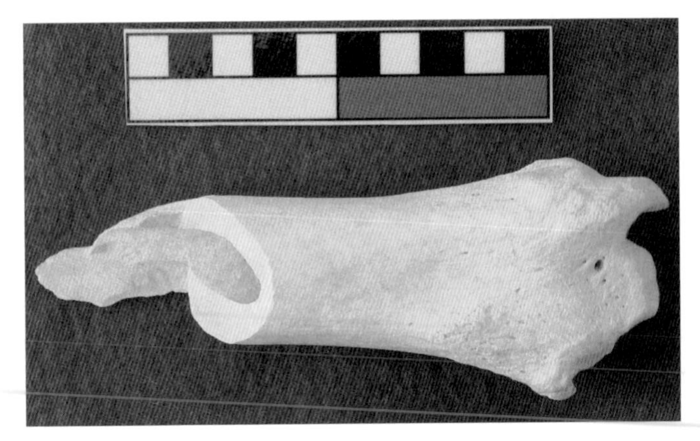

75 *A cattle tibia exhibiting a fresh bone fracture, that is suggestive of marrow extraction. Scale is 10cm long*

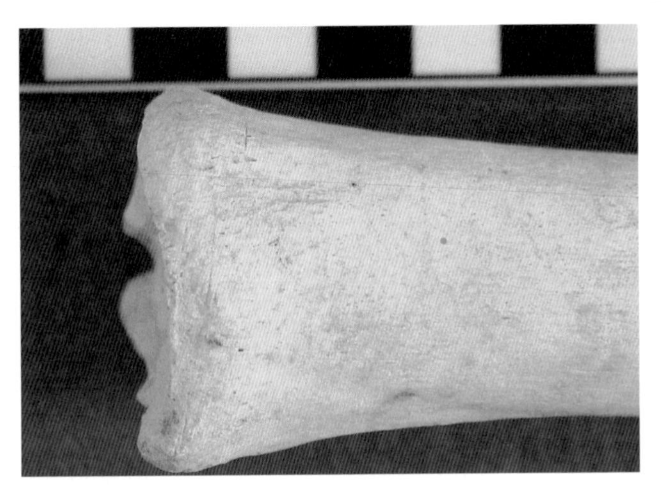

76 *Cut marks around the proximal end of a horse metatarsal from Iron Age Oram's Arbour, Winchester. Scale bar in 1cm divisions*

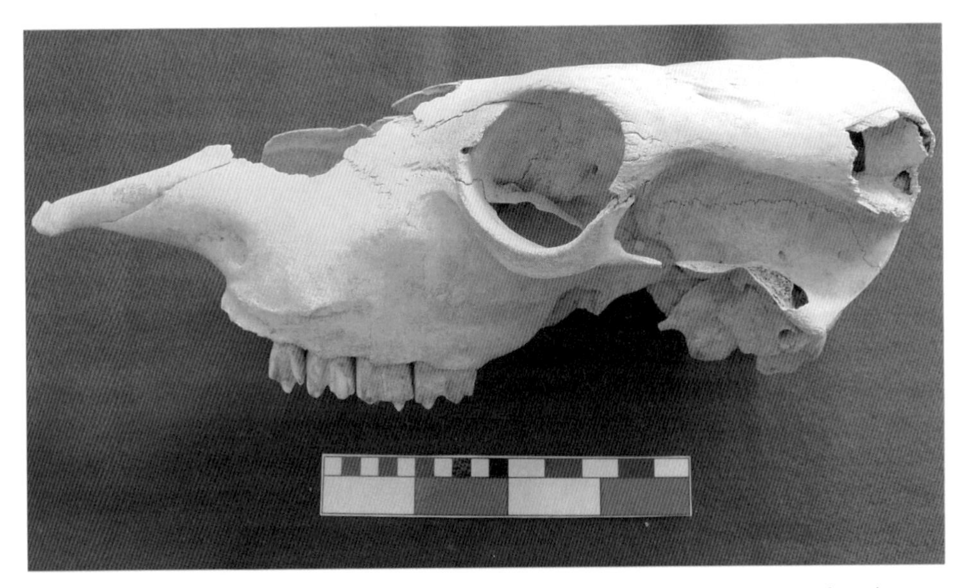

77 *Cattle skull from Oram's Arbour showing chop marks where the horn core has been removed. Scale is 20cm long*

charring, waterlogging (which we also covered in section 2), mineralisation and desiccation. Of these, attention has been overwhelmingly focused on charred plant macro-fossils as these are ubiquitous finds on most archaeological sites. They are also the category of plant macrofossil that best survives in those areas of world which saw the origins of agriculture, namely the Near East and central America, and the Bronze Age agricultural intensification (the Near East and eastern Mediterranean margins). For each preservation circumstance we examined the types of contexts in which macrofossils survive, the importance of understanding the taphonomy of the deposit, methods of extraction and analysis and, most importantly, interpretatory frameworks.

After considering the botanical data we turned to that of zoology. Given the importance of vertebrates, and in particular mammals, to human subsistence we concentrated on animal bones (although we examined marine molluscs briefly for their importance as a dietary supplement). Animal bones survive well on the majority of archaeological sites that do not have acid subsoils and are usually recovered in a similar way to artefacts, i.e. by hand excavation. We examined how animal bones are identified, to both part and side of the body, and then quantified in order to examine the importance of individual species to subsistence. We also examined why it is important to estimate the age and sex of the vertebrates that are found on archaeological sites, and how this is achieved in practice.

Biomolecular studies are a relatively new, but a rapidly expanding, area of environmental archaeology, that is rapidly evolving into a separate sub-discipline of archaeology. We examined, albeit briefly, how residues can be chemically examined from ceramic vessels to determine what liquids they once held and how morphological evidence from starch grains allows us to reconstruct ancient brewing practice. Isotope studies are an even more recent development, and – using the maxim of 'you are what you eat' – have enabled researchers to determine the relative contribution of marine and terrestrial foods for prehistoric peoples using ratios of carbon and nitrogen isotopes. Finally we looked at the 'new kid on the block' as far as bioarchaeology is concerned: the analysis of DNA and its contribution to examining the domestication process. We saw how studies are in their infancy, but that already we now know that domesticated maize was bred from a single population of teosinte in central Mexico and that the first wheat domesticate (einkorn) originated in south-east Turkey.

In the last part of the section we returned to the subjects of production and consumption with which we began. We looked at the stages by which a cereal crop is produced, examining for each what the archaeological and bioarchaeological signal might be, and what this might tell us of past economies and subsistence. We concentrated in particular on processing stages that come with and follow the harvest, and how specific suites of plant macro-remains can be indicative of how much processing had been carried out prior to crop storage.

We then gave vertebrates the same treatment, examining how herds are reared and how this depends upon what their intended use is. We looked at how the bone record can inform us both of what this usage was, how carcasses have been dismembered depending on use and how we can recognise signs of animal husbandry for milk, wool and hide from the bone data.

Throughout this and the previous sections we have taken a functionalist approach. Plants and animals are indicative of an environment that people exploited, or plants and animals that were consumed or otherwise *used* by people. However not all bioremains recovered from archaeological sites are simply evidence of past environments (section 2) or economies (section 3). There is a growing realisation among archaeologists that the plants are deposited and animals killed and used for reasons other than subsistence, and which may not be typical of the surrounding environment. The study of these is the subject of our next section.

SECTION 4

ENVIRONMENTAL ARCHAEOLOGY AND IDEOLOGY

Integrating the natural environment with social ideology and economy

The interpretation of bioarchaeological data usually ends with an economic or environmental reconstruction that is based on function. This is partly because many environmental archaeologists believe economic questions are more approachable than those which relate to ideology – or as it is more commonly termed in archaeology: ritual. By ritual we refer to actions that are associated consciously or subconsciously with symbolic meaning, for example shaking a stranger's right hand because it was once their sword hand, and could not be used to simultaneously run you through.

The rise of post-processualist theory has prompted non-functionalist interpretations in archaeology, but these interpretations are just as applicable to biological remains as they are to artefacts and archaeological features. Therefore, in this section we consider how the data of environmental archaeology have been used in the interpretation of past ritual, how rituals may have come into being and how they relate to economic practice.

Ideology, economics and attitudes

A division is often made between economy and ideology, but as any good sociologist will tell you they are in truth inseparable. Future archaeologists digging up the remains of our meals might recover chicken femurs, ribs of pigs or sheep's scapulas – drumsticks, pork ribs and shoulders of lamb. They might find charred remains of Solanaceae tubers, legumes and baked wheat flour – chips, beans and toast. Such foods differ considerably from what we find archaeologically. Our own rubbish rarely contains the skulls of animals, cereal grains and chaff. Rather it reflects the types of foods we like and the ways we choose to exchange and consume them (**78**).

The deposition of our 'food-waste' is also different. People in the past put 'waste' on middens to be spread on the fields or disposed of it in 'special' ways. Today much of our waste lies buried in landfills.

Our own society's ideology affects consumption through social values, resulting in for example vegetarianism or the purchase of organic products. It acts through religion, for instance Muslims abstaining from pork and alcohol. Through new trends, many of our parents and grandparents ate very different

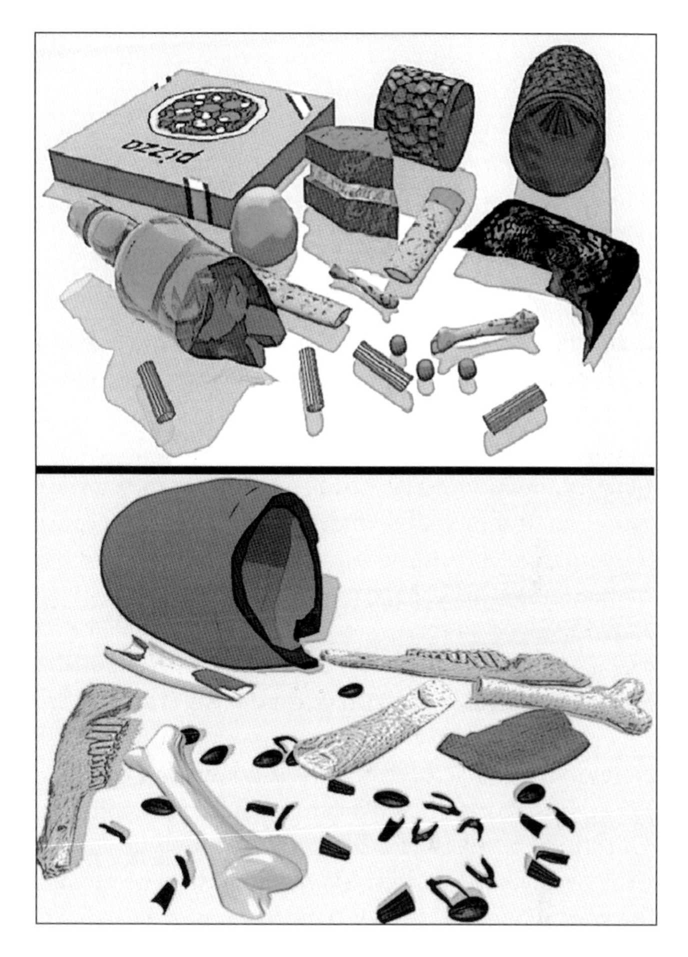

78 *Modern versus archaeological waste*

foods to those we consume today. Lastly, while many of us eat meat, our psychological distance from its production means few of us could stomach working in an abattoir. Much of this reflects attitudes embedded within our society. We make personal choices, but these choices are tailored by how we relate our personal experiences to the ideological concepts or values of our society. We also have ideological beliefs that shape our attitudes towards the environment. We keep pets, we value conservation of beautiful natural places, we watch and play games within sports 'fields', we grow flowers in our gardens, we teach our children to read using animals and plants, we watch cartoon animals on television.

If ideology plays such an important role in our own attitudes towards food and the environment then this is likely to have been true of people in the past. When we look at biological evidence recovered from an archaeological site it is probably revealing much more than just what people ate or the environment in which they lived.

If we aim to use environmental archaeology to understand the relationship between humans and environment in the past then we need to understand how past peoples might have thought about their environment. For literate societies views of this type are sometimes articulated through text, but for most past cultures we have nothing but the archaeological record, including ecofacts, to help us.

As a start to exploring the topic we will examine the fundamental aspect of ideology, namely religion, and how perceptions of the environment embodied within hunter-gatherer belief systems became transformed into the religions of early civilisations. We will then look at how elements of the environment become embodied within ritual, myth and economic practice. We follow this by discussing the symbolic role that biological remains had within past societies, including the nature of ritual offerings and 'deposits' as seen from both the ethnographic and archaeological record. Finally, we discuss how ideology has led to the formulation of taboos and how these are manifested in the bioarchaeological record.

From natural spirits to gods in the environment

We are all familiar with the various ways in which today we use elements of the environment to attribute certain characteristics to people. She was as beautiful as a rose. She's been a right cow. He behaved like a monkey. He's a real weed. They followed like sheep. This use of environmental associations to describe and classify social concepts seems straightforward. Yet as the sociologist Herbert Mead stated, such associations become symbolised within social ideology through language.

The work by anthropologists and sociologists such as Claude Lévi-Strauss and Emile Durkheim has shown that the way in which people use plants and animals to think can be extremely complex. Environmental elements are not only symbolised in people, but become incorporated into myths, creation stories and cosmologies.

Animism and totemism

One such group of beliefs is collectively called **animism**. This term covers those belief structures that attribute souls and spirits to particular aspects of the environment. By possessing a soul, environmental phenomena (including not only animals and plants but also aspects of the physical environment such as the moon, sun, Earth, stars, volcanoes, rivers, and even climatological phenomena, such as thunder, storms and the wind,) are brought to life or 'personified'. They are seen as having emotions, the ability to communicate and a 'social' order. Animist societies see the environment (including humans) as formed from a great many reciprocal relationships. Each element is dependent upon another; each knows its place and performs its reciprocal duties.

In contrast to animism, **totemism** uses environmental phenomena to identify and hence symbolise individuals and social groups. Such symbols are used to construct social ties between living people, and between them and their ancestors. Totemism shapes lives through structuring an individual's identity, both in their social world and with their natural environment.

Animism and totemism, through the association of environmental phenomena with gods and creation myths, have been seen as the basis of all religions. Through ritual these symbolic associations are brought into every aspect of social life. To paraphrase Catherine Bell, rituals like myths impose an order on the world that simultaneously refers to the origin of that order. They then shape people's experience of, and dispositions towards that order.

The main ritual associated with animism is **shamanism**. The shaman provides a means of assessing the souls of environmental phenomena, the link between the living and spiritual world. Shamans communicate with the spirit world by entering trances in which the spirit exits its body to seek guidance from other spirits. In these journeys it may travel upon the earth, to the sky, or, subterranean, underwater, or ancestral worlds. Journeys may be made to seek knowledge of better hunting grounds, better weather or to make medicinal cures effective. The shaman may guide the spirits of the dead on their journey into the afterlife or seek guidance from ancestral spirits. To undertake this journey the spirit of the shaman often enters the body of an animal. As the shaman takes animal form so may other spirits of environmental phenomena adopt animal form. The shaman may therefore become a bear, tiger, walrus or jaguar, seeking assistance from other spirits perhaps themselves in such a form.

The shaman's role is as much about ritual and performance as anything else. During trances the shaman will 'act' the roles of the spirits that are encountered on the 'journey'. Often they undergo physical transformations to aid this spiritual transformation, wearing costumes consisting of the pelts of animals, antlers, horns, skulls and/or branches. Shamanism, and by association animism, is probably as old as human culture, a point testified by finds in Palaeolithic contexts. 'Modern' rock art often depicts these transformations. One from South African !San shows an antelope-human shaman in a trance that has induced a nose-bleed. An almost identical depiction of a half-human half-bison is seen from the Palaeolithic cave, Trois Frères in France (**79**). The lion-headed man carved in mammoth ivory from Hohlenstein-Stadel cave, south Germany (*c.*34,000-30,000 BP.) might also represent such beliefs.

Evidence for shamanism might also come from the British Mesolithic. At Star Carr (*c.*10,750-10,400 BP) fragment of deer skull with the antlers attached had holes bored through it, suggesting they had been used as a headdress. This might have been used as a hunting aid as is suggested from ethnographic

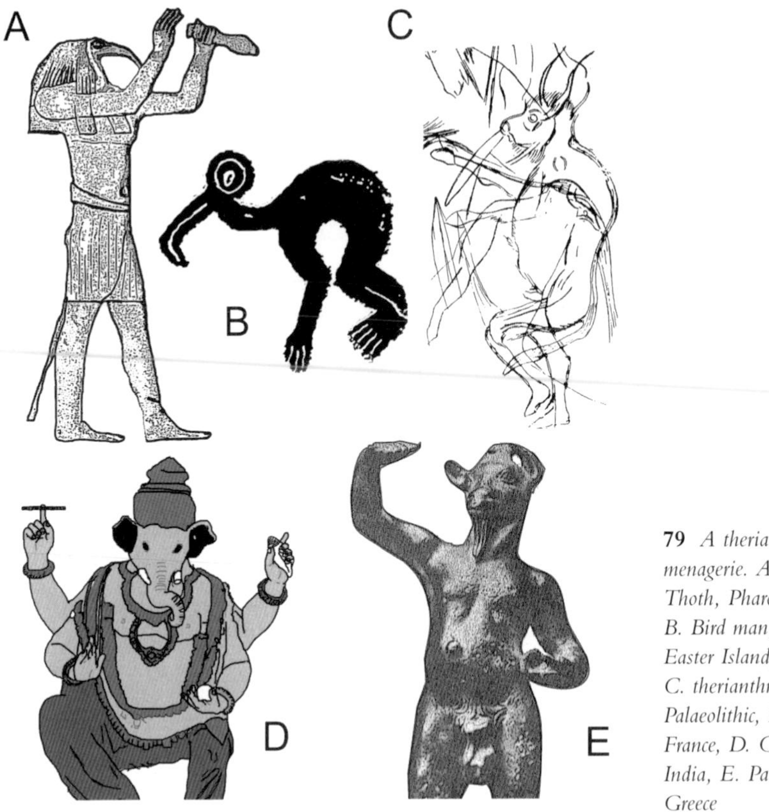

79 *A therianthrope menagerie. A. Ibis-headed Thoth, Pharonic Egypt, B. Bird man, c.AD 900, Easter Island, C. therianthrope, Upper Palaeolithic, Les Trois Frères, France, D. Ganesh, modern India, E. Pan, Classical Greece*

accounts of the Karok Indians of California, but equally plausible is the suggestion that it represents part of a shaman's costume (Grahame Clark found a similar depiction of a shaman from an eighteenth-century Russian engraving, wearing almost identical head gear (**80**)). Such costumes are also used within many societies to denote important persons such as chiefs.

The legacy of animism and shamanism is often clear to see in the religions of later societies. Within agricultural societies many aspects of shamanism become reversed. Rather than entering a trance the 'shaman' become possessed, allowing spirits to speak through them, as seen for example in the oracle priestess at Delphi. Similarly gods in early religions transform into animal disguise to visit the world of humans rather than the shaman coming to them. Lastly, we see how the 'souls' of associated environmental phenomena become unified or personified as a single spirit being, deity or god. The associated environmental phenomena then become their totems.

In this way the gods of the Old World: the Greek Zeus, Roman Jupiter, Scandinavian Thor, Slavic Perun, Lithuanian Perkunas, Indic Parjanyas and Teutonic Dunar all share similar characteristics. They are principal gods of the sky, rain, fertility, thunder and oak trees. As shamans were once depicted as animals, so divine beings and even kings become depicted as animals, therianthropes (half-animal half-human beings) or in human form alongside animal totems (**79**). Demeter, the Greek corn goddess is sometimes portrayed with a horse's head, Dionysus, the fertility-nature god as a goat or faun or sometimes a bull; these were also the animal forms they assumed. The Norse god Loki could transform into many different animals including a seal and a hawk, while the 'Celtic' god Cernunnos is often depicted looking shaman-esque with an antler headdress.

The Hittites of Anatolia (*c.*1700-1200 BC) had bull-headed 'gods' who supported the sky. More frequently gods stood on the back of their animal symbols, as was common in Mesopotamia. The Ziggurat of Ur was dedicated to the city's main god Nana, the moon god. His symbols were a recumbent moon, the bull, lion-dragon or antelope. Enki or Ea the water god was depicted with many horns and water containing fish running from him.

Ancient Egypt too had many animal-headed gods. Geb, god of the earth, was shown as goose-headed man or as man with a goose, sometimes on his head. His sister and wife Nut represented the sky, and was sometimes shown as a cow with stars on her underbelly. Their son Osiris was the fertility god, associated with the Nile floods. Perhaps best known is Horus, the falcon-headed man. The falcon was the symbol of kingship and the myth of Horus tells of Horus as the first king.

Such representations are known beyond the Old World. The Olmec people of Central America (*c.*1500 to 600 BC) had numerous carvings of half-human half-jaguar beings, sometimes with rain emanating from their mouths,

80 *An eighteenth-century engraving of a shaman from Tungusic, Russia*

interpreted as rain gods. It is thought such symbols emerged in shamanism to be moulded by Olmec kings into symbols of power, prestige and control.

Evidence for totemism in past civilisations is limited. Between *c*.3000 and 2500 BC in both Sumer and Egypt we find symbols for individual cities that might have evolved from totemic symbols of clans or tribes. The city of Larsa, in Southern Iraq, was symbolised by an altar with a sun; other cities had birds or snakes on altars. Such symbols seem to correspond to the main god of the city. So at Larsa we find Utu, the sun god; at Eridu, Enki the water god; at Ur, Nana the moon god. In Egypt at the same period something similar may have existed. The Tjehenu palette depicts seven fortified 'towns' (**81**). Each 'town' has an animal or 'plant' within it and a different animal on the corner clasping a hoe. While the scene has been interpreted as the animals attacking the towns it would seem more likely that they represent totemic emblems of towns and the names of individual rulers or gods. Certainly early Egyptian Kings had such animal names. One such King was the Scorpion King who is depicted with a hoe on a ceremonial mace head.

Myths, metaphors and natural rhythms

The interaction of humans with the natural environment is often symbolised through a combination of ritual, religious belief and myth. It is not unusual for celestial cycles, myths and environmental symbols to become embodied in the use and structure of space and may even govern the movement of people through it. Such interactions may also develop into special treatment of certain spaces and places.

81 *The Tjehenu palette (pre-Dynastic Egypt – found in Libya)*

Sacred places and spaces

The Pawnee tribe of North America associated the wolf and bobcat with the southeast and southwest respectively, the mountain lion and bear with the northeast and northwest. Each animal was associated with bringing a coloured maize into the world; red, white, black or yellow; and one of four tree species; box elder, poplar, elm or willow. The Pawnee lodge was constructed around four central posts, each post carved from one of the four tree species chosen according to its position. It was then painted in its symbolic 'maize' colour.

For many totemic societies the occupant's clan and totem might dictate the position of their house within the settlement. For example, the Australian aborigines of Cherbourg, Queensland laid the dead facing east or west depending on their moiety.

In the ancient world many structures also seem to pay attention to features of the natural environment. It was once thought that the southeast orientation of prehistoric British roundhouses was to keep the door away from northerly and westerly winds. That such attention to orientation is seen all across Britain, including for non-domestic structures, suggests rather a symbolic connection to the winter equinox sunrise.

Recently, there have been a number of works exploring the relationship of archaeological monuments to the natural environment. Richard Bradley has suggested Scottish cairns were orientated southwest towards the summer moon. Many also showed selective use of geological materials. On some the south-west faces included substantial amounts of quartz that would have reflected the midsummer moonlight.

Aspects of nature symbolised within both ancient and modern societies are often environment interfaces – where, for example, land meets sky, water meets land or 'upper world' meets 'lower world'. Chris Scarre has noted that passage graves in northwest Europe were deliberately placed on the coast or in inland positions visible from the sea. Such positions can be seen as a reflection of such beliefs, the division between land and sea symbolising the division between living and spiritual worlds.

The deposition of sacred objects within watery environments is characteristic of several ancient European ideologies. Examples include the swords placed by Bronze Age peoples in the River Thames or the Gunderstrup cauldron deposited in the peat-bogs of Denmark some time in the Iron Age.

Ritualised shafts and wells are also reasonably common in Europe. Frequently they contain waterlogged deposits, thus providing a unique source of biological data. At some point in the late Bronze Age at Wilsford near Stonehenge in southern England a 30m deep shaft was dug into the chalk. As we saw in section 2, biological evidence showed it to be located in cleared grassland which was used for pasture. Finds of wooden bucket bases and numerous pieces of rope suggested it was used as a well. However, the presence of carved bone pins, animal bones, pottery and beads suggested a ritual purpose. The two views have been seen as conflicting, but evidence from elsewhere in Europe clearly demonstrates that wells (used primarily for drawing water) were also seen as sacred places. One of many examples comes from Berlin-Lichterfelde, Germany where 100 small pots were found in the well. In the pots were the remains of willow catkins, lime flowers, grain, orache and aromatic herbs. Lossow, a defended site in eastern Germany dated to the Halstatt period (*c.*850-500 BC), had over 50 wells. One contained the remains of a whole stag, another a human sacrifice and others, further articulated remains of animals and humans.

Beliefs such as these were not confined to Europe alone. At Hattusas, the Hittite capital, a shrine was found whose floor was sunk one and half metres below that of the surrounding rooms. Layers of mud and sand indicated that water flowed within the room, with an outlet at one end. Within the sand and mud were 'ritual' deposits, including piles of shells and votive pots.

Two of Britain's most famous Bronze Age sites, Flag Fen and Seahenge, both in East Anglia, reveal similar ideological practices. At Flag Fen, around 1000 BC a huge platform the size of a football pitch was built, linked to the land by two alignments of posts strung out across the fen. In the water around it were large numbers of what appeared to be votive deposits; pottery, tools, metal, weapons and animals, including a dog with a stake through it and an alder wheel. The distribution of finds also suggested structure in their deposition.

Seahenge was a ring of posts that stood just out to sea, long buried by sand and peat. It came to the attention of archaeologists after storms had exposed much of it, threatening its future survival (**82**). Much of the environmental context of the site had long disappeared, but it is certain that even in the past it would have stood where land meets sea. The circle consisted of some 55 oak posts forming a continuous ring wall. In the centre was the base of a large upturned oak, inserted so that the roots formed an altar of sorts. Around the 'altar' were the remains of mistletoe ropes, which were perhaps used to bring the oak to the ring. We do not know what ceremonies were performed her, but the exhuming of the dead and/or sacrificial offerings would seem most probable. The position of the site and the presence of the up-turned oak could symbolise the meeting of worlds – the 'living-world' and the subterranean, watery, spiritual world. Other evidence gives us even more clues as to the nature of the site. Dendrochronology revealed that the trees were all felled between April and June in the years 2050 and/or 2049 BC. This short period of construction must have involved a great many people. This is borne out by the cut marks on the wood revealing that some 50 to 60 different axes had been used.

Caves also formed the focus of past 'interface ideologies'. Their importance to Upper Palaeolithic societies is obvious from cave art, but they also featured in later cultures. Sir Arthur Evans, for example noted that the Bronze Age Minoans used caves for ritual purposes including the deposition of animal and human figurines.

82 *Seahenge reconstructed*

Gratifying the spirits of nature

Animist beliefs often encompass elements of reciprocal exchange between humans and the spirits of the natural world. Hunters will treat their prey with absolute respect and conduct certain proper procedures to ensure that they have completed their obligations to the animals' soul. The Cree Indians believed animals presented themselves as 'willing' the hunter to take their life. Only when eaten was the animal's soul released to become re-fleshed. The importance of respect is that if the soul is displeased, future animals will no longer present themselves to the hunter. For many hunters such respect might also involve the begging of forgiveness or over-acted expressions of guilt and sorrow for the taking of the animal's life. Respect must be shown to the animal when it is killed and it must be butchered in the 'correct' way. The meat and other organs must be shared out, sometimes with the wider community. Specific parts sometime go to specific persons, while the bones must then be disposed of or treated in a fitting manner. Such treatment sometimes involves burial, such as was practised by Alaskan hunters with the bones of beavers and sable which they protected for a year before disposal. To paraphrase Timothy Ingold, the animal is welcomed into the human community following a successful hunt. The soul leaves 'satisfied' following the 'correct' disposal of the bones. Only with the correct procedure is the cycle permitted to continue.

Some hunters select bones or organs thought to attract the spirits or contain the souls of hunted animals. Fishermen of the Timor-laut Islands in the Indian Archipelago prayed to turtle skulls kept under their houses so that their souls might entice others to be caught. Hunters of the Ewe of Togoland, Ghana erected shrines of branches on which they hung lower jawbones of their prey. Before hunting they went to the shrine offering food and wine to the souls of the deceased animals so that they may tell other game to be taken by the hunter. Hunters of Poso, Indonesia hung jawbones of pigs and deer near the fire, so that the souls might attract living animals to the hunter.

Natural rhythms

For many societies life and death, celestial rhythms and seasonal cycles in nature are all embodied in ritual and myth. Natural cycles that dictate the availability of wild plants, mating and migration of animals, or times of sowing and harvesting become themselves symbolised through association with natural phenomena. For example, in ancient Greece the festival of Proacturia before Arcturus marked the beginning of ploughing. Arcturus is from the Greek *arktos* or bear referring to the bright star of the same name, while the festival took place when the star was high in the night sky. The Osage tribe of North America associated the flowering of Blazing Star, *Lancinaria pycnostachya*, with the end of bison hunting, the ripening of maize and the beginning of harvesting.

Through the association of natural phenomena with gods, myths are constructed that encompass the natural cycle. These then become embedded in human ideology through ritual. The rituals then shape the activities themselves, and, therefore, the activities become synonymous with the acting out of the myth. Nowhere can these associations be more clearly seen than in James George Frazier's *The Golden Bough*. Frazier could see the same rituals and myths repeated time and time again in both the records of classical civilisations and observation of tradiitonal farming commodities. The basis of the myth is that the principle character is taken to the underworld and later reborn, a sequence of events that represents the growth of the crop. The events that befall the character are related to events in the agricultural cycle. For example, the Greek goddess Persephone is often depicted as having sprouting cereals emanating from her body. In the myth Persephone is carried to the underworld by Hades to be his bride. Demeter, her mother, learns from the sun of Persephone's whereabouts and appeals to Zeus. Zeus decrees that Persephone should spend a third of the year with Hades in winter below ground, returning in spring with the flowers to be with her mother. Apollo undergoes a similar cycle. On his birth Aphrodite entrusts Apollo to Persephone, who refuses to return him. Zeus again decrees that he should spend part of the year above ground, part below.

In ancient Syria and Babylon the story is retold, only with Tammuz or Dumuzi as the principal character. In various versions of the myth, Dumuzi becomes trapped in the underworld, sometimes with his lover Ishtar. The lovers are then sprinkled with water and escape. In another myth Dumuzi is killed and his bones ground like grain and scattered to the wind. In another, Ishtar is held captive and once a year Dumuzi takes her place so she can return to the earth.

The story of the Egyptian god Osiris, the symbol of fertility and the Nile floods follows the same pattern. Osiris is tricked by Seth (his brother) into getting into a large chest. Isis recovers his body only to lose it to Seth. Seth cuts the body into pieces scattering the parts along the Nile. The body is finally recovered and reborn although he can no longer leave the world of the dead.

Offerings to, and celebrations of, the gods

Ritual or rubbish?

For a long time the possibility of ritual taking place in European prehistoric settlements was barely considered by archaeologists. Deposits of ash, bone and pottery were almost always dismissed as 'rubbish'. Gradually excavations began to reveal patterns that suggested something more was occurring. Since the beginnings of post-processual archaeology the importance of ritual in shaping the archaeological record has slowly come to the fore.

One particular aspect to emerge int he intervening period is the idea that few deposits are simply rubbish, discarded in a haphazard manner. Rather deposition is frequently structured according to ideological factors. This is not to say that 'rubbish' or middens did not exist, but that rules often exist that may govern how 'rubbish' is disposed of, such as those already discussed for hunter-gatherers. Henrietta Moore and Ian Hodder have also drawn attention to the fact that in some societies, ash from hearths is treated differently with respect to gender, being deposited in specific areas of the settlement. Here we describe some of the multitude of evidence for ritual sacrifice and votive offering of animal and plants that may be represented in the bioarchaeological record.

Ritual slaughters, animal sacrifice and ancient feasts

The slaughtering of animals is an important example of where ideology meets economy. A distinction is sometimes drawn between **ritualised slaughter** and **animal sacrifice**. In the former the animal is ritually killed but then consumed, in the latter the whole, or part of the 'sacrifice' is destroyed without being eaten. The difference is a subtle one and here 'sacrifice' is used to describe both.

There are many scenarios in which animal sacrifices take place. Important sacrifices may form part of annual festivals, but sacrifice is also a feature of 'everyday' living. We are familiar with the Hebrew and Islamic law that a priest or suitably trained person kills all the animals consumed, thereby producing Halal or kosher meat. This practice can be viewed as a type of sacrifice. These practices are by no means unusual and for many societies all animals are sacrificed in a similar way. We must not then view sacrifice as always something 'special' or out of the ordinary.

One of the most important aspects of sacrifice is that of sharing. Evan-Pritchard noted that the Neur and Dinka of Africa sacrificed and shared all their cattle. As the hunted animal is shared with the community, so is the sacrificed animal. Sacrifice also involves sharing the animal with the spiritual world. Through sacrifice the living community is unified with the spiritual community. For the Bara of Madagascar not only were large numbers of cattle slaughtered and consumed at funerals, but cattle sacrifice in general was seen as a means of communicating with dead ancestors. At such sacrifices they also believed that the head should be buried in the earth to give fertility to the crops.

Sacrifice also features in hierarchical relationships built around reciprocity. An important chief may hold regular **feasts** ensuring the support of both people and gods. In ancient civilisations the offering of animals by individuals to be sacrificed at the temple can be seen as part of such beliefs. Temple texts at Ur in present day Iraq tell of the sacrifice of some 350,000 sheep and goats and 35,000 cattle during a single year of the reign of Shulgi (2093-2046 BC) alone. These were brought to Ur as tribute from many cities, some as far away as Byblos on the Lebanese coast. The large numbers of animals killed in this manner have been taken to indicate that the main supply of meat to the city came primarily through ritual sacrifice in the temple.

In ancient Greece animals sacrificed by individuals for feasts, or by temple priests might eventually all be sold in the Greek market place or *agora*. The best descriptions of sacrifice in ancient Greece relate to the more important festivals. The Panathenaia at Athens and Hyakinthia at Sparta involved the sacrifice of large numbers of cattle to feed the city's citizens. The animals were ceremonially slaughtered with a knife concealed in a basket of grain. Water was sprinkled on the animal's head, and then some grain and a few of its hairs thrown onto the altar fire. The throat was then cut, with the head turned skywards so that the blood might spurt forth towards the sky. The butchering and distribution of the carcasses followed similarly strict rules. The animal was skinned and gutted, the hindquarters separated, the meat removed and the bones placed on the altar where they were burned or sacrificed to the gods. Indeed burnt sacrifices were commonplace offerings to the classical gods of ancient Greece. These offerings often involved specific parts of the carcass including the hind-limbs, tail, horn cores, skulls or internal organs. The remainder of the animal was then consumed. Zoorchaeological evidence for such practices from temples and sanctuaries is numerous and has been recounted by Valasia Isaakidou, Paul Halstead, Jack Davis and Sharon Stocker.

Sometimes variations in offerings are evident. For example, the sanctuary of Kabeiroi near Thebes and Zeytin Tepe contained numerous astragali (ankle/foot bones), many of which were unburnt. Such practices probably have a longer antiquity, perhaps dating back to Mycenaean times (*c.*1200 BC). In Homer's *Odyssey*, Odysseus's son, Telemachus, is said to have given burnt offerings of bones stripped of meat during his visit to King Nestor of Pylos (*Odyssey* iii, 447–63). Wall paintings from the palace at Nestor depict animals being slaughtered while during excavations at Mycenae have been found, which are numerous burnt bones, also interpreted as sacrificial animals.

Walter Burkett has argued that the roots of sacrifice in classical societies lie in the animist beliefs and practices of hunter-gathers. In modern hunter-gather communities he noted that certain parts of the animal were treated in special ways, something also true of the Palaeolithic evidence. These parts are those from which the animal was believed to regenerate or in which the soul was believed to dwell, Burkett proposed that in the sacrificial traditions of Classical Greece the parts offered up to the gods to be sacrificed in the truest sense followed these 'rules'. This is not to say that ancient Greeks necessarily perceived these parts of the animals as still housing the soul. It might simply have been the correct thing to do, a reciprocal exchange between humans and gods.

The ancient Egyptians also sacrificed animals. Depictions show cattle, sometimes with garlands of lotus flowers and fruits led to the slaughter. A common Egyptian depiction is the removal of the front-right foreleg by knife, the prized *khepesh* cut (**83**). This cut has its own hieroglyph used to denote its presence on offering tables or altars. The depiction is also associated with the constellation we know as the plough. We will return to this subject in the last part of this section.

Archaeological evidence from elsewhere in the Near East reveals similar practices perhaps even resulting from Egyptian influence. At the Iron Age site of Tel Qiri, near Hazorea, Israel, Simon Davies recorded large numbers of sheep or goat right fore-limbs, cut from young animals, that were associated with shrines or cult places.

The Etruscans were heavily influenced by the Greeks and had similar sacrificial traditions. Burnt bones of piglets, in addition to other animals, have been recovered from an Etruscan temple at Cerveteri. Another Etruscan custom was the inspection of the liver. As the largest organ, it was believed to hold the soul and hence a means of communicating with the gods. Individual parts of the liver were believed to hold messages from particular Gods to be read like tea leaves. Models of sheep's livers in terracotta, and in one case of Bronze, are found incised with these gods names. The Chinese Bronze Age also saw similar practices that involved the 'reading' of the cracks on burnt scapulae of cattle and sheep by oracles, which were then inscribed with both question and 'reading'. Such practices may have originated in the Neolithic

83 *The Khepesh cut*

as piles of burnt pig and deer scapulae have been found on sites of this period in China.

Returning to northern Europe and continuing with the pig theme, Celtic myths tell of the importance of these animals in forming a social bond between living and spirit worlds. The pig was believed to regenerate, cycling from the spirit world to the world of the living then back again. The killing and eating of the pig was then believed to please gods and ancestors by replenishing the stocks of pigs within the spirit world.

Zooarchaeological evidence from British Neolithic causewayed enclosures, such as Abingdon and Windmill Hill indicate that Neolithic communities engaged in pig feasting. What is interesting is that these sites also have evidence for the 'ritual' deposition of animal bones in their surrounding ditches. Sometimes the ditches even appear to have been cleaned out, perhaps to make room for more offerings. Such deposits are especially prominent in the terminals of ditches, such as the bundles of cattle ribs at Eton in East Anglia. At Windmill Hill the evidence suggests that animal remains may even have been bagged. Many pig bones from these sites have signs of burning, sometimes taken to indicate they were spit-roasted. However, as Julian Thomas has noted, many deposits from causewayed enclosures show signs of having been deliberately burned following deposition, so that such evidence may not relate to cooking at all.

The excavations at the Iron Age hillfort at Danebury in southern England revealed large numbers of pits with the same types of deposits repeated again and again. These consisted of skulls and mandibles, mainly of horse, with rounded 'balls' of chalk next to them. Whole articulated skeletons and limbs of domestic animals, along with birds, mainly crows and ravens, were also found. Those of articulated animals sometimes suggested total 'sacrifice' with no trace of butchery marks. Studies of other British prehistoric sites have revealed similar patterns.

Animal sacrifices are also found in the archaeology of the New World. The jaguar in particular was revered as a creature that was able to cross from the world of the living to the world of the dead. Its symbol was thought to offer protection and was therefore used in the houses of the Maya elite. Excavations by William Fash at an altar dedicated to the Kings of the royal dynasty at Copan revealed that in AD 775 King Yax Pac had 15 jaguars sacrificed and buried within a crypt, one for each ancestral king. A jaguar was also excavated from beneath the Great Temple at the Aztec capital of Tenochtitlan, a green stone ball having been placed within its mouth. Offerings to Tlaloc, the Aztec God of the rain and water, including large numbers of bones of fish as well as other marine animals were also found beneath the temple.

The last category of sacrifice that we shall look at is that of people. Direct archaeological evidence for human sacrifice is often difficult to ascertain, but in some cases no other explanation could possibly account for the nature of what we find.

'Bog-body' is the general term applied to those remains of humans found in peatbogs. These environments preserve skin and hair, but not bone. The majority of known bog-bodies come from Denmark although some are known from England, Ireland, northern Germany and the Netherlands. It was once thought that the people deposited in bogs were criminals, but much of the evidence now suggests otherwise. Examination of the bodies showed young disabled people to be disproportionately represented. Many had undergone elaborate deaths and their very presence within watery peat bogs also points towards ritual sacrifices. Many analyses have been conducted on these bodies including on their stomach contents. Frequently these reveal evidence for the consumption of a basic gruel, bread or porridge that was weed-infested and chaff-ridden. This was used to suggest that such meals were only suitable for ill-treated outcasts, but other evidence contradicts this. The skin on the hands of the bog-bodies suggests that they carried out little manual labour, while because they are generally above average height, they may have received a more nutritional diet than most.

The poor quality of the food found in bog-body stomachs should probably be taken as the normal diet for Iron Age peoples. Finds of prehistoric bread, such as that recovered from Glastonbury Lake Village, have been examined and reveal that the flour still contained large quantities of weed seeds and chaff. Therefore, these people seem to have been well treated before their sacrifice.

For a possible parallel we might turn to Greek texts and ethnographic accounts of human sacrifice. The Athenians kept people for human sacrifice, the *pharmakoi,* a group of outcasts deemed not useful to society and used for periodic human sacrifice when the city was experiencing problems. Records of Pawnee sacrifices indicate that such persons, like hunted animals, were also exceptionally well treated prior to their sacrifice.

Spirits in the corn

Although in this discussion of rituals concerning the exploitation of plant foods we shall only be discussing agricultural societies, the collection of plant foods from the wild may also be steeped in ritual. It is first worth making a point that to our knowledge seems to have largely gone unnoticed in studies of agricultural customs. That is how many of the attitudes and beliefs surrounding arable customs would seem to mirror those already discussed for animals. There is perhaps a reason for this. A key difference between those who cultivate cereal crops and those who gather wild fruits or seeds is connected with the destruction of the vegetation. When gathering the seed from wild plants the collector is clearly not killing the plant in the same way that the farmer is when he puts sickle to stem. While we may only speculate as to how the transition to agriculture was incorporated into existing animist-type beliefs, similarities to the ideological transition from hunting to animal sacrifice are also clearly evident in agricultural customs. Firstly, rather than the animal's soul being embodied in a particular bone or organ we have the animalised or personified crop as a **corn-spirit**, present usually in the last sheaf. Secondly, are the expressions of sorrow and guilt among peoples at the killing of the corn-spirit. As certain bones are sacrificed or specially treated, so then is the chosen representation of the corn-spirit.

The arable cycle has for many societies become embodied within mythology. Elements of these myths also relate to the actions of sowing, harvesting and processing. In the Egyptian myth, Osiris' body is broken up in the same way that the crop is cut and threshed. His body is then scattered along the Nile, representing sowing or winnowing. In one version Isis is even said to have collected his limbs in an agricultural sieve. We can then see how the myths are constructed as metaphors for the human actions. The harvesters and threshers act out the myth, in part through performing agricultural tasks, in part by performing certain rituals. Through acting out the ritual they are completing both the agricultural cycle and the mythological cycle. Elements of these myths were elaborated upon to a greater or lesser extent. Often they became incorporated into 'games' played in the field or on the threshing floor. In state religions they became elaborate ceremonies involving statues and sacrifices that began to have increasingly less to do with agriculture other than their timing. For rural communities they transformed into traditions and customs with the original myth and meaning having been long forgotten.

Harvesters often avoid being the ones to cut or thresh the last sheaf. In this respect harvesting and threshing become akin to the killing of the corn-spirit, the sending of Persephone back to the underworld, or the slaying of Osiris.

The ancient Egyptians are even recorded by Greek writers as making a great show of lamenting and calling upon Isis to come to Osiris' aid while harvesting. In European cultures it was common to ridicule the person who killed the corn-spirit.

The embodiment of the corn-spirit in human form by making an effigy from the last sheaf is commonly recorded. The European corn-dolly or the Arabic 'old-man' can be related to the embodiment of the corn-spirit as various deities. Sometimes it is embodied as the equivalents of mother and daughter, from the Scottish barley-mother and barley-daughter, and to the Malaysian's 'rice-mother' and 'rice-child'. Animal representations of the corn-spirit include horses, pigs, bulls, cows, goats, geese, quails, cocks, wolves, dogs, foxes, cats and hares. The animal is sometimes modelled from a sheaf, or carved in wood to be carried back to the settlement with the harvested crop. Sometimes the last sheaf is just named, for example, the 'Oats-Goat', 'Barley-Bull', 'Wheat-Wolf' and so forth. The effigy might then be carried to a neighbouring farm where threshing continues, for the spirit is thought to be present in the crop as a whole rather than just that of an individual farmer's field. Like animal sacrifice, harvesting and threshing rituals serve to bond the community. As with animal sacrifice, the sacrifice of the corn-spirit too completes the cycle so that it may be 're-born'.

Sacrifices may be real or just token. In Sweden the grain from the last sheaf was once made into bread that was shaped like a girl or pig to be eaten at particular times. Animal sacrifices, however, were once common practice in France where a live goat or cow, decorated in ribbons and garlands, was sacrificed and its body roasted for the harvest feast. Such sacrifices were usually made after harvest, but sometimes they also accompanied ploughing, sowing and the winter or spring solstices. In Rome, it was during ploughing and sowing that a horse was sacrificed, its head strung with loaves. Its tail, symbolising an ear of corn, was cut off and taken to the king's house along with its head.

In ancient Egypt each region sacrificed different animals during harvest and sowing. In Memphis it was the bull, while for many others it was the pig, which was sacrificed to Osiris. The festival of Osiris was celebrated as the floodwaters receded and the time for sowing approached. It involved a ceremony that acted out the myth, removing a statue of Osiris that was buried for three days and three nights to miraculously re-appear. Even in modern times some Egyptian farmers bury the last sheaf, 'the old-man', so that he may rise from the dead.

Equally important as the last sheaf, for many cultures, are first – fruit ceremonies. As with animal sacrifice first fruit ceremonies are associated with social bonding and the sharing of the harvest. In chiefdoms and big-man communities it may be associated with tribute and redistribution. Both Congo and Tonga farmers would offer their first fruits to a priest before they themselves partook of the rest of the harvest. In Greece we can see an extension of these systems in which 'first-fruits' were given as a form of tribute by members of the Athenian Empire. The exact amount to be offered was recorded on an inscribed stone at Eleusis, along with orders for three vast subterranean granaries where the grain was to be stored. The 'best' of the grain was offered in sacrifice to Demeter, Persephone, Hades and Athena. The rest was sold on the market and the money used to buy votive offerings with inscriptions noting who gave them.

While many of these rituals might seem to leave little trace in the bioarchaeological record, environmental archaeologists should at least appreciate the wider significance of their data beyond how many seeds or bones were recovered from a site. In section 3 we saw how differences in storage practice could be related to differences in agricultural organisation, especially that relating to the harvesting and processing of the crop prior to its storage. We also emphasised the importance of the timing of harvest and sowing, while organising enough people to do these tasks was paramount to the success of the community. Ritual within arable societies then plays a significant role in instigating the sharing of responsibility and labour at those times when it is needed, as well as the sharing of the products of that labour.

Votive offerings

Catherine Bell has highlighted the connection between the method of sacrifice or treatment of the dead, and the beliefs of the people involved. Those who burn sacrifices wish them to go to gods that dwell in the sky, as seen in many of the rituals of the ancient Hebrews, Greeks and Romans. Those who wished offerings to go to the subterranean spiritual world deposit offerings within water, caverns and deep pits, as seen with many 'Celtic' beliefs. Those who leave sacrifices or bodies to the natural elements to become 'de-fleshed' (**excarnation**) believe that the spirits reside in the natural environment, as is common within many hunter-gatherer groups, and for which we have evidence a in Neolithic Britain.

In some cases the ritual nature of biological deposits is obvious. Eridu, an ancient city in Iraq and contemporary with Ur is famous for a series of temples dedicated to the god Enki. Enki, as we saw earlier, was the god of fresh water and was often depicted together with fish. The recovery of large numbers of blackened bones of fish from one level of the city can then be interpreted as offerings, similar to those offered to Tlaloc discussed earlier.

Between 700 BC-300 BC Egypt was to see a rise in the prominence of animal cults. Mummified rams sacred to the god Khnum were recovered at Elephantine, while crocodiles, sacred to Sobek, were found at Kom Ombo and Faiyum, while clutches of crocodile eggs were also found at the latter site. One particularly prominent site is at Tunah-Al-Gabel, the Greek Hermopolis. Here many underground corridors were discovered with galleries full of pottery vessels that contained the mummified bodies of baboons and ibises sacred to the god Thoth. A giant artificial lake with palm trees for the sacred ibises to live in had even been created at the site. Similar subterranean galleries were excavated at Sakkara. The frequent finds of eggs even suggested that the ibises were bred here. Texts indicate some 60,000 ibises to have been present at any one time in the temple. From the burials alone it seems probable that some 10,000 were slaughtered each year. The records also indicate that only one burial took place each year, suggesting a single mass slaughter.

Ritual offering pits of burnt sacrifices were a conspicuous feature of the Indus civilisation. Excavations at the Harrapan city of Kalibangan revealed mud-brick platforms with rows of pits or 'fire-altars'. The pits were lined with kiln-fired bricks and contained bones of cattle and antlers suggesting animal sacrifices. Such sacrifices are also represented in Indus seals depicting individuals kneeling before deities with animals behind them, we which were probably, brought as an offering for sacrifice (**84**).

The oracle bones of the Shang dynasty also provide evidence of burning rituals in ancient China to worship the hills, rivers and earth. Evidence for

84 *An Indus seal depicting sacrifice*

these events has been forthcoming from excavations at the ancient city of Sanxingdui. In pits dated to between 1300 and 1200 BC animal bones, elephant tusks, ceramic and bronze items were found, including bronze bird figurines, all of which were burned in one mass sacrificial fire. For ancient northern Europe of course we have a number of sacrifices, such as the dog with the stake through it from Flag Fen, or the bog-bodies that by virtue of their deposition within watery-environments, may also be seen as votive offerings.

Plant foods are less commonly recovered as votive offerings, unless they are burned or placed in waterlogged environments. Remains of burnt food offerings have been recorded in Roman Europe. At Pompeii, excavations of gardens dating from the second century BC to the volcanic eruption of AD 79, revealed pits in which burnt offerings to household gods had been placed. Analysis of this material by Mark Robinson revealed remains of pine nuts, dates, olives, figs, grapes, walnuts and hazelnuts, all of which are rarely recorded within normal charred assemblages. Barbara Zach found similar remains at Roman Mainz in western Germany. Here they were dedicated to the worship of Isis or Magna Mater, the Roman equivalent of the Greek mother goddess Cybele. It is interesting to note that a comparison with offerings from other sites associated with Mithraism, Fortuna and Mercury, generally revealed little difference in which plant foods were offered to the different gods.

Grave goods and burial customs

Given that animals and plants may be endowed with spiritual souls it is unsurprising that the treatment of human dead is undertaken with similar deference to belief and ideology. For many people human souls may journey to different worlds or take on different shapes or attributes after death. Animals, especially birds and fish, are often favourites for 'housing' human souls or transferring them between worlds.

As with the treatment of animals, the treatment of the human dead is often also tied to the economic prosperity of the community. In the world's oldest known villages it was once common to bury the dead within or close to granaries, a practice still seen in the settlements of the Mura-Urza of Cameroon. At Jerf-el-Ahmar, Syria (11200-9000 BC) a building thought to be a granary contained human remains, especially parts of skulls. A similar situation is seen at Tell Ramad I (6500-6000 BC) where lime plaster 'bins' thought to be for food storage were butted by a deposit of human skulls.

At other early agricultural sites it was common to bury ancestors within or close to the house as is seen at Abu Hureyra, Syria and Jericho, while at the city of Ur, ancestral shrines continued to be housed under rich private houses. At Çatalhöyük excarnated human remains were wrapped in textiles and placed under benches and sleeping platforms within the houses. Male burials were separated from female and child burials, with the former in the east, the latter two in the northwest of the house.

Many burials have ecofacts that are buried with them, such as animals or their bones that accompany artefacts and personal belongings. These may be remains of funeral feasts, food to sustain the individual on their journey to or within the afterlife or offerings to buy their way in. A further possibility is that the souls of animals might lead the dead on their journey into the afterlife. So it is that van Gennep in *The Rites of Passage* tells us that the Assam of India butchered and ate a dog, goat and pig at funerals. The animals' souls then guided the soul of the dead to the afterworld, although it is unclear whether their bones were also buried in the grave of the deceased.

From the earliest times animals sometimes accompanied human burials. At Skateholm on the tip of southern Sweden a Mesolithic cemetery numerous dog burials were found. Sometimes whole skeletons were buried with human inhumations, sometimes separately, sometimes just certain bones. In these cases the dogs might represent personal items, pets or hunting companions to accompany the soul of the deceased in the afterlife. A more unusual symbolic discovery comes from a Mesolithic burial of a woman and newborn child at Vedbaek, Denmark. The woman probably wore a garment with pendants of pig and deer teeth that are commonplace in other Mesolithic burials. The child, however, appeared to have been laid upon the wing of a swan during the burial.

Another unusual occurrence is reported from several Neolithic lakeside rock tombs in northern Italy. At Monte Covolo it was established that bodies were excarnated, placed in a mortuary house, then taken out and burned along with offerings of food, then reburied. The examination of these assemblages by Sue Colledge revealed that alongside the remains of domesticated cereal, many more unusual items were found, including pistachio nuts and fruits, carbonised pips of grape, fig, strawberry, elder, wild service, dogwood and cornel. Going back to the statement of Catherine Bell earlier in which disposal of the dead is seen to be reflective of where the spiritual world lay, the inhabitants of Monte Covolo would appear to be hedging their bets.

In the ancient cities of Mesopotamia the souls of the dead were thought to continue living in the underworld so that food offerings had to be regularly made to them at the house shrine. Such offerings were often burned upon an altar, the smoke exiting the house through a chimney. Sometimes the dead were also buried with joints of meat to sustain them in the afterlife.

A common occurrence in Egyptian tombs is that of 'Corn-Osiris' figurines similar to those used during sowing. One such hollow wooden figure found in the tomb of Tutankhamun (*c.*1327 BC) had been lined with linen and filled with grain and wet Nile silts so that the grain had partially germinated. Such figurines then tie the agricultural cycle to that of the Osiris myth and rebirth of the deceased in the afterlife. Desiccated offerings of plant foods are a frequent feature of Egyptian burials. Many plant offerings including figs, dates, grain, sesame, and wreaths of lotus, olive, willow, cornflowers, mayweeds and various fruits were found in Tutankhamun's tomb. In ancient Rome garlands of flowers were also scattered liberally over the dead, around the interior of tombs, or over the funeral urns. The flowers were often associated with, and dedicated to, individual (usually female) deities, such as poppies to Ceres and anemones to Venus.

As archaeologists we might ask about the antiquity of the association of flowers and funerals. Examination of pollen samples might well reveal the presence of certain plants, but this can only be confidently asserted if comparable 'control' samples are taken from contemporary layers, both outside and within the tomb. The earliest evidence claimed for the use of flowers in burial comes from the cave of Shanidar in the Zagros Mountains of Iran, dating to sometime between 120,000-50,000 BP. Here pollen samples from a Neanderthal burial had high concentrations of certain pollen types only within the vicinity of the body itself. The excavator concluded that the burial had been laid out on a bed of wild flowers, which included yarrow, hyacinth, cornflower, ragwort, groundsel, hollyhock and ramose branches. The claim has been challenged by the suggestion that the pollen was introduced by burrowing mammals, or even the clothes of excavators, certainly nothing unusual stands out within the assemblage, and most are representative of a common weedy flora.

Other investigations have revealed finds of special clothing. Claw and tail bones of a leopard found with a Neanderthal burial from Hortus in southern France have been interpreted as part of a costume, or possibly that the individual was buried wearing skin. Even cremations may reveal such findings. At Welwyn Garden City, north of London, a rich burial of an important individual dated to around 50-10 BC was discovered. The body had been cremated but amongst the human bones were found burnt phalanges (the foot and hand bones) of a bear suggesting that the individual was cremated in a bearskin. Finds from British Neolithic barrows have also demonstrated a clear association of cattle with incarnation. Human bodies had either been wrapped in the hide or alternatively that hides were hung over the mortuary house together with cattle skulls.

Some of the most unusual burial deposits investigated by environmental archaeologists come from the American mid-west. Here from the beginning of the first millennium BC, large earthworks and conical shaped mounds were built in which the dead were buried. Excavations of these Middle Woodland period mounds often reveal the use made of soils from many different regions and habitats, including alluvial soils. Animal effigies often accompanied the burials suggesting the mounds were deliberately structured. Burials from Moundville cemeteries have animal parts commonly placed in the burials. These may be bones of deer, but often claws of birds and turtle shells appear, while in the Later Woodland period the mounds themselves begin to take the shape and form of birds, bears, snakes and lizards. All these elements are certainly symbolic and probably relate to creation stories and origins myths.

Shrines and foundations deposits

Common features of many sites are deposits of artefactual or biological material incorporated into building foundations. While such deposits often consist of animal figurines made of clay, earth, and wax or carved stone, they sometimes incorporate animals or plants, or consist of real animals and plants. In the excavations of the high priestess of Nana's residence, which dates to the final days of Ur (*c.*668-627 BC), Leonard Wooley found numerous figurines of unbaked clay loosely moulded into animals and therianthropes under the floors. Bones of small animals and birds, along with grains of barley preserved through mineralisation, were pressed into these figurines. In Neolithic Europe such foundation deposits occasionally consisted of real animals. At Parva in Romania a raised platform had several cattle skulls buried underneath while deposits of the Tisza culture of Hungary included dog skulls, bull horns and antlers.

As we have already seen bones of a jaguar were incorporated in the foundations of the Great Temple at Tenochtitlan while in ancient Egypt the *khepesh* cut or its effigy sometimes served a similar purpose. It is not, however, just animals that are thus incorporated. In exceptional circumstances plant remains may be preserved, most notably in desiccated environments. So it was that excavations of an eighteenth-dynasty temple (*c.*1500 BC) at Semna, Nubia revealed the seeds of watermelons, buried in the foundations.

At other sites it is not so much foundation deposits but rather the building itself that incorporates such remains. The site of Çatalhöyük is famous for its bull shrines. In the rooms of the numerous houses were the horns of cattle, especially those of the wild aurochs, plastered in clay and set upon walls and benches (**85**). The symbolism of the vulture was also a common occurrence in shrines at the site, and perhaps relates to the excarnation of human bodies. However, it is not only Çatalhöyük that contains such remains. At many similarly dated Near Eastern sites plastered heads of bulls, rams and stags built around real skulls were commonplace. Often, they too were placed on low benches, or as at Ganj Dareh in Iran two wild 'goat' skulls were set one above the other in a niche. At Mureybit in Syria rooms interpreted as possible shrines had carnivore jaws embedded into the walls, in others, bullhorns were used.

The ideology behind such deposits is probably varied. It may be that animal spirits help protect the household or may relate back to animism, ensuring good hunting or rather prosperity in the keeping of domestic animals. It is often suggested that human figurines, including mother goddesses within houses, pertain to similar aspects, the success of the household and its fertility

85 *Reconstruction of a house shrine at Neolithic Çatalhöyük*

in terms of both having children and good crops. Of course such meanings have been lost, the practices now being nothing more than tradition.

Consuming ideologies

Consumption, like production, is also deeply embedded within social ideology, and plays a key role in formulating social relationships. In unison with ideology, consumption, together with production, defines the identity of every individual society. How and what an individual produces and consumes is shaped by their age, sex, caste, clan, tribe, status and religion.

The Indian caste system has rules prohibiting and prescribing the consumption of certain foods and adoption of certain occupations, something Lévi-Strauss saw as emerging from totemism. In early industrial Europe individuals of certain classes would live in set places, undertake certain professions and eat certain foods. The ability to control limited resources is often given as the reason for the elite consuming 'better' foods. Only the bourgeoisie might be able to import exotic fruits and spices. Foods, however, may become

86 *Safe from lunch. Swans may feature only on the dining tables of the Queen and King's College, Cambridge*

associated with the elite for other reasons. In England swans are no longer exotic nor particularly rare (**86**). Yet only the Queen and certain Colleges in Cambridge can serve swan meat or swan eggs to their guests.

For hunter-gatherers and in traditional agricultural communities, the allocation of work is largely determined by age and sex. For example, in hunter-gatherer communities where plant foods form the staple diet, women are generally responsible for their collection.

As well as distinguishing individuals from other individuals within the same community, consumption and production may distinguish whole cultures from one another. The French are French because they eat snails, frog legs and horsemeat; in Eastern Asia people eat with chopsticks; the Egyptians are renowned for fine linens; New Zealanders and Australians for raising sheep. In the past the nature of sacrifices and harvesting customs may also have formed part of such social identities.

Food and crops are much more than just something to eat; both their production and consumption help form important social ties. Recent archaeological studies have begun to concentrate on this aspect, tying changes of foods in the archaeobotanical and zooarchaeological record with changes in pottery types. In investigating the effect of Romanisation, for instance, Karen Meadows highlighted simultaneous changes in both butchery practices and pottery styles on some sites in the Thames Valley, that could be related to changes in food fashion, particularly the way in which food was cooked and was intended to be served.

Taboos

Taboos are an important part of the formation of social identity. They may apply to all individuals in a society or only certain groups. They may only apply at certain times during an individual's life or perhaps may only be broken under certain circumstances. Many taboos prohibit the eating and killing of certain animals. Taboos against eating domestic pets are commonplace simply because it is considered akin to cannibalism. The Fang tribe of Gabon even avoid eating mice caught in their house because they are thought of as household members.

One of the most common taboos is against the killing and eating of totemic symbols. Mainly this forbids the killing and eating of one's own totem. For some this may make little difference, for instance those whose totem is the bear or wolf. For others it may be more significant; Omaha clans whose symbol is the elk will not eat male elks, those of the clan symbolised by the red maize, will not eat red maize and so on. For tribes in Bengal the consumption of any plant or animal that was any clan's totem was taboo. Taboos against 60 to as many as 300 species of plants and animals have been recorded for these peoples! Two of the most common reasons given for taboos are that the animal is sacred, as with Hinduism and cattle, or because the animal is seen as unclean and will have undesirable consequences (rendering the consumer unclean), as with Islam, Judaism and pork.

Sometimes the reasons associated with the animal being sacred in some way are combined with undesirable consequences. The ancient Syrians thought fish sacred and believed that eating them brought about stomach ulcers and feet swelling. Tribes in southern Nigeria did not eat fish because they believe that their own souls resided within fish. To kill a fish would result in a death of one of their own.

Taboos against the pig extend back into ancient times. Ancient Egyptians reviled the pig in Herodotus' day (*c*.500-400 BC). It was after all, the form Seth took to injure Horus, or in Greek-influenced Asia Minor the killer of Apollo. However, both of these reasons are much later than the earliest evidence for the existence of the pig taboo. For example, at the onset of the Egyptian and Mesopotamian civilisations from the late fourth millennium BC onwards pigs are rarely depicted in official art. Nevertheless zooarchaeological evidence suggests pigs were commonly killed for food in the Near East, although this practice became scarcer in the second millennium BC.

Herodotus says pigs were still kept in Egypt in Classical times for the sole purpose of sacrifice, although the swineherds lived in separate villages and could not enter a temple. In these sacrifices to Osiris and the moon, pigs were removed by the swineherd in the former and eaten in the latter.

Many theories have been forwarded to explain how the pig taboo came into existence. Marvin Harris that suggested pigs were inefficient to rear, while Mary Douglas suggests that because they have hooves but do not eatgrass, they did not fit ancient Hebrew classification systems that put hoofed animals eating grass and leaves together. Neither explanation is satisfactory. Pigs continued to be kept in the Near East but were simply not eaten, while Douglas's explanation is not applicable to other cultures with pork taboos. Frazier suggested that pigs were once sacred, to be eaten only at certain times. To break these taboos would then render one unclean and later this uncleanness was to become associated with the pig itself. Roman writers were certainly unclear on the issue. Lucian (second century AD) noted that at Hierapolis in Syria, pigs were kept, but neither eaten nor sacrificed, although he could not decide whether it was because they were sacred or unclean. Plutarch (*c.*50-120 AD) had a similar dilemma concerning the Jewish attitude towards pigs, for while they could not eat pigs why should this prohibit them from killing them?

At Iron Age sites in Britain bones of fish, birds and other wild animals are rare, although some are commonly recovered from both earlier and later periods. This absence can be considered in relation to Caesar's observations of Late Iron Age people in Britain – 'hares, fowl, and geese they think unlawful to eat, but rear them for pleasure and amusement' (*De Bello Gallica* V.15.1). The zooarchaeological evidence generally supports these remarks, but raises the possibility that these animals were more than just taboo. Wild animal bones are, however, sometimes present on archaeological sites in spite of Caesar's assertions. For such finds it is important to establish the presence of butchery marks to determine that they were eaten rather than dying natural deaths. Bones of domestic fowl are more commonly found in later Iron Age deposits in Britain and their numbers increase into the Roman period. A bone of mallard duck with butchery marks from later Iron Age Danebury might then suggest the lapsing of the taboo. However, many earlier Iron Age finds of wild animals are in deposits of ritual significance. For example, a single pit at the hilltop enclosure of Winklebury in southern England contained bones of a red deer along with twelve foxes, while the Lindow man bog body from Cheshire had an armband made of fox fur.

Finds of wild animals on Iron Age sites of continental Europe have frequently been interpreted in relation to the fur trade, such as a single deposit rich in the bones of fox and badger from the *oppida* at Villeneuve-Saint-Germain, France. However, as Miranda Green argues, the rarity and special treatment of such finds casts doubt on such functional explanations. A similar 'trade' argument was made for finds from the enclosure at Haddenham on the fen margins of Cambridgeshire (*c.*150-0 BC). The excavations revealed bones of beaver, eggshells of swan, feathers and bones of large birds, pelicans, cranes and/or swan. It was suggested that they related to trading networks, but as the excavator, Chris Evans, noted this would seem to conflict with the pottery evidence that suggested nothing but local wares. Although it is too early to draw conclusions there exists a possibility that these animals were taboo and that this site had a ritual significance. The unusual nature of the deposits, the position of the site itself close to the fen edge and the association of the animals with water would all have similar connotations.

An association of taboos with water symbolism might also account for the general absence of fishbone from most Iron Age sites in Britain. One of the few exceptions is Glastonbury Lake Village, where fish have been found along with bones of other wetland species. We might well ask how old this taboo could be in Britain. Fish remains are reported from British Neolithic sites, although they are uncommon. A pit at Coneybury contained trout along with beaver, although given the presence of some forty pots in the feature the pit was certainly a 'special deposit'. Similarly chemical analysis of residues in Neolithic pots from Runnymede, Surrey revealed the use of fish.

A bear by no other name

To conclude let us draw these comments to how nature becomes symbolised in language. The names of several animals in northern Europe can be seen to differ from the more commonly known Indo-European names for such animals. The Indo-European root for bear is **rtko-**, giving us *arktos* in Greek, and *ursus* in Latin. However, in the languages of Northern European countries we find many different derivatives from the root **bher-** or brown**.** Bear literally means 'brown one'. Such names are known as **taboo-variants**, alternative names because the real names cannot be spoken. The true name of the bear is also not spoken by the Khanti of Western Siberia, because they regard the bear or 'bear-spirit' as sacred – the protector of people from illness, the reconciler of disputes and instigator of successful moose hunts. Such variants are also common in northern European cultures for wolves, foxes, beavers, weasels and stoats. That it is these same animals that are found only on occasion on archaeological sites, but often in 'ritual contexts', demonstrates perhaps the antiquity of these taboo names. If the real name of the bear became taboo for similar reasons, then it is the greatest of ironies that while the real animal itself must now be protected, its symbol and taboo-name still perform their protective roles (**87**).

87 *Instigator of moose hunts? A modern representation of the bear as a cuddly toy – a far cry from earlier views on the animal*

SECTION 5

ENVIRONMENTAL ARCHAEOLOGY:

THEORY WITHOUT THEORISTS

> The alienation of archaeological science, and with it environmental archaeology from the rest of archaeology has had a number of undesirable effects . . . It is paradoxical that the diffidence that recent trends in archaeological theory have shown towards science has not led to the questioning of the methods borrowed from the natural sciences, but rather to their isolation and consequently to their uncritical acceptance.
>
> Umberto Albarella, *Environmental Archaeology: meaning and purpose* (2000) 7

Environmental archaeology as a science

There is little doubt that environmental archaeology has problems with recent archaeological theory. The contentious issues originate as a result of the development of environmental archaeology within **New Archaeology**. New Archaeology (including **processual archaeology**) was a theoretical movement beginning in the early 1960s one aim of which was to bring scientific vigour into the discipline of archaeology. Environmental archaeology with its roots firmly situated within the biological sciences naturally flourished in this new theoretical framework. Environmental archaeology enjoyed a prominent place in the hearts of New Archaeologists. Then in the 1980s, like the appearance of a wicked stepmother, post-processual archaeology arrived (post-processual archaeology is discussed in detail later in this section). The affections shown towards environmental archaeology was lost and seemingly transferred to the new theories that showed little respect or regard to science. Arguably this has left environmental archaeologists stranded, retreating into a shell against a tirade of criticism that most do not fully understand. The aim of this section is to examine the theoretical basis from which environmental archaeology developed, take a look as some of the criticisms levelled by post-processual theorists at environmental archaeology, and finally explore theoretical frameworks that would help environmental archaeology to maintain its relevance in the archaeology of the twenty-first century.

Incommensurability is a useful concept to help explain why post-processual and processual archaeologists do not always see eye to eye. Incommensurability is exactly the sort of thing that occurs when two academic groups with different aims, objectives and methods talk completely over one another's heads without either understanding what the other is discussing. To give an example, a scientist looking at problems in recent African history may conclude that it is the things that science studies, such as disease and the deterioration of land, that are the main reasons for poverty. However, a social historian looking at the same scenario may instead conclude that war, corruption and the bad advice from western governments in the affairs of African states are to blame. Both answers may be equally valid, but the point is that each group is asking different questions, using different data and different methods. Therefore each may conclude that it has the 'right' or 'truthful' answer. Incommensurability occurs where neither side can understand the other's objectives, methods or the often jargon-riddled language that describes these. Such a situation certainly describes the relationship between environmental archaeology and post-processual theory. Many environmental archaeologists do not understand post-processualist aims, objectives and language, while many post-processual archaeologists do not understand enough of archaeological science to be able to criticise it.

Further considerations on the definition of environmental archaeology?

Before proceeding further we need to address three relatively simple questions that lie at the heart of the issues surrounding incommensurability.

What is environmental archaeology?
What does it aim to achieve?
What are its methods?

In section 1 we examined in a broad sense the meaning of environmental archaeology to those who practise it. As we saw, environmental archaeology encompasses a broad range of meanings. It is usually taken to mean the study of 'natural' biological or mineral remains, the ecofacts. These include among other things animal and plant remains, sediments and soils. The environmental archaeologist specifically studies ecofacts that have been modified by humans.

In the 1950s television game show *Animal, Vegetable or Mineral?* which featured the famous archaeologist Mortimer Wheeler, panellists would guess what an object was, beginning with the question 'animal, vegetable or mineral?' This is logical as all material culture manufactured by humans begins life as part of the 'natural' environment. The problem is in deciding when the ecofact studied by environmental archaeologists becomes the artefact (including structures and features) studied by what we might term 'conventional' archaeologists. A simple answer might be when it is turned into something for human use. However, such a definition is problematic as ploughed soils for example are modified by people and clearly have a human use. Similarly cereal crops, when harvested and processed, are destined for human use, as are domesticated animals, while a forest clearing detected by pollen or molluscan analysis might have been made for human use.

The issue becomes clearer still when we explore the aims of environmental archaeology, which as we saw in section 1 broadly covers the relationship of humans with the natural environment. In its simplest form environmental archaeology is seen as dealing only with environmental reconstruction. Many environmental archaeologists (e.g. Karl Butzer, Rosemary Luff and Peter Rowley-Conwy) have fought against this as the sole aim. The main environmental archaeology textbooks continue to deal mostly with environmental reconstruction, while similarly the fifth chapter in Colin Renfrew and Paul Bahn's standard textbook, *Archaeology: Theories, Methods and Practice* (2000), entitled 'What was the environment? Environmental archaeology', speaks volumes about the continuing perception of environmental archaeology within wider archaeological circles.

The problem is: where should the study of the relationship between humans and environment stop? Why is the transformation of biological and geological material (soils-fields, animals-bones, plants-cereals or wood) into food or fuel dealt with by environmental archaeologists, while the transformation of clay into pottery, trees into houses, wool into textiles is often dealt with by conventional archaeologists? In this way environmental archaeologists examine only one aspect of a set of related activities. By way of an example, an archaeo botanist will study the transformation of the biological aspects of agriculture, the grain, chaff and weeds (see section 3), while an archaeologist studies the tools of this transformation, for instance the sickle and quern. What then justifies environmental archaeology as a separate discipline?

We might argue that the division is an historical one as we saw in section 1. Throughout the history of archaeology specialists have been drawn in from other disciplines to study particular classes of material. The divisions are not then necessarily beneficial to studying past activities or societies, rather they arise from the nature of the material itself. Environmental archaeologists were drawn from other disciplines to classify and study a body of material that was not easily classified by traditional archaeological methods. Soils, sediments, animal bones and plant remains are difficult to classify in the same way as artefacts and tend to be categorised using specialist terms (e.g. Latin names for biological organisms) taken from the natural sciences. Pottery, house types, tools etc. are classified using terminology, taken from the common language, that either describes their function, for example a pot, an axe, a house, a ditch, or their relation to archaeological sites e.g. Peterborough ware, Clovis points. People in the past, however, may have classified soils, animal bones and plants using a variety of terms and concepts now lost to us.

This leads nicely to the last question, and one which we have covered in much greater detail in section 1. How do environmental archaeologists investigate their material? Environmental archaeologists use techniques borrowed from the natural sciences to study 'ecofacts' and the humanly induced processes that have transformed them. These techniques were, however, designed for examining 'natural' phenomena. As environmental archaeologists are interested in the relationship between humans and the environment, the question posed by Umberto Albarella (quoted above) still remains. How appropriate are the techniques and approaches of 'natural' sciences for examining human-environment relationships? One criticism of science in archaeology is that it only asks scientific questions, uses scientific techniques and hence comes up with scientific answers, but is rather less good at addressing social issues.

Paradigms in archaeology

In order to examine how and why theoretical perspectives that are of relevance to environmental archaeology have changed we need to take a brief look at the work of Thomas Kuhn. Kuhn, a philosopher of science, suggested in the early 1960s that science operated within 'intellectual climates', or what he called **paradigms**. Paradigms are systems of knowledge that encompass scientific theories, favoured beliefs about the world, socially accepted methods and objectives; in short, the social elements that accompany the formulation of scientific theories. Kuhn realised that past scientific theories which appear wrong today were once 'correct'. In other words, past scientists had solved accepted puzzles within their current system of knowledge and as they saw them. Despite paradigms shifting, Kuhn thought that in general science progressed.

Kuhn outlined the history of science as one in which new scientific theories arose through revolutionary changes. For long periods scientists conducted their work within accepted paradigms, what Kuhn called **normal science**. Every so often a scientific revolution occurred in which old paradigms were replaced by new ones. Therefore, rather than being truly objective, science operated within socially accepted frameworks that defined objectivity. Kuhn's work was not easily accepted by the scientific community. Karl Popper and other leading philosophers of science organised a conference in 1965 with the sole aim of debunking Kuhn. By the end of the meeting it was clear they had failed and Kuhn's work is now regarded as a landmark of scientific thought. Indeed Paul Feyerabend took Kuhn's ideas one stage further in suggesting that science was entirely socially – rather than scientifically – constructed.

To understand the concepts we have outlined here is to understand part of the current debate within archaeology over the role of science. As New archaeology developed in the 1960s and 1970s, Colin Renfrew and other advocates of the new ideas argued that New Archaeology represented a Kuhn-style paradigm shift. They saw New Archaeology as beginning a period of unified normal science in which new theories would be developed. However, as Matthew Johnson has pointed out, the irony is that while Renfrew used the work of Kuhn to legitimise New Archaeology, Kuhn's ideas concerning shifting paradigms can also be used to undermine it. In other words, paradigms will inevitably shift once more with, as arguably they have to lead to another orthodoxy: post-processual archaeology.

The creation and shaping of paradigms

Archaeological theory is sometimes seen as being born with New Archaeology. However, before the 1960s there were a number of significant shifts in perceptions as to how the past should be viewed, that might be seen as paradigm shifts in themselves. In the following text we will take a detailed look at the development of archaeological theory as it relates to environmental archaeology, to the point at which New Archaeology had become generally accepted.

Theory before the 'New' theory

The concept of social evolution is tied to an earlier form of biological evolution than that proposed by Darwin and known as **Lamarckism**. J.B. Lamarck (1744-1829) proposed that organisms 'adapted' to an environment (e.g. giraffes got longer necks through stretching to reach leaves) and passed on these 'acquired' characteristics to their offspring. This was a very different model to that proposed by Darwin, which instead stated that in certain environments some characteristics might favour the reproduction and survival of certain individuals more than others. These characteristics were then more likely to be passed to the next generation (giraffes with longer necks have a better chance of reproducing). According to Darwin then, organisms did not adapt to exploit the environment in an optimal way. In fact, in Darwinian evolution animals may even de-evolve characteristics that later become their undoing; Pacific Island birds such as the kiwi becoming flightless then falling victim to cats and rats introduced by humans. As a result of both Lamarck's and Darwin's ideas social evolutionism became the scientific paradigm of the late nineteenth century. Theories concerned everything from the evolution of law proposed by Sir Henry James Sumner Maine, to the evolution of knowledge proposed by August Comte. In archaeology it was a neighbour of Darwin, John Lubbock, later Lord Avebury, who brought such theories to the fore.

Historicism, materialism and functionalism

The first half of the twentieth century saw many theoretical developments in both sociology and anthropology. Three strands of thought adopted from these disciplines, were to become particularly influential in New Archaeology. They were historicism, materialism and functionalism.

As long ago as the seventeenth century Sir John Locke (1632-1704) had suggested that science could be used to answer social questions, although he doubted whether it would ever produce the scientific laws that characterised the natural sciences. The success of Darwin's evolutionary theory, however, brought such hope. Auguste Comte (1798-1857), the accepted founder of sociology, had died only a couple of years before the publication of *On the Origin of Species*, yet he had already proposed that it was only a matter of time until scientific laws were discovered for human society. Such a belief is known as **historicism**, in other words the idea that human society is subject to scientific laws that may be deduced through positivist scientific enquiry.

Among the important early sociologists few are better known than Karl Marx (1818-83) and Friedrich Engels (1820-95). Marx and Engels have often been accused of historicism, although Marx was keen to distance himself from it. However, it is with them that we find our second influential theoretical strand, **materialism**. Materialism embodies theories that stress the importance of how people satisfy their material needs (as opposed to their ideological needs) as the primary means of understanding human societies. Marx and Engels saw the ways in which humans organised the provision of their material needs as the main factor governing social organisation and the development of human societies. They proposed that human society had 'evolved' through a series of stages. Each stage was characterised by a different type of economic organisation.

Economy was also to play an important part in the theories of Emile Durkheim (1858-1917). Developing the ideas of Herbert Spencer (1820-1903), Durkheim saw society as an organism with all the various parts functioning to keep the whole going. His ideas were key to the development of what became known as **functionalism**. However, functionalism is more commonly associated with the anthropologists Arthur Radcliffe-Brown and Bronislaw Malinowski. Following the analogy between organisms in nature and society put forward by Durkheim, Radcliffe-Brown suggested that cultural elements functioned to maintain the balance and integrity of social structure as a whole. In his scheme of things customs, traditions, institutions, and ideologies were all interrelated and functioned to serve the persistence of each other. Malinowski's functionalist theories were somewhat different, especially in the role that materialism played. For Marx and Engels 'needs' were associated with socially or culturally determined actions. For Malinowski and later economic anthropologists 'needs' largely constituted people's biological requirements. Culture then replaced natural selection as 'a more efficient and better founded way of satisfying the innate biological desires of man' (Malinowski 1947, 33). As far as Malinowski was concerned the natural selection of Darwin had been superseded by the adaptation found in Lamarckian evolution.

A fundamental aspect of Malinowski's work is that the cultural elements of our lives exist to satisfy our biological needs. However, there are obvious problems with such a model. For example, in the western world we can eat almost any food that we can afford. If culture simply reflects our need to eat then there would be no need to eat luxury items such as caviar. Such arguments return us to the incommensurability that we examined in the beginning of this section. For instance, Person A says 'we eat because we need to eat' and Person B says 'but eating is clearly more than a biological activity'. Person A may be right, but sees different things and asks different questions to Person B. Person B knows that people eat for biological reasons (this is obvious to us all; what Person B is interested in is why people eat different things and if this tells us something more interesting than just that all things are biological.

While Malinowski explained human culture in terms of 'improved' biological adaptation he did not suggest why human culture should change or vary. It was left to others to explain cultural change, and in particular the logical next step to the argument, that adaptation was environmentally determined.

Behaviourism

By stressing the biological element Malinowski's theories were treating humans and animals as alike. This point can be linked to an approach in psychology that was becoming increasingly popular in the middle of the twentieth century, **behaviourism**. Behaviourism saw human action as being determined by innate biological responses to environmental stimuli. It began in the early twentieth century where it was heralded as the 'scientific approach' to understanding the human mind. Most readers will have heard of Pavlov's dog. Pavlov was a Russian physiologist (he did not like to think of himself as a psychologist) who laboured on the problem of the workings of the human mind. In an experiment he rang a bell while feeding a dog. In time the dog began to salivate on just hearing the ringing bell. This was known as conditioning, in that the dog's behaviour was conditioned by external stimuli (in this case the ringing bell). Clearly it is a long way from ringing a bell to make a dog salivate to suggesting that people's actions, indeed culture, could be caused by such environmental stimuli. However, if we replace salivating (the dog's conditioned behavioural response) with culture and the ringing bell (the external stimulus) with the 'natural environment' then we have an environmentally determined culture. Culture in this view is the stimulated 'natural' response for humans to the environment. A paradigm was beginning to emerge, within which people were considered predictable – in that they all acted to satisfy their biological needs. However, unlike animals, humans had culture to help them satisfy such needs. Culture, therefore, began to be viewed as functioning to aid people in extracting their biological needs from the environment.

If we accept these behavioural views we may logically ask: what determines the way in which people exploit the environment? Will humans naturally follow one method (a natural stimulus), of exploiting the environment? In other words, is there a universal way of exploiting the environment, or are there a number of different ways and only chance or prevailing ideology will decide which way people choose?

Common sense from our own life experiences would tell us that there are many ways of doing things and that we do not always choose the same one as other people. However, as this was a scientific paradigm following social-evolutionary principles, scientists opted for (Lamarckian) **adaptation**. Animals had the best ways to exploit their ecological niches, long necks for giraffes, long beaks on humming birds. Humans and human culture would logically find the best or most efficient way of exploiting the environment.

Hawkes' Ladder

A further step towards New Archaeology came not from social scientists, ecologists or anthropologists, but instead from an archaeologist, in the form of a highly influential article published by Christopher Hawkes in 1954. Hawkes suggested the existence of a scale of difficulty in interpreting archaeological data, which became known as **Hawkes' Ladder of inference**. At the bottom of the ladder, being the easiest to interpret, he put technology. Above this were activities associated with, in ascending order, economics, social organisation, and politics, which could be reconstructed with increasing levels of difficulty. At the top of the ladder were past ideologies, the interpretation of which he deemed almost impossible. We might think that Hawkes' ladder reflected what archaeologists had discovered, or indeed been interested in based on the prior years of research. However, this was not the case as economy, for example, had not been a major focus of previous archaeological research. Indeed, it was only a couple of years earlier that Grahame Clark had proposed that archaeology should concentrate on just such matters. Perhaps more truthfully Hawkes' ladder was the product of a new archaeological paradigm that had begun to put the importance of studying ideological questions below economic ones. In Clark's work science was certainly beginning to play an important role and therefore Hawkes' ladder may say more about what science was perceived to be good at discovering rather than archaeological inference *per se*.

A scientific archaeology

Cultural ecology and cultural materialism

The early 1960s saw renewed interest in social evolutionary theory. For example, Leslie White saw technologies and cultures developing as a means of exploiting the environment in a more energy efficient way. White, along with a group of economic anthropologists known as the **formal economists** argued for an idea we have previously discussed under behaviourism, that humans adapted to find a universal best way of doing things. An important distinction should be made here. While some theories stressed the importance of culture in creating and maintaining efficiency, others (notably those developed by economic anthropologists) argued that it was individuals that acted efficiently, rather than society as a whole. White suggested that cultural change was driven by new technologies that more efficiently satisfied biological needs. Complexity and progression occurred as societies increased their ability to capture and use energy more efficiently through technology. White was known as a cultural materialist insofar ashe believed that culture emerged as a means of satisfying material needs. Although White's theories played down the influence of adaptation (he did not see the development of new technologies as being inevitable) they still suggest that elements of culture functioned as a means of society adapting to the natural environment in a more efficient way.

White's contemporary Julian Steward on the other hand saw technological advance as the inevitable adaptation of culture to particular environments. If Steward's line of reasoning is followed, differences in environment explain variations in human culture – an argument that would later be seized upon by some New Archaeologists. The other factor that Steward drew upon was population dynamics. These he suggested operated in tandem with technology through what he termed **push-pull mechanisms**. At the end of the Weichselian agriculture was developed in order to feed people as Near Eastern environments dried up and wild resources diminished. This led to population increase, promoting further technological advance in the form of irrigation systems. In this way environmental change was ultimately responsible for technological advance. As we saw in section 1, Steward's approach was known as **cultural ecology**. Its main concern was with how cultural change arose from human adaptation to living in specific environments. Cultural adaptation in this context was not through increased efficiency but through technological advance.

While White was the architect of the term cultural materialism, it was to become synonymous with the name Marvin Harris. From 1964 onwards Harris developed an elaborate theory that he claimed to be subject to the same objective scientific methods as Darwin's theory of evolution. The theory proposed that cultural traits (as opposed to technologies) arose as a means by which people culturally adapted to exploit their particular environment more efficiently. For example, Harris proposed that the pig taboos operating in the Near East occurred as the environment became drier. Increasing aridity led to a loss of woodland and therefore the husbandry of pigs became inefficient and wasteful. The energy return from raising pigs was less than that put into feeding them. Taboos against the eating of pigs, therefore, served as a cultural adaptation allowing the more efficient exploitation of the new arid environment.

Ester Boserup's work of the same period is a good example of the difference between the Harris and Steward approaches, although she was closer to the latter. Boserup argued that as populations increased in size technological advances would have to be made to find means of extracting more food from the environment. She did not argue that such means would be more energy efficient. In fact she advocated the idea that more, rather than less, energy might be needed. A classic example of this line of thinking came from the study of the !Kung bushmen of the Kalahari desert of southern Africa undertaken by Lee and Devore. The study of these hunter-gatherers revealed that they had more leisure time than agriculturalists did, raising the question as to why hunter-gathers turned to agriculture. Of course a follower of Harris might argue that this is precisely why the !Kung remained hunter-gatherers rather than turning to farming.

New Archaeology – using 'science' to see the past

Whether the archaeologists of today agree or disagree with every aspect of New Archaeology it is nevertheless true that it created an atmosphere which through its promotion of interest in the economy and environment led to the formulation of environmental archaeology. It is therefore worth spending some time considering the ideas that it promoted. New Archaeology itself grew out of the dissatisfaction of younger archaeologists with the state of archaeological theory in the early 1960s. Three in particular, Lewis Binford, Colin Renfrew and David Clarke had firm ideas of the direction in which archaeology should head. They wanted archaeology to become more scientific, to tackle the big questions, to scientific laws governing the development and structure of societies. During the 1960s and 1970s many aspects of functionalism, materialism and historicism were to become prevalent in archaeology. In particular cultural materialism and cultural ecology came to dominate archaeological science. While few environmental archaeologists would acknowledge these works as a direct influence, much of their work nevertheless continues in this tradition.

Of the theories to be adopted by New Archaeology few show the influence of the new approaches then current in anthropology more than what we may group under the heading of optimal location models. **Optimal foraging theory** proposed that hunter-gatherers would set up base camps according to energy expenditure within hunting and gathering territories. Energy expenditure included the distance travelled to hunting and foraging grounds and the weight of the targeted food resource. Hunters and foragers would travel far to acquire a resource providing high-energy returns, such as animals. However, when foraging for low-energy and/or bulky foods, such as tubers, nuts and shellfish, much shorter distances would be travelled each day. The location of the base camp would reflect the optimal place from which all subsistence needs could be met with the minimal of energy expended. It therefore follows that the base camp will be moved when people can no longer exploit the remaining resources in an optimal manner. Importantly, this optimal behaviour was not something that was calculated each time by individuals. In accordance with Malinowski's ideas of cultural adaptation it was suggested that such behaviour developed over many generations, eventually become embedded within cultural practices. In respect of the last Tim Ingold has pointed out that if the environment changes then because exploitation patterns are supposedly culturally embedded over such long times they will be slow to alter and therefore foraging will become less than optimal!

Another optimal location model was **site catchment analysis,** developed by Eric Higgs mainly to examine populations of sedentary agriculturalists. As with optimal foraging theory, site catchment analysis proposed that people located settlements in a rational way that aimed to maximise the returns from the surrounding environment. By mapping potential resources around an archaeological site, the optimal location of pasture or arable field could be plotted and the ways in which people exploited the environment could then be mapped out.

Binford's middle range theory

Certainly, one of the most famous of the New Archaeologists was Lewis Binford. Elements of cultural ecology and cultural materialism were unified in his work. He saw cultural adaptation as the exploitation of the environment by the most efficient means. Social phenomena were therefore responses either to the natural environment and/or to adjacent competing cultures. Remember the question we asked earlier; why, if people act to exploit the environment in the best way, should culture ever change? The answer to this according to Binford was that without environmental change all societies would have remained the same. The implication is that 'environment' takes on an all-important role in cultural change.

Malinowski's influence is clear to see in Binford's definition of culture as 'man's extrasomatic means of adaptation'. Binford argued that an animal is biologically adapted to its environment through its body (somatic). Polar bears have bodies capable of withstanding extreme cold and powerful jaws to tear through seal flesh. Seals have flippers for swimming. Human culture was hence derived from man's intellectual or 'extrasomatic' adaptation to his environment. The Inuit become culturally adapted to the environment in that they make snow-shoes, hunt with bows and arrows, wear fur coats and live in ice-built houses called igloos.

In the 1980s Binford modified his previous ideas by adopting a term borrowed from sociology, **middle range theory**. It is within his middle range theory that we see the emphasis on cultural adaptation through increased efficiency. middle range theories are those ideas that link archaeological data to the understanding of past cultural systems. For Binford such theories were developed through ethnography, or the study of present day 'primitive' societies. In ethnographic studies of these we can observe people in the present growing and processing crops, butchering animals or building houses using traditional methods. We may then examine the remains resulting from these activities in detail. When we find the same evidence archaeologically to that observed ethnographically we can thereby determine the past activity. In other words middle range theory relies on the uniformitarian concepts we introduced in section 1.

While this helped in deciphering past human actions, Binford extended such theories to the understanding and interpreting of actions within a social context. One observation Binford made from ethnographic work with the hunter-gatherer Nunamiut of Alaska is that only the larger bones carrying most meat are carried back from kill sites to the main camp site (home base). From this observation Binford developed a theory: where hominids are killing animals and bringing meat to a home base we would expect to find only large bones. Conversely, where they are scavenging carcasses, such sites would contain only smaller bones.

The emphasis on efficiency should be noted in Binford's ideas. The assumption of the **universality of efficiency** is that the most efficient way of hunting is to butcher animals and bring back the bones bearing the most meat. This argument is emphasised in his latest work, *Constructing Frames of Reference: An Analytical method for Archaeological Theory Building using Ethnographic Data Sets,* (2001), where he generalises that 'human actors…attempt to maximise their vital security in any habitat, limited only by their capacities and means' (p.41). Even if this is true it still leaves a lot of room for variation. A key capacity affecting behaviour is knowledge. We may only act in ways that our knowledge permits even if we are capable of putting current knowledge to new uses.

Human ecology as a framework for archaeology

As was the case with cultural materialism, **human ecology** also had its roots in North American sociology. The approach was brought to archaeology by Karl Butzer in a seminal work *Archaeology as human ecology* (1982), which is still a core textbook for the majority of students studying the archaeological sciences at the present day. The similarity of this work to those already discussed is apparent in its opening pages. Butzer emphasises '*adaptation* (specifically as a strategy for survival) and *adaptability* (as the capacity of the cultural system to adjust)'. He goes on to state that 'These concepts, as defined in cultural rather than biological terms . . . are at the heart of the human ecosystem.' (p.11), the latter is a concept that we examined in section 1.

While Butzer concedes that few societies display optimal adaptation, his view of what he terms cultural factors is that they prevent optimality. Butzer did not consider in detail what optimal behaviour might be, but he assumed that it fulfils biological needs in the most efficient way possible. As optimal behaviour has the aim of maximising the chances of attaining cultural goals, then it is difficult to see how cultural factors could possibly get in the way. If all cultural factors arise as a human means of adapting to enable a better (presumably optimal) strategy for survival, then how can cultural factors simultaneously prevent such optimal adaptation? As Tim Ingold has shown this paradox is a common trait to be found in cultural ecology theories.

There are many other theories that come under the umbrella of New Archaeology, but the selection we have discussed were chosen as those most relevant to environmental archaeology. The key point is that, although most practitioners may not realise it, environmental archaeology is the child of New Archaeology, and therefore the theoretical viewpoints of processual archaeology underpin it. We began by raising some minor criticisms of environmental archaeology. Now we will get to the crux of the argument: what did post-processualist archaeologists find so controversial about science in archaeology?

Science or science fiction?

The most serious objection to science in archaeology for post-processual archaeology was that for the most part it was less science and more science fiction, this comparison being made by Julian Thomas in an article we have already examined. In the early 1970s Ian Hodder, like many others in archaeology, were building models to test various archaeological hypotheses – an approach fundamental to all science. When he began to delve deeper into 'scientific' practice in archaeology he found that he often ended up just confirming what he set out to prove. Constructing a hypothesis around archaeological data would seem logical and straightforward. The problem was that if he set out to prove a realistic alternative hypothesis, the data could prove this as well.

Another line of attack came from Michael Shanks and Christopher Tilley who stated that archaeological science is based on modern economics. It therefore created a past that reflected the attitudes to be found in modern Western societies. They argued that people in the past did not necessarily plan to maximise their economic gains in the way people of our own society do. We have discussed such arguments in greater detail in section 3.

How serious are these allegations and does environmental archaeology really simply create myths based on supposedly scientific data? To address these questions we have written two entirely fictitious site reports using environmental archaeological data. It is likely that the reader might recognise in the examples of *Excavations at Donny Brook Farm and Griffin Hill* parallels in the environmental archaeological literature. It is not our aim in doing this to undermine all scientific reasoning that is found in environmental archaeology. The examples given are exaggerated and certainly provide less detailed reasoning than is to be found within much of the literature. Nevertheless, they serve to show the potential dangers of purely scientific approaches within archaeology, especially with regard to the lack of plurality in interpretation currently voiced. In light of this it is then left to the reader to consider the merits of any particular environmental archaeology study that we consider elsewhere in this book.

Donny Brook Farm

The excavation of the earliest occupation phase at Donny Brook Farm revealed bones of cattle, sheep and pig with the last predominating. The age profile of the cattle and sheep showed a high proportion of animals aged two to four years. In the latest phase the number of pig bones declined, while the assemblage of cattle and sheep bones contained a greater proportion of animals aged over five years than before.

A hypothesis was constructed suggesting that pigs were initially kept in woodland and, along with sheep and cattle, slaughtered for their meat. The change in the age profile suggested to the zooarchaeologist that cattle and sheep were increasingly being kept for milk and wool rather than meat. A decline in tree pollen dated to the later phase of the site was interpreted as indicating woodland clearance. The clearance of woodland was seen as making way for further pasture and arable. As pigs are often associated with woodland their decline was suggested as being related to this new regime.

The charred plant remains from Donny Brook Farm also revealed changes in subsistence. In the earliest phase emmer wheat dominated over naked wheat and naked barley. The weed flora indicated a high number of annual species, especially those of the Chenopodietea. In the later phase hulled barley replaced naked barley and spelt replaced emmer and free-threshing wheat. The weed flora changed to one high in seeds of perennial species and annuals of the Secalietea.

Modern varieties of emmer wheat are mainly spring-sown and grow best on light soils. Spelt wheat grows on a much wider range of soils. The archaeobotanist concluded that the change from emmer to spelt might indicate the extension of agriculture onto more marginal soils. This was correlated with archaeological evidence for numerous drainage ditches dated to this phase. A change in climate towards wetter conditions was seen as the primary cause of the abandonment of free-threshing wheat and naked barley. Both are more susceptible to fungal attack in wet conditions than hulled barley or wheat. Modern varieties of hulled barley also respond better to manuring. The change in the animal economy, it was concluded, may have led to increased availability of manure and hence hulled barley being favoured. It was also noted that barley makes a good fodder crop.

Species of the Chenopodietea have been associated with spring sowing, the Secalietea with autumn-sowing. The increase in the Secalietea was taken as indicating a change to the latter regime. As autumn sowing allows for a longer period of growth, autumn sown cereals often give better yields. Perennial species are known to be more common where a system of leys (short periods of grassland) is practised. It was suggested that the change might be associated with a more intensive animal economy combined with a more intensive field rotation system.

The general conclusion of the environmental archaeological report from Donny Brook Farm was that the changes in the environmental data demonstrated the intensification and extension of agricultural practice.

All these conclusions may seem reasonable enough, but do they reflect or prove what really happened in the past? We will examine this by reversing the archaeozoological and charred plant macro-evidence, then reinterpreting them.

Griffin Hill

In the earliest phase of the Griffin Hill site the archaeozoologist found that bones of sheep and cattle were dominated by individuals aged over five years. In the later phase bones of pigs increased, while the age profile showed an increase in the number of younger individuals of sheep and cattle.

Spelt and hulled barley dominated the earliest phase, while the weed flora contained seeds of perennial species and annuals of the Secalietea. Later emmer wheat, free-threshing wheat and naked barley replaced spelt and hulled barley. Seeds of annual weeds, especially those of the Chenopodietea, came to dominate the weed flora in the later phase.

The zooarchaeologist concluded that in the earliest phases animal husbandry was less intensive and mainly for meat. The change that led to higher numbers of juveniles in the bone record was interpreted as being indicative of a more intensive milk economy in which young males were slaughtered for meat. The increase in pigs was associated with their usefulness in the settlement, eating waste that is effectively transformed into meat.

Both free-threshing wheat and naked barley are easier to process and cost less to transport. Free-threshing wheat can also give very high yields with manuring and weeding. The change to these crops was seen as intensifying production for exchange. The switch from spelt to emmer was associated with archaeological evidence for improved drainage as modern varieties of emmer are recorded as performing well on drier soils. Annual weed species are associated with higher levels of soil disturbance and continuous cultivation. Species of the Chenopodietea, unlike the Secalietea have been associated with more intensive cultivation of highly fertile soils. It was proposed that such a change was linked through manuring to the intensification of the animal economy. The general conclusion drawn by the specialist in the environmental report from Griffin Farm was that agricultural practices had intensified.

From totally conflicting evidence it is possible to draw very similar conclusions by just selecting those attributes that fit the hypothesis that people were attempting to maximise efficiency. None of the information given here is unusual. Every attribute or interpretation can be found in the archaeological, anthropological, and ecological literature, although some of the interpretations, for example associations of emmer wheat with soil types and sowing times have been disputed. The general point is that we may argue conflicting hypotheses with relative ease using biological data and drawing only on the archaeological data that fit our argument.

Neither hypothesis is necessarily wrong, indeed both may be valid; the interpretations are suggestions rather than proof of the causes responsible for change. Without this proof they are, however, just stories no more correct than an alternative set of stories. More worryingly they are stories that only look at the data from one angle. They do not take account of other possible interpretations. There is then reason to suspect that Hodder, Shanks, Tilley and Thomas may have a case in their criticism of environmental archaeology.

Shaping a lock to fit a key

Four main problems may be identified with scientific interpretation in environmental archaeology.

1. It generalises that human action and culture are directed towards increasing efficiency.
2. It only defines efficiency within the framework of existing knowledge.
3. Any change seen in the biological record is only interpreted as functioning to increase efficiency. Therefore, only economic hypotheses are created to fit the data. Alternative, perhaps equally valid, hypotheses go unexplored.
4. The interpretation of data rarely goes beyond efficiency as an explanation to understand the evidence in a social context.

With regard to these it is worth considering the distinction between how we as archaeologists may interpret something in the past compared to how people in the past may have interpreted it. A functionalist explanation of a feature or artefact may indeed coincide with some elements of the purpose, intention and meaning that past peoples had attributed to it. For instance, we may interpret a pit as a well for obtaining water. This may encompass some of its intended purpose and meaning in the past. However, much criticism has been directed towards purely functionalist explanations in archaeology. One accusation is that such simple explanations mask deeper meanings that objects may have had beyond that of their immediate function. A well may be for drawing water, but it may also have had a symbolic meaning that may explain the presence of other artefacts found in association with it. Further to this is the interpretation of an artefact, feature or identified process within its social context. Different types of crops may be cultivated in many different ways, different animals may be reared and utilised according to varied practices. But how are these differences to be interpreted from archaeological data if they are largely determined by past **social values** reflecting past ideologies and symbolism?

It is at this point that functionalist arguments begin to run into difficulties. At best they may just be too simplistic. For instance we may hypothesise that a cereal crop was chosen for planting because it produces better yields. This social value makes sense within our society as grain can be sold for a profit. The more grain that is produced the greater the profit. However, we must ask how and why, if increased production was also the aim of past peoples, they profitted if selling the grain was not possible? How did the desire and the means to increase production for past people relate to their social context? It is not enough just to say it was more efficient. This brings us to a further point. At worst functionalist arguments may be incorrect about what the archaeological phenomena we are observing meant to people in the past. A crop may simply have been chosen for production because it was required in what had become the most fashionable dish.

The sociologist Robert Merton pointed out that it is possible that some cultural traits may have no social function or purpose. Merton asks about things that might be seen as negative cultural traits. Do wars really have a function? According to some archaeological theories, a cause of war is scarce resources. Johnston and Earle in *The evolution of human societies* (1987) explain fighting between the Yanomamo of Venezuela as related to 'scarce supplies of dietary protein'. Would such a functionalist explanation provide an acceptable account of Nazi atrocities or the bombing of Hiroshima/Nagasaki? It might be argued that elements of the war were caused by economic depression or economic aims, but this still does not explain everything that occurred. This case illustrates a trait of science, to reduce highly emotional ideologically fuelled events to cold hard 'facts'.

It is perhaps time to consider the place of science in our own society. Jean-François Lyotard wrote about changes in the technologically advanced Western nations occurring after the Second World War. He argued that within our society knowledge had become tied to economic production. In turn, the main goal of society had become directed towards efficiency, in particular economic efficiency. Science that once sought knowledge for its own worth had turned into an economic and political tool. Through such associations the drive for increased knowledge, as realised through science, had become intertwined with concepts of increased economic productivity. An example of this relationship comes from the importance placed on research into 'natural' medicines in 'primitive' groups with the aim of marketing new drugs. Through the interjection of science the knowledge of hunter-gatherers becomes a sellable commodity in Western economies.

The trouble with science is that it can only study certain things. Things that can be measured or calculated, in terms of length, weight, energy, cost, money or physical reaction. Science cannot measure goodness, fairness, aesthetics, fashion, morality, beauty, spiritual righteousness, bravery, loyalty or emotional responses. Lyotard pointed out that science has been seen as a dark force. For example, the weapons of mass destruction that characterised World wars in the twentieth century did little for the reputation of science. More recently people have been concerned with genetic crops and other 'scientific' farming practices. All these practices are conducted to increase efficiency and are fuelled by an objective science that only fulfils the objectives of Western economies. Here lies the heart of the problem. Charles Lyell in his description of uniformitarianism said 'the present is the key to the past'. In other words that the same processes we observe today also occurred in the past, a perception that has served environmental archaeology well (with some modifications), shown and discussed in sections 1 to 3. However, as Matthew Johnson states while this is all very well for natural or physical processes, it might be less applicable for the mental or emotional processes of the human mind.

The gradual introduction of science in archaeology, beginning in the early 1950s, appears to have imposed the social values of our own society onto people in the past, the same social values that Lyotard saw as gaining prominence in our own society from 1950 onwards. In recent years these same social values have been increasingly questioned by our own society (hence the new paradigm with which this section is written.

Turning middle range theory upon itself

Robert Merton, the original architect of middle range theory, used it to explain how individual action related to grand social theory. For example, middle range theories might be used to explain how the actions of certain individuals are shaped by their societies' goals and how their actions are conducted with respect to the social values and ethics held by their society. In modern Western societies economic success is a major goal. The legitimate socially accepted (and expected) means of attaining economic success is through hard work within the confines of social and ethical law. Merton then uses middle range theory to explain suicide and drug abuse as a rejection of both the goals of society and the means, when the goals seem no longer attainable. Crime is a way of attaining societies goal by rejecting the legitimate means. For others such as policemen, administrators and politicians upholding or defining the accepted means becomes their goal.

Arguably we might see the application of science within archaeology as – even if subconsciously – directed at using the past to justify the means and goals of our own society. Science in archaeology has continually given the impression that people's activities in the past were also orientated towards the goal of economic success. It also justifies the means. By interpreting all past activities as succeeding in attaining increased efficiency, by our own definition, it gives a modern scientific rationale to people in the past, in other words, while we use science to calculate success and efficiency, people in the past were also capable of doing so, using methods that we can recognise.

What we have done then is to use middle range theory in its original sense to explain the actions of Binford and the New Archaeologists. The fact New Archaeology only exists to justify the goals and uphold the means of our own society. The use of science in environmental archaeology then appears to have redefined the past lock to fit the key of present social values, rather than trying to unlock past social values. There is an even greater irony to this argument. Science is often upheld not only as the supreme tool by which we can better our understanding of the past, but perhaps more widely as the tool by which our society can better its future and become even more efficient. Yet it is also accepted that people in the past did not practise the science that we currently recognise. This implies that if people in the past still acted efficiently then they must have had other means, equally as successful as modern science, with which to gauge their own efficiency. If this is true then clearly other ways of thinking exist that are equally capable of producing the same explanations as science. By their own arguments those, such as Marvin Harris, who claim that science is the ultimate tool for understanding human culture end up by implying that any other means of thinking are likely to be equally valid!

Taking the theme of turning academic theories on themselves further we can consider how appropriate science is for explaining what archaeologists do. Modern archaeology is not driven by biological or materialist needs (if it was there would be no archaeologists, teachers and nurses, and many more well-paid accountants and lawyers) and is only understandable as something we put a social value on. Considering that what many people in the present do (e.g. archaeologists) is beyond science, biological and materialist explanations then we cannot by reason expect science and materialist explanations alone to explain people's actions in the past. Even the funding of archaeological science and what does or what does not get published is down not to scientific rationale but to social values. Thus it might be cynically argued that the success of New Archaeology was not down to good science but rather a reflection of the high social value put on science by society in the 1960s and 1970s and hence the funding and publishing of scientific work.

Lessons in efficiency and social values

While archaeological scientists seldom hypothesise that people may have acted inefficiently in the past, there are a large number of examples from both history and ethnography where precisely this can be argued. William Durham tells us of how the cultural practice of eating dead relatives, adopted for purely ideological reasons (as a sign of respect!) almost ended up wiping out some populations in South-east Asia. This was because a form of the degenerative brain disease scrapies was continually being transmitted through the practice.

We can also take the example of the taboo against eating pigs in the Near East, which Harris explained in terms of the inefficiency of raising pigs in less forested environments. To begin with it should be noted that the taboo is against eating pigs not keeping them. The Greek author Eudoxus certainly describes Egyptians keeping pigs around 350 BC. The animals were apparently allowed to eat rubbish and wander the fields after sowing, trampling in seed-grain, but were not eaten. Eudoxus came to the conclusion that pigs were still kept in spite of the taboo because of the pig's usefulness in agriculture. Contrary to the views of Marvin Harris, then, Eudoxus assumes that pigs were still kept because reasons of efficiency prevailed over the cultural taboo.

Tim Ingold has argued that individual hunters are often not slave to any larger scheme such as optimal foraging theory. In what are often unpredictable environments the hunter, rather than setting out to maximise his return, acts according to individual experience and personal whim. Given that we ourselves do not live the lives of those we study or have their knowledge and experience, it is highly debatable how capable we are of predicting what might have been efficient or suitable behaviour in the past. However, clearly both hunters and gatherers do need to act in ways that provide a reasonable degree of success. After all there is little point carrying out an action if it does not provide a good chance of achieving a favourable outcome. Nevertheless, in a resource-rich environment would people really need to act with maximum efficiency? Tim Ingold surmises that culture provides the means and tools of thinking and acting out hunting and foraging but how this is put into action is often down to the individual.

Even if we accept that people in the past wished to act efficiently, we still have to ask how they came by and measured efficiency in order to calculate whether it was worth them changing their practices. Binford states that efficiency is governed by people's capacity and means. In short, this implies that people can only choose to act efficiently within their knowledge of what efficient action is and how they may attain it. Even with this knowledge people may still choose to reject more efficient methods, as we have seen in section 3 in the case of Jethro Tull's seed drill. We must also consider that even in the past people might have acted in ways they perceived as efficient but which turned out not to be. An historic example of a perceived improvement in efficiency that went badly wrong was the provisioning of African farmers with tractors that soon became abandoned when they broke down because parts could not be obtained.

The majority of social evolutionary theories, as well as those associated with New Archaeology, rely on the existence of a single law governing the nature of all human behaviour (such a law might be that all people consciously or subconsciously strive to optimise and increase production). As behaviour is governed by knowledge that itself is gained through experience, then such a law must be able to predict how new knowledge is formed.

The problem of tying knowledge in with historicism

The argument that there can be no unified theory governing knowledge is exactly what the sociologist Max Weber argued for in 1922. However, it was Karl Popper who really put the problem into perspective. In the *Poverty of historicism* (1957), Popper argued that knowledge plays an extremely important (perhaps even the greatest) role in the way societies are structured. The construction of new knowledge is, therefore, an important aspect governing the ways in which societies change. Such knowledge may be technological. For example, our own society has been transformed dramatically through the invention of electricity, the aeroplane, motor car, and silicon chip. Knowledge may even be religious. The history of the World is dominated by the influence of the major religions, Judaism, Buddhism, Christianity, Hinduism and Islam. To discover laws concerning the nature and development of society would then be to have laws regarding the formulation of knowledge. If such a law could be formulated we would be able to know right now what we should only be able to know in the future, something which Popper states is clearly impossible. We may discern patterns in history but can we truly find laws that indicate patterns in social 'development' were inevitable?

This idea is similar to that of Robin Collingwood and Wilhelm Dilthy concerning historical explanation. **Historical explanation** sees each historical event as unique in the reasons for the event occurring, the form the event takes and the effect it has. This leads to the proposition that we can only understand and should only study the past in relationship to itself and its own past. This is an idea that even some Darwinian evolutionists would agree with. These conclusions are very similar to those reached by Franz Boas, the very man whose work was rejected by the New Archaeologists. Franz Boas was what we call a **cultural relativist**. Scientific positivists do not like relativists much. Relativists often concern themselves with what are known as **hermeneutics**. According to Boas material culture can only be understood through understanding ideology. Yet according to Gordon Childe past ideologies can only be understood through material culture. Hermeneutic circles, then, are problems in which a solution to a question can only be found by finding the solution first, clearly impossible!

The future of environmental archaeology – in theory

A relatively bleak picture has been painted of the theoretical perspective that underpins environmental archaeology. Is environmental archaeology really built upon foundations of shifting sand? After all if we accept, as many post-processual archaeologists do, that archaeology is inevitably caught up in hermeneutics then it is difficult to see how we are to understand anything about the past from material remains. However, to take a more positive view, we believe, as we hope we have demonstrated in the previous sections, that science does have, and has made an important contribution to archaeology. What we need to agree on is what science can tell us and to accept what it cannot. If we are to carry on digging up and recording archaeological data we need to have faith that this activity has some purpose.

Perhaps the most important theories that can bear upon these issues come not from anthropology or archaeology, but from the work of the sociologist Max Weber. Weber drew attention to the difference between what he termed **social facts** and **social values**. He saw these as dividing science and politics. Science is good at establishing facts, but that science itself is politically determined according to contemporary social values. In archaeology it is the same. Science is generally good at establishing what we may term 'archaeological facts' but cannot establish 'past social values'. If archaeological scientists can recognise and distinguish between the two in their archaeological interpretations then environmental archaeology will enjoy a bright future even alongside post-processual archaeology. This may sound straightforward but even interpreting a dog-gnawed bone as someone's rubbish is imposing a social value on how that person perceived it!

Let us look pragmatically at what we believe environmental archaeology does well. At the most basic empirical scientific level we may compare different data sets and note that one is different from the other. As we saw in section 1, Michael Schiffer has demonstrated that we may compare our evidence to present naturally and culturally induced processes and deduce which are present and how they interacted to form the archaeological record.

Throughout this book we have been relatively uncritical of the use of science in environmental reconstruction. This is because we believe this to be a strong point of science in environmental archaeology. Nevertheless, we stated at the beginning of this section that environmental archaeology should be more than simple environmental reconstruction.

This brings us to a further strength of science and environmental archaeology. Namely that as a tool for interpreting past human actions in terms of basic physical processes it has undoubtedly made a significant contribution to our knowledge of the physics involved in past activities. Both Schiffer and Binford have suggested that we can look at ecofactual remains and see how they compare to what we know from ethnographic evidence. Similarly scientific approaches have proven invaluable in interpreting how past crops were processed, how old an animal was when it died, what species it was, whether the marks upon the bones were cut marks or what contents a vessel last held. The strengths of science lie, therefore, in interpreting past human action in terms of its effect on physical processes.

Problems begin to arise when interpretations are taken further. What science cannot tell us is the purpose of past actions. Science must be neutral and value-free. However much any of the theories outlined earlier try to convince us that they are scientific they all fail by this one definition. Neither past human actions nor our interpretation of them can ever be value-free. The moment we make the assumption that people act in ways that are optimal or to maximise efficiency we are making a value judgement about what we consider optimal or efficient behaviour to be. As a means of finding explanations archaeologists should turn to those disciplines that are better equipped to examine what are essentially social issues, in particular sociology, history, philosophy, cognitive psychology and anthropology. In other words, unless environmental archaeologists simply want to be technicians whose data are interpreted by others, they must be versed in the theories of the social sciences.

Using social values to go beyond environmental reconstruction

Interpreting what past actions meant to past peoples and how they related to a social context is difficult, and we must accept the validity of multiple interpretations to explain the same phenomena. As we previously discussed, actions, in as far as they represent people's intentions, will always relate to their knowledge, experience and social values. Both the sociologist Herbert Mead and the psychologist Jean Piaget suggest that individuals act according to their experiences, as gained from growing up within a certain society and culture. These experiences will then greatly influence aspects of their social values and knowledge, and through these reinforce elements of their society's social structure. It is their own unique personal experience and ways of interpreting these experiences within the context of their current knowledge, that lead to every individual reformulating that knowledge in new ways and being capable of acting relatively 'independently' of their social structure (**agency**).

To understand the relationship of actions, knowledge and social values to social contexts in the past, archaeologists must compare archaeological data to more than just ethnographic evidence. It is necessary to begin to try to understand such evidence within its own 'historical' or archaeological context. Peoples of a single culture do not define themselves by elements of their own culture alone. As we saw in section 4, they define their identity in relationship to the world around them. This will include how elements of their own culture (this might be anything from how food is prepared or how a house constructed to marriage customs, social histories or ancestral myths) are different from those of their neighbours.

We can then begin to define and build up a picture of past cultures by comparing the actions that we have interpreted from archaeological evidence to similar actions as seen from neighbouring regions. Any action in the past was performed means that it had some social value attached to it, even if we do not know what that value was. Where differences are seen between the actions of past peoples, either between adjacent cultures or from one generation to the next, we may assume that social values were also different.

There has been a tendency with scientific evolutionary approaches to see past developments as a somehow inevitable consequence that accumulates in our own society. What we must remember is that past societies were shaped by their own histories not their future histories. When we look for a single point in time and space that marks the origin of agriculture or the beginning of civilisation, we will always be deceiving ourselves, as the peoples involved in this will still have acted according to previous experiences built up over millennia. To look at any moment in time is akin to staring up at the night sky. We may see a single picture, but it is made up of many stars, the light from each one originating at a different moment in time.

By studying the past in its own historical and social context what we are doing is building up **relational analogies**, comparing data pertaining to aspects of past societies to their contemporaries or other similarly dated past societies. As Ian Hodder argues relational analogies are more appropriate than **formal analogies**. This is simply because in relational analogies we are comparing evidence pertaining to past knowledge and social values to other contemporary evidence, while formal analogies only compare such evidence to our present knowledge. Many environmental archaeologists have tended to take the 'formal approach', applying no more than the interpretation of a comparable ethnographic site.

We might still wish to look at the archaeological record to understand the importance of technological advances, intensification and improvements in efficiency, all of which are still (in spite of all that has gone before) perfectly valuable concepts. But these issues are examined more objectively taking into account the social context in which they occur and the role of both knowledge and social values in such transitions.

Phenomenology

One issue that we have not yet drawn on is the relationship between environmental reconstruction and past ideologies. **Phenomenology** is the understanding of archaeological monuments through experiencing them, or in other words: what we feel the meaning of such structures was for those who built them, by walking through the monument at the present day. Similar experiential approaches have been used in computer modelling, virtual reality, Geographical Information Systems and computer reconstructions to investigate the archaeological site in its landscape. All these approaches would be strengthened by understanding how that relationship and perhaps the social meaning of the monument would alter with the changing environment. It is worth spending a little more time discussing phenomenology as this is an area in which environmental archaeology could have a useful input. The major proponent of the approach is Christopher Tilley who has 'experienced' monuments such as the Dorset *cursus* (a long banked Neolithic 'trackway'), by walking along, around, and within, them. Such experiences are shaped by the monument's relationship to its 'natural environment' and archaeological landscape. The monument is experienced and understood by taking in its setting and combining this experience with the stories, legends and myths that surround it. Can running water be heard? What parts of the landscape are difficult to cross? What other archaeological features of the landscape are prominent? What symbolism might be attached to these features?

From what we have previously seen in sections 1 and 2 it is not hard to see why the approach is much criticised. Firstly, it is often assumed that the landscape today is similar to that in the past. The romanticism and mysticism of visiting long-deserted 'lost cities', overgrown in vegetation in South American rainforests, may provoke experiences similar to those of the early mid nineteenth-century pioneers John Lloyd Stephens and Fredrick

Catherwood, but these are of course not how they were experienced by those who actually lived in them.

In Europe many sites such as the Dorset *cursus*, are situated today within cleared landscapes of rolling arable fields and grasslands. These are landscapes that have been shaped by millennia of farming. Using pollen and molluscan evidence Michael Allen has described the view along the same *cursus* in the Neolithic period. Rather than the open fields of today, Allen's alter-ego (the persona of a Middle Neolithic woman buried nearby) could see 'the grassy ditches and banks of the *Cursus* sweeping down in front of her as a short-grassy ride, cutting through the open patchwork of . . . birch and hazel shrubs and long tussocky grassland . . . down across the Allen valley, with its rushed-edged stream and damper grassland snaking next to it and then lifted itself across the cliff . . . and processed up the gentle slope to the skyline . . .' (M.J. Allen, 2002, 61-63).

Another feature Tilley discusses is Chesil Beach in Dorset, which he suggests was an important element in determining the location of the early to middle Neolithic causewayed camp at Maiden Castle. Chesil Beach is a long narrow ridge composed of rounded pebbles, joined to the land at either end with sea on both sides. It was formed when pebbles and sand were deposited by long-shore drifts forming a spit. The spit extends for some 8 miles across the sea to the immediate west of the modern town of Weymouth. However, if the palaeoenvironmental evidence is examined it becomes clear that Chesil Beach had a very different form in the Neolithic – if it existed at all – than it has at the present day. Spits such as Chesil only form during lengthy periods of sea level standstill. The last sea level standstill was around 3000 BC; prior to this the area would have been coastal marshes. This is confirmed by the examination of peat layers under the present day beach that date from around 8000 BC when sea level was some 20m lower than at present to 3000 BC when it was 3m lower.

These examples are not given to undermine the potential of Tilley's approaches. Rather, they serve as an indication of the importance of environmental archaeology in our understanding of past archaeological landscapes. This is especially true of some Neolithic sites constructed when landscapes were still wooded. At Coneybury, near Stonehenge, Neolithic people built a henge consisting of a circular ditch, bank and post ring. Buried soils under the monument bank did not contain a turf layer, indicative of grassland. Rather together with molluscan evidence it showed the monument to be constructed within a recently made clearing. An experience of this monument would have been very different to that of many later Neolithic rings, such as the later phases of Stonehenge itself, constructed in relatively open landscapes. Perhaps to the people who built it, it was more reflective and symbolic of the clearings where their Mesolithic ancestors had once gathered.

As with the criticisms of scientific approaches made earlier, phenomenology also suffers in its relationship to how individual knowledge shapes experience.

People in the past obviously had very different experiences and knowledge to us. For example, Stonehenge must have seemed particularly awe-inspiring to people who had no experience of substantial architecture. Further we do not know how or if the experience of monuments was socially constructed. Perhaps people did not take Sunday strolls down the *cursus*, but rather drove noisy parades of cattle down its length, cheered by others who watched from the banks.

While these particular problems with phenomenology lie outside the realm of environmental archaeology (perhaps even archaeology!), locating a site within its reconstructed past landscape will at least provide an impression as to how past landscapes might have shaped past experience. Tilley has quite rightly stated that past peoples would have created myths around and marvelled at natural phenomena. As such, natural phenomena should be taken into account when interpreting archaeological sites.

Beyond science to a real world

As undergraduates we were both brought up on a diet of Karl Butzer and taught that archaeology should be a science, something that appeared to make perfect sense at the time. With the benefit of hindsight we have come to realise that there are, and always will be elements of archaeology (including parts of environmental archaeology) that are simply beyond scientific interpretation and that archaeological science is often neither truly objective or indeed reflective of past people's lives. Yet we think that only through the use of science within archaeology can we carry out any interpretation of the archaeological record. The environmental archaeology that originated in New Archaeology, has brought about great benefits. It has produced a new body of data that has allowed an insight into many unexplained aspects of past societies. It has been able to enlighten us about the many ways in which humans have influenced the environment.

In our opinion the acceptance of the limitations of science (that it cannot reveal to us an objective truth concerning past social values) would not weaken the discipline, but rather strengthen it by allowing a more varied and broader range of questions to be asked of the data it produces. Often our interpretations of the past are more limited by the questions we ask than by the nature of the environmental evidence itself. Part of this limitation is the tendency of environmental archaeology to look at ecofacts in isolation from artefacts (including structures and features). In limiting interpretation to biological evidence alone environmental archaeology rarely understands the environment and subsistence beyond this one component. The categories we use to divide up specialists in archaeology are determined by the nature of the archaeological material. They do not necessarily form a unified category of data that allow us to understand and interpret any single aspect of people's lives in the past. In fact, having read many environmental archaeological reports one might be forgiven for thinking that past peoples ate charred cereals and buried the bones of skeletal sheep, while swimming in a sea of molluscs and pollen.

We must accept that we can never escape social values. However, we feel there is nothing wrong with making assumptions about past social values as long as we are aware of our own influence on such interpretations. After all archaeology would be extremely dull if it were limited to table after table of scientific data alone!

SECTION 6

PLANNING, INTERPRETING AND WRITING ENVIRONMENTAL ARCHAEOLOGY

Introduction

In earlier sections of this book we have discussed the principles that underpin twenty-first-century environmental archaeology, the methodologies by which sub-fossil remains are extracted from the soil or sediment matrix, subsequently identified and interpreted, and the way in which geological materials are examined. In an ideal world all the classes of remains that we have reviewed would survive in easily interpreted contexts from all archaeological sites, while budgetary and time constraints would not be a consideration in their investigation. In the real world of course, this ideal is seldom even approached and choices have to be made about what it is in theory, possible to investigate (i.e. what survives) and based on that, what we choose to examine. We have looked at the first of these considerations in previous sections of this book, but here we deal with the latter, the vexed issue of choice. Later on in this section we go on to look at how environmental archaeological data are integrated to provide detailed interpretations of past human situations, before moving on to examine how environmental archaeology is written. In doing this we have taken a case study approach.

Planning environmental archaeology

As we have seen as the result of differential taphonomic factors, preservation of biological remains varies from site to site, and whereas many categories may survive on wetland sites for example, only one or two classes may survive on or in a decalcified sand dune. The one exception of course is geoarchaeology as sediment provides the matrix for almost all archaeological sites, and soils are also present on many. Nevertheless, selecting which techniques to use on any given site is not simply a case of assessing whether particular sub-fossil remains or sediment sequences/soil profiles exist on that site and then employing the relevant specialist. Techniques used should depend on the archaeological questions asked of a site rather than the mere presence of a particular sub-fossil organism. In other words, investigations should be led by questions – usually in the case of present day archaeological investigations set down in a **research** or **project design**. This document is written long before fieldwork begins, hopefully with contributions from environmental archaeologists. For example, questions asked of an investigation of a Bronze Age riverside settlement may largely concern the subsistence economy of the inhabitants. Even though aquatic organisms such as ostracods may be present in the sediments that comprise the site, these will provide no indication of what people ate or how they processed food remains, and relatively little idea of where those food resources came from. Therefore, if the questions asked of the site were only economic, there would be no justification in examining the ostracod remains even if they survived in perfect condition and huge numbers. Rather attention (and of course money) would be far better focused in studying the plant remains and animal bones that would provide the best evidence. Of course such a restricted research agenda is extremely unlikely and in this case it is more plausible that our putative prehistorians would also be interested in the surrounding environment, including that of the nearby river, that was once inhabited by the ostracods now incorporated in the site matrix. However, there is a further consideration before the time-consuming process of sampling the sediments and extracting the ostracod tests commences. This is whether, despite their good state of preservation and abundance, ostracods provide the best indicator given the unique circumstances of the site of past river conditions. It is for example possible that diatoms, or in estuarine or coastal conditions, Foraminifera, may provide better a idea of the aquatic environment at the time the site was occupied.

A second reason for not simply working on all categories of sub-fossil evidence that survive on a site. As we saw in section 1 some contexts may be so heavily bioturbated, so far removed from their point of origin or so subject to post-depositional modification, that the included bio-remains do not provide reliable indications of the past situation – no matter how well preserved they are. For example, mollusc shells are frequent finds in archaeological sites on calcareous geologies, and occur in many types of context. However, as was discussed in section 2, there are problems in interpreting shells from pit fills, particularly with regard to their origin. Even if it were possible to separate autochthonous (shells from snails living within the pit) and allochthonous (shells derived from outside the pit) components, the results from the analysis will only relate to an environment immediately around the pit. Interpretations based on the results would also be subject to numerous caveats. For instance, can it be determined if the pit was deliberately filled or did it gradually fill up from erosion of the contemporary ground surface? If the former, what was the origin of the infilling sediment: the contemporary soil or another archaeological context? If the first of, was the whole soil profile removed, or just the upper part? If the latter, can the original context be identified, and how did it form? Rarely can these questions be answered with any reliability, and consequently mollusc analysis of pit fills is, or rather should be rare practice. This may seem a particularly extreme example, but it does emphasise the point that a clear idea of site formation, taphonomic and post-depositional processes must be obtained before deciding on which bioarchaeological techniques to employ, and from what contexts. It is for this reason that a geoarchaeological input is vital in designing all archaeological fieldwork projects if for no other reason than understanding how these processes functioned.

Despite what we have said about the careful selection of techniques to specifically answer questions posed of a site by using two or more techniques which appear to address the same aspect of past environments, a more detailed reconstruction is often possible. The use of more than one category of biological remains also enables conclusions to be checked, origin of material to be better investigated and apparently unusual findings independently verified. The advantages of multiple proxies to investigate a single aspect of a past environment are of course offset by increases in cost. Therefore, whether to proceed in this direction must be decided on a site by site basis after weighing the advantages offered against extra resources expended. One site where multiple data sources have proved particularly useful is in the investigation of the Dover Bronze Age boat. This apparently sea-going vessel, dated to around 1600 BC, was found in thick silts and **tufas** (calcareous sediments resulting from the re-deposition of carbonate in shallow freshwater conditions) during a road and sewerage construction scheme in 1991. In addition to a detailed examination of the sediments surrounding the boat (**88**) in both the field and laboratory, analysis was carried out of other remains that might determine the environment in which the boat had been abandoned. These included molluscs, plant macro-remains, insects, ostracods and diatoms (pollen was also examined to provide a regional environmental reconstruction). Diatom, ostracod and mollusc data indicate that, despite the position in which the Dover Boat was found, only some 100m from the sea front, the boat had originally been placed in a freshwater creek that was not subject to any tidal influence. The molluscan data suggests that water in the creek was shallow, carbonate-rich and fast-moving (**46**), while plant macrofossils indicate that the side of the creek was inhabitated by rushes. Outside the creek on dry land, mollusc and pollen evidence suggests the landscape was open, while the insect evidence indicates it was comprised of poor-quality grazing. The evidence of the boat structure itself suggests that it was deliberately scuttled in its place of rest, while domestic rubbish had been deposited both within, and around it, including bones of cattle, sheep/goat, pig, red deer and dog, as well as burnt hazel nuts and pottery. The River Dour, in which the boat was found, would not have been navigable for such a large craft, and, therefore, the impression is that the boat was dragged along the river channel specifically to be disposed of. In this case the biological and artefactual evidence have informed us in great detail about the circumstances surrounding a single event, which may have taken no more than a single day. They also tell us of environments that both preceded and post-dated that event, although there is not enough space to recount them here.

88 *The Dover Bronze Age boat in a partially excavated state. Note the dark silt and clay deposits which have been deposited inside the boat. The tufa deposits on which the boat was abandoned can also be seen.* Photograph courtesy of the Canterbury Archaeological Trust

As a further aid in considering what environmental archaeology sampling strategy to employ, the questions given below are posed for the budding project manager's consideration.

Question and *example answers*:

What is the geology of the site (solid and superficial)?
– *Chalk (calcareous limestone) covered by a 0.3-0.4m thick, calcareous rendzina soil.*
What sub-fossil biological remains have the potential of surviving on the site?
– *Vertebrate bone, mollusc shells (terrestrial and marine), charred plant remains (including wood charcoal).*
What questions are being considered in the investigation of the site?
– *Was this site a primary producer of plant and animal food, or a secondary consumer?*
What sub-fossil biological remains that are likely to survive on the site have the potential to answer these questions?

– Vertebrate bone, charred plant macro-remains (not wood charcoal)

What are the archaeological contexts present on the site in which these biological remains are most likely to survive and which will allow the most reliable interpretation of results?

– Storage pits and pits used for rubbish disposal. The latter offer some interpretative problems relating to the original use of the pit (if re-used) and the source of the rubbish, but are the most likely to contain waste indicative of processing activities. Before field sampling is undertaken it must be determined how each separate pit fill accumulated, and whether by deliberate deposition or erosion of pit sides and contemporary soil.

Is it likely that post-deposition agents of disturbance will have so impacted on these contexts that the interpretation of included sub-fossil biological remains is likely to be unreliable?

– No. The top most fills of the pits may have been subject to plough damage and caution should be exercised here, while worm activity is likely to have caused minor movement of sub-fossil remains. Otherwise the remains are unlikely to have been disturbed.

What environmental archaeological strategy should therefore be adopted?

– Pit fills to be separately bulk sampled by fill context for plant macro-remains and the samples processed using the flotation technique. All unsampled sediment from each fill context to be dry sieved on site to recover vertebrate bones.

The key point that we hope to have emphasised in this discussion is that a great deal of thought must be given with regard to the techniques of environmental archaeology that should be employed in any project and how they will be utilised. These decisions must be made before the first spade – or more usually, mechanical excavator bucket – hits the earth. Fortunately the wisdom of such an approach has been realised by those who manage archaeology – particularly in Britain and North America – and few archaeological investigations go ahead in these areas that are not initiated by a full and well-argued project design, which is frequently subject to external expert scrutiny. Hopefully, the days where numerous environmental archaeology techniques are 'thrown' at an archaeological problem in the hope that at least one will resolve it, have passed.

Before we move on to consider how environmental archaeology should be written we will take a detailed look at one case study which turns everything that we have so far examined in this section on its head. The case study concerns Amarna in Egypt, where due to exceptional climatic conditions and cultural history, there has been almost total preservation of biological remains. In such unusual circumstances almost any archaeological question can be asked of the data, while cultural phenomena are relatively easily examined from a combination of the biological and artefactual remains.

Tell el-Amarna The Workmen's Village – a case study of total biological preservation

The ancient city of Amarna in central Egypt is unique in many respects. The particularly dry conditions prevailing on the site have preserved desiccated remains thereby providing an excellent source of biological data not normally found on archaeological sites. As well as the biological data there is also a multiplicity of archaeological, written and artistic sources relating to economic practices with which the biological record can be compared. Finally there is the unique history of Amarna itself, inextricably linked with one of the most controversial figures of ancient Egypt; Akhenaten. Traditionally, there were many gods in ancient Egypt, each requiring a host of priests to take charge of the temples erected in their honour. However, on accession, the 18th dynasty Pharaoh Amenhotep IV decreed the worship of just one, Aten, the sun god. Around 1341 BC in Amenhotep's IV fourth year of reign he founded a new city at el-Amarna half-way between the traditional capitals of Thebes and Memphis, dedicated specifically to Aten. He called it Akhetaten 'the horizon (or seat) of the Aten' and changed his own name to that of Akhenaten 'spirit of the Aten'.

Akhenaten's rule lasted approximately 17 years, but the imposition of the 'new religion' amongst people who had dedicated their lives to the worship of many gods, made it an unpopular one. Although the city was occupied for some years after his death until the succession of the boy pharaoh, Tutankhamun (possibly Akhenaten's son), it was abandoned shortly thereafter. The unpopular nature of the 'heretic' pharaoh provoked those rulers who followed to destroy the city and with it all memory of Akhenaten. The total length of occupation of the site is therefore 20-22 years. This means that at Amarna we have the unprecedented preservation of an Egyptian city captured in a frozen moment of time that spans little more than a single human generation.

Amarna lies some 300km south of Cairo, between the modern villages of El-Till and el-Hagg Qandil, on the eastern bank of the Nile (**89a**). The 're-discovery' of the city in 1887 led to a series of high-profile excavations that continue to this day. The area around the city has also yielded other important archaeological discoveries, including the tombs of Akhenaten, his wife Nefertiti and an entourage of family and administrators (**89b**). Lying between city and tombs is the settlement of the small community who worked on the construction of the royal tombs and which is called the 'Workmen's Village'. While many excavations had been conducted at Amarna almost no environmental archaeology had ever been carried out until the 1980s when Barry Kemp began his excavations.

As with many archaeological sites in Egypt that are situated on the desert fringe beyond the edges of the Nile cultivation, Amarna contains rich assemblages of desiccated organic remains. The abandonment and short period of occupation meant that deposits were largely undisturbed by later inhabitants.

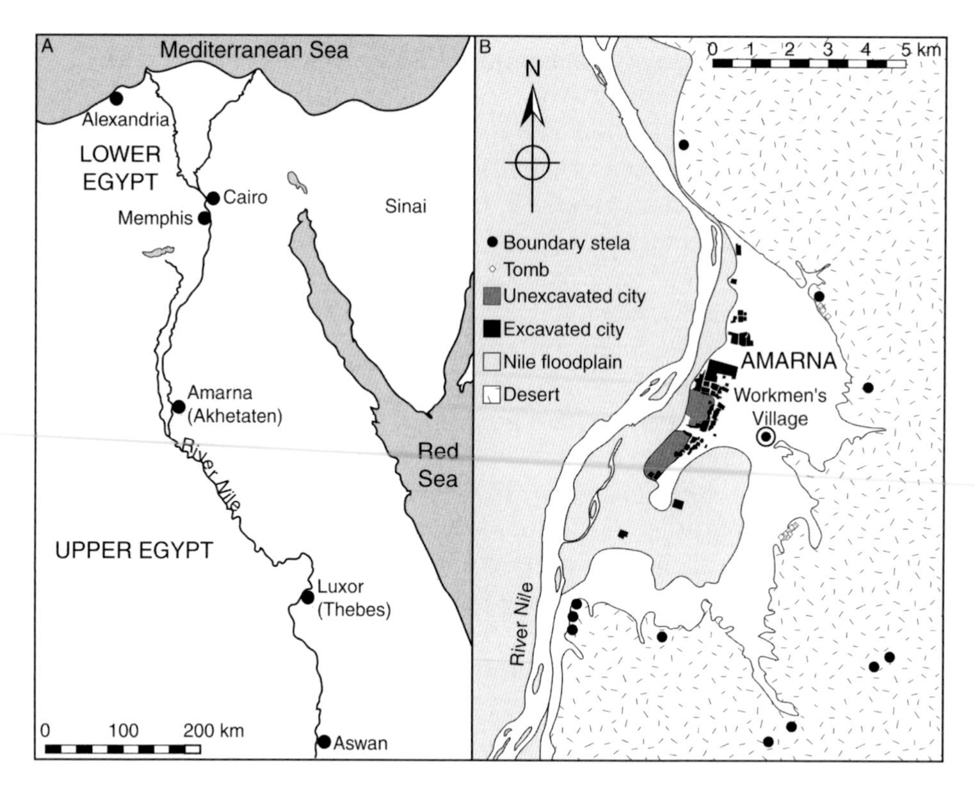

89 *Amarna in Egypt (a) and the relationship of the Workmen's Village to the city (b)*

Indeed, the 100 years of excavation by archaeologists has probably been the greatest source of post-depositional disturbance to the site since the fifteenth millennium BC. Biological preservation at the Workmen's Village was especially good, although desiccated material also occurred in the city itself, and at a nearby palace at Kom-el Nana. Preservation by desiccation means that in addition to cereal remains, other organic remains have survived, including a variety of animal and plant remains, along with objects crafted from them such as textiles, tools, basketry, shoes and woodwork. Animal bones often had skin, fur, hooves, hair and tendons still attached. Feathers, parasite eggs and insects also survived. Dung was recovered, mainly from herbivores and therefore as much botanical in composition as it was zoological in origin.

The representation of skeletal remains and butchery patterns has been studied by Howard Heckler and Rosemary Luff. They revealed not only the range of animals exploited, but also provided insight into the exchange of select joints of meat, and the methods employed in the processing of animal carcasses. The insect remains were studied by Eva Panagiotakopulu and Paul Buckland, while C. Donald examined the eggs of parasites found within animal dung. All these provided further information helping us construct a more detailed picture of life at Amarna. The plant remains were worked on by a number of authors. Jane Renfrew undertook an initial study of the material from the village. Delwen Samuel conducted later work on the relationship between cereal processing waste and quern and mortar emplacements. The investigation of further remains was carried out by one of the authors (Chris Stevens) together with Alan Clapham. Gillian Vogelsang-Eastwood and Barry Kemp have conducted work on the textiles.

The ancient Egyptian economy was very different to our own. It was largely governed and controlled by the state through intense and meticulous forms of taxation. In return the Egyptian state kept huge stores of grain to deal with fluctuations in supply and to maintain the non-farming elements of the system. At Amarna we have evidence of the non-farming elements of the community, the servants and administrators, the government officials, policemen, military men and priests, the workers within the temple bakeries and slaughter yards, and of course, the workmen in the village and their families. However, while biological remains exist to be compared for the two main classes of Egyptians, those who dwelt in the city and the specialist workmen in the village, settlements of the common Egyptian farming peasantry are all but unknown. Indeed, the best archaeological evidence we have of these providers of Akhetaten is the result of their labour, the remains of the crops they grew and the animals they raised. For the remaining parts of this case study we will focus on the way that biological evidence from the Amarna Workmen's village has been used to interpret the archaeology of the site.

The Workmen's Village

The Workmen's Village is so named because it was believed to have housed those who worked on the tombs including both miners and skilled artisans. It is possible that at the end of Amarna's occupation, when work on the royal tombs ceased, it was used for housing the local police and tomb guards.

The village lay 2km from the Nile, and a relatively short walk from the outskirts of the city (**89b**). The absence of water meant that most provisions are thought to have been transported from the city to the village. It is believed that the provisions entered the village from the south-west. A trail of broken sherds delineated the route these provisions came by following the edge of a shallow wadi bed. The provisions were possibly offloaded at a site (X2) lying between the wadi and the edge of the village, distinguished by a large amount of broken pottery waste. For what reason produce was off-loaded here rather than at the village itself is somewhat of a mystery. Possibly the inhabitants of the village were kept in isolation from those delivering the goods. Alternatively it would allow scribes to keep a tighter control on goods entering and leaving the village.

There are almost no plants growing in or around the remains at the present day except in the wadis. The closest source of water, both today and in Pharaonic times, is within the meandering branches of the Royal and Great wadi that run either side of the village, to the north and south respectively.

Today these shallow beds are almost always dry and are replenished only for a short period every few or more years by the rains that run off the eastern desert plateau.

Modern surveys of these wadis have revealed that many of the plants growing today are the same as those found in archaeological samples dating to the reign of Akhenaten. The thorny violet-flowered tumbleweed, *Zilla spinosa* which grows in the desert close to the south tombs, would have been a common spectacle in the second millennium BC. The mauve flower of *Fagonia arabica*, the yellow and white of *Zygophyllum* sp., the thorny thistle-like heads and pale blue flowers of *Echinops spinosissimus* and white flowers of *Heliotropium bacciferum* would all have been familiar to the past inhabitants of the village, just as they are to the inhabitants of the two Nile villages today.

While not a part of the natural environment, cereal chaff must have covered almost every surface of the village as almost every sample contained such remains. Today chaff of free-threshing wheat, barley and sorghum is blown across the ruins of the main city, gathering in the corners of the mud-brick foundations (**90**). The distribution of ancient cereal finds may then tell us little of where processing activities were conducted without further corrobative evidence. That they are present at all on the site, however, suggests that more cereal processing was conducted within the ancient Village than is carried out in the two modern villages today.

90 *A mud brick structure in the Workmen's Village with accumulated cereal chaff in the lead of its walls*

There was also a greater variety of animal life in and around the village than today according to the preserved faunal assemblages. The Nile fox we may assume was a common scavenger around the village. Smaller animals also appear to be present on the site based on the presence of their bones, such as the Egyptian gerbil. The dung of these creatures also survived including miniature pellets of gerbil or mouse and larger pellets, presumably of rats. In addition, there was a variety of insects some of which are local to the desert, others that appear to have been brought to the village with cereals and other crops. We will now look at a number of contexts where biological remains have particularly aided interpretation.

Buildings at the village margin

A group of buildings at the edge of the village (X1) were amongst the first excavated by Barry Kemp. The function of these buildings is somewhat of an enigma. Some have evidence for fine crafts, such as bead-making. Others contained large numbers of horn cores that had had the outer horn removed. As the removal of horn is a particularly smelly task, Rosemary Luff has suggested that the buildings were conveniently located away from the village. However, taking into account the position of the buildings at the village entrance they have also been assumed to have had an administrative function, perhaps the checking or temporary storage of goods entering and leaving the village. Two aspects of biological evidence point to this possible use. Large numbers of goat droppings indicate that goats were periodically kept in the building, perhaps as they entered the village. Secondly the buildings produced the only real evidence for the presence of whole spikelets of emmer wheat. Assuming ancient wheats were transported and stored in spikelet form then the evidence suggests that it may have served as a temporary store. Numerous rodent droppings of a size indicating that suggests that they are more likely to be rat than gerbil. Mice were also found amongst the spikelets suggesting a problem with vermin. Flax capsule fragments were also frequent compared with elsewhere, suggesting that it too may have been kept here. It is then possible that the buildings served for the storage of crops entering the village before their distribution to the villagers. Of course, it might be that the spikelets were food for the goats. However, an examination of goat dung undertaken by G. Hosey did not reveal the presence of cereals. What it did reveal was stems of darnel or rye grass, rushes and date leaves. The presence of rush is of some interest. Rushes would have only grown next to the Nile. While it is possible that fodder was brought to the village, no evidence for rush was found. This suggests that unless such vegetation had been cut especially for fodder, the goats had been feeding within the immediate vicinity of the Nile and its floodplain. As it can take up to two or three days for goats to completely digest their food these data suggest that the goats were brought to the Workmen's Village from such a location.

A third possible function is that the buildings housed guards for the site. The fact that the site contained the largest number of dog bones was seen by Howard Heckler as in keeping with this role. Of course, animals are not always buried where they lived, but it is possible that the village guards disposed of the bodies of such dogs in the immediate vicinity of their post. It is possible then that the goats were kept by the guards and the grain was among their rations.

Middens

Although no doubt used for obtaining stone, the village quarry eventually became the depository for the vast amounts of waste the village produced. The quarry samples were amongst the first to be examined by Jane Renfrew. They proved a good choice, providing evidence for almost every species that is to be found anywhere else within the village. The deep stratigraphy and continuous burial and sealing of older deposits also meant that some of the best-preserved organic material came from the quarries. Whole fruits of sycamore figs, garlic bulbs, onion skin, pomegranate pips, olive stones and almonds are among just a few of the finds from the midden. Fruits of balanatites (*Balantites aegyptica*) were also present in large numbers. The kernel of this plant produces a precious oil used in perfume and Jane Renfrew notes that they appeared to have been opened at one end to extract this part. Other finds included dôm palm; the outer fruit can be chewed, the inner kernel used as a form of vegetable ivory for crafts. Charred cereal remains were particularly prominent, especially the many thousands of glumes suggestive of the waste from cereal processing burned on the hearths of the villager's houses and perhaps within the chapels.

The waste within the quarry must have come from almost every part of the village. So while revealing a large variety of economic activities they tell us nothing of where these were carried out. One exception is the large quantity of animal dung present within the quarry, both goat droppings and other dung that almost certainly came from a set of pens located just to one side of the quarry. Jane Renfrew noted that many of the cereal spikelets were mashed up and covered in dung as if they had been fed to animals and passed through them almost intact.

Animal pens

Animal pens are an unusual feature of the Amarna Workmen's Village, which are not found at other contemporary settlements. The floors of the pens contained large amounts of animal dung, while the entrances were both low and narrow, suggesting they were for goats and/or pigs. An examination of all the available evidence by Barry Kemp strongly indictates the latter. Evidence of this was the close resemblance of the archaeological structures to modern traditional pig rearing facilities. The narrow doorway and the presence of low poles are common features of sties today. The placing of low poles some 20-30cm above the ground keeps the piglets confined, but allows the mother to step over it. The vast quantities of animal dung within the pens were evidently not derived from goat, the intact fragments morphologically resembling those of modern pigs. Finally H.M. Appleyard identified hairs and bristles in the pens as being from pigs.

Internal analysis of the dung revealed eggs of the tapeworm *ascaris,* a parasite of both humans and pigs. The presence of dung elsewhere in the village probably indicates that other pigs were allowed to roam more freely eating whatever waste was left lying around and that the pens were primarily used for breeding. Eva Panagiotakopulu also identified insect remains from pig coprolites, specifically *Palorusratze burgi,* the small-eyed flour beetle, a common pest within granaries, mills and flour. Such insects undoubtedly came from the city granaries, brought in with the crops and eaten with them by the pigs.

The samples taken from the pig pens were rich in organic material and provided good evidence for a whole range of foodstuffs consumed by the pigs. The large number of mashed-up spikelets of emmer and barley palaes and lemmas showed that the pigs had been fed on cereal remains in large quantities. Occasionally high amounts of barley rachises and culm nodes were found which indicate that barley may have been brought to the village as ears for use as pig feed. In fact, the pigs appear to have been eating everything that did not go into the village dump. In other words the pigs acted as an organic rubbish disposal facility.

Workmen's houses

The houses of the workmen, some 73 in total, were contained within a square high-walled area known as the 'Walled Village'. Only a single narrow entrance permitted entry to this area. Samples from the entrance were rich in goat dung, and as the south wall of the village would have provided the main source of shade within the village it is possible that animals were tethered or left to roam around the entrance.

The workmen's houses were built side by side and line both sides of five narrow streets. Evidence suggests that some, if not all, were two stories high. The ground floors of these houses often had up to three rooms. From the biological remains recovered from the houses and their immediate environs the predominant cereals utilised by the villagers were emmer wheat and hulled six-row barley. The quantity of cereals brought to the village must have been vast. Emmer requires further processing to remove the glumes. That almost every single sample contains their chaff, both desiccated and charred, indicates that they were brought to the village as spikelets, and only then processed to remove the grain.

The site provided an excellent example and at the same time a warning, of the bias of charred plant macro-remain assemblages towards cereal agriculture. Many species whose parts were readily preserved in the desiccated samples, including those such as dôm palm and date that are highly robust, were rarely found in a charred state. Yet charred remains of cereals were also present in almost every sample. Such high quantities of charred material would be surprising on European sites that were occupied for such a short period of time. However, absence of many post-depositional processes associated with soil formation at Amarna also meant that charred material was exceptionally well preserved. It was no surprise then that the charred seeds of wild species interpreted as weeds of the crop, were present in greatest quantities where desiccated cereals were also most numerous. These charred remains enabled the life-history of the crop to be traced, from its sowing in the flood-inundated fields adjacent to the Nile to the houses of the workmen in the village. The growing season for cereals in Egypt is limited not by cold winters as it is in more northerly latitudes, but by low rainfall. Today the Aswan Dam and intensive irrigation mean that crops can be grown throughout much of the year. In ancient Egypt, however, the agricultural cycle was slave to the Nile floods. Only after the flood would the farmers begin to sow the crops. Curiously from depictions of such scenes in Egyptian tombs it would appear that fields were tilled after sowing rather than before. Such a sequence is, however, logical. Ards do not turn the soil, but merely break it up, so that rather than being buried the grain would be trampled under the oxen's feet. The ard would aerate the soil, something that would certainly be needed after the flood. After these activities the crop could be left with weeds and crops growing together in the field ready for harvest at the end of the winter (**91**).

The most predominant weeds of ancient New Kingdom fields appear to have been the wild grasses; rye grass, darnel and canary grass. Other species were also found such as dock and, less commonly, clover and purselane. Occasionally seeds of sedges or other species of wetlands are present. Presumably these seeds come from plants growing in the wetter parts of fields from which the floodwaters were slow to recede. Darnel is thought to be indicative of broadcast sowing. Having a short seed-bank it needs to be resown with the crop by virtue of its large seeds. This accords with tomb scenes in which crops are sown by the broadcast method. Indeed tomb scenes can help us better understand the Egyptian harvest generally. One example is the tomb of Menna in the Valley of the Kings, dating to *c.*1380 BC. Menna was by all accounts a high-ranking agricultural inspector employed by Akhenaten's father. In scenes in his tomb we find the tax assessor determining the yield of the crop before harvest. Alongside this harvesters are shown cutting the corn with sickles.

The fact that seeds of non-twining weeds are present in the Amarna cereal crops suggests that cereals were harvested by sickle. The height of the crops at harvest can also be determined from the weed flora. Purselane and clover are low-growing plants, nestling in the shaded spaces between cereal plants, flowering only some 30-40cm above the field's surface. The presence of these

91 *A tomb painting from Tunah el Gebel, showing Egyptian harvesting practice in post-Pharaonic times*

plants at the Workmen's Village implies that the harvesters bent down to cut the straw low on its stem, inadvertently passing the sickle through the stems of these weeds as well. So it was that when the sheaves were gathered to take to the threshing floor, these plants came with them. As noted earlier, emmer wheat produces ears at many different heights. It is notable that in Menna's tomb this same phenomenon can be noted, some ears scarcely rising above the tax official's knees! Although the scale of the figures is not necessarily to be taken literally, the image clearly demonstrates variation in the height of ears.

Having reconstructed both sowing and harvesting we can now turn to the products. The charred remains from the workmen's village were dominated by glumes and a few larger spikelet- or grain-sized weed seeds, predominantly of rye-grass. This indicates that the grain had probably been transported to the village as semi-cleaned spikelets, meaning that crops were threshed, raked, winnowed and coarse-sieved before transportation. The general lack of smaller weed seeds probably indicates that the spikelets had been at least partially fine-sieved too. Such activities were probably carried out in the fields or the surrounding agricultural villages before the grain was transported to the city or Workmen's Village. Scenes from Menna's tomb also show these processing activities. Firstly, cattle can be seen seemingly dancing amongst the ears. What is happening here is that the cattle are being used to trample the grain so threshing it beneath their feet. In African villages today one can still occasionally see an almost identical scene, with villagers and cattle flocking together in a sea of cereals (usually sorghum rather than wheat), their heads only just visible above the surface. In front of the cattle, labourers winnow the crop by throwing it up into the air. The biological record then provides us with evidence that agrees with the scenes from the tomb, although the exact methods of threshing and winnowing can never be seen in the charred record.

The grain arriving at Amarna itself no doubt came by boat, arriving at the city docks to be further transported to the city granaries. From here it was brought to the Workmen's Village to be stored, probably by donkeys along the Wadi's edge. Textual evidence from the Village at Deir el-Medina in the Valley of the Kings recounts how each workman received about 310 litres of emmer and 115 litres of barley a month. This would be ample enough to provide food for a family of eight to ten individuals; bearing in mind that children would need much less. Given that the textual evidence at Deir el-Medina refers to a census of 13 houses in which none have over five members and half are occupied by single men, surplus rations might easily have been accumulated. Some of these rations should have been used as fodder, although it may be noted that barley brought to the village for this purpose may have arrived as partially threshed ears rather than semi-cleaned grain.

A survey of the position of mortar emplacements, presumably used for processing the grain in the village, carried out by Delwen Samuel, revealed that these structures were present in individual houses and often located in small rooms or courtyards at the front of the building. These mortars are often wrongly thought to be for crushing grain whereas they would have been used for dehusking hulled wheats and perhaps hulled barley as well. Large amounts of glume chaff were recorded around them by Samuel, a testimony to the last time they were used. Following de-husking in the mortars, the glumes were removed by further sieving and winnowing. A sieve suitable for such purposes was recovered from a probable kitchen at 12 Gate Street by Peet and Wooley. It was some 21cm in diameter and made entirely of palm fibre.

After processing, the clean grain was milled in each household on saddle querns. Such querns had special emplacements, and a survey of their distribution showed them to be similarly located to the mortars in individual houses. Most houses also contained ovens. Kemp records these as more commonly located at the rear of the house or in a few cases perhaps even on the upper storey. Bread may have been made on trays, in moulds or by sticking the flattened dough to the side of the oven. Although not every house had an oven, a mortar or a quern – in which case sharing of facilities may have taken place – the overall impression is of payment for work being made to the workmen in agricultural produce to individual houses, rather than to the community as a whole.

Many of the houses also yielded evidence for textile manufacture. The preservation conditions meant that alongside the textiles themselves, wooden spinning whorls were commonplace. Stone socket blocks were also found. These are limestone blocks with square holes thought to be for the placing of upright looms. These, like the mortars and querns were found in several of the workmen's houses. While woollen textiles were recovered from the village they were fewer in number compared to those of linen. This may, however, be a product of preservation rather than a true reflection of relative proportions, and in this respect it is notable that the city produced even fewer finds of woollen textiles. A few textiles spun from long, dark-brown goat hairs were also found. The textiles would have been used for clothes, cloth bags, and also for household furnishings, cushions, carpet, bed covers and curtains. Other finds indicate that at night the inhabitants would have lit lamps consisting of twists of woven flax soaked in oil.

Chapels

Chapels were found to the north-east and south-east of the Walled Village. One role of the chapel would have been the dedication and preparation of dedications to the Aten. For example, pottery stands with small bowls were a common feature of the chapels and were probably made for the burning of incense and food offerings. Analysis of the oil used in the lamp wicks using gas chromatography and mass spectrometry by Maragaret Serpico, revealed the presence of pistacia resin, probably terebinth. Another feature of the chapels that probably relates to the rituals performed within them is the high number of leaves recovered in samples. Fragments of leaves were relatively uncommon finds in the samples from the Walled Village and the animal pens, and given the smaller amount of botanical material from the chapels we can suggest that they probably represent the remnants of wreaths and garlands. Flower and leaf garlands are also a common feature in Egyptian tombs. Tutankhamun was wearing a garland of olive leaves, and other garlands of olive and willow leaves were also found within his tomb. Wreaths from other sites reveal the presence of flowers, especially those of Chrysanthemum-type plants, such flower heads were also quite a common feature of the chapel samples.

The chapels were also host to activities that may have been both ritual and economic. For example the slaughter and butchery of pigs appears to have been conducted in the chapels. In pre-industrial societies pigs are usually slaughtered by slitting their throats then hoisting them up to drain the blood and stop the meat from spoiling. We cannot be sure that the ancient Egyptians treated pigs in the same way, but Kemp suggests that the presence of gypsum in distinct areas would provide suitable surfaces for these practices. After slaughter the pig's hair is removed by either boiling water or by covering the skin with chaff and setting light to it, after which it is scraped off. Boiling water removes the hair with the root while burning removes the hair without the root. Large numbers of pig hairs were found in the chapels. Under microscopic analysis they were seen to still have the root attached suggesting to Kemp that boiling water rather than burning was used to remove them.

Analysis of the pig bones by Howard Heckler showed that they were generally slaughtered in their first to second year, while studies of the butchery marks by Rosemary Luff indicated that a skilled butcher had dismembered them using a knife. As was seen in section 3, studies of butchery marks can be used to suggest whether the meat had been removed from carcasses while they were still fresh. The evidence from the Amarna chapels seems to indicate that the joints from pigs were stored perhaps by salting, smoking or drying. No signs of heating were visible on the bones, suggesting that the meat was probably removed from the bone before it was cooked; pictorial evidence most commonly shows the boiling of meat.

The chapels also contained ovens. In one of the chapels it was noted that two different types of oven existed, one for baking bread and the other for firing clay objects. The ancient Egyptians sometimes baked bread using clay moulds, the ancient equivalent of today's bread tins. However, an important distinction exists. Today's bread tins can be used repeatedly, yet experiments by Paul Nicholson and Delwen Samuel suggest that clay moulds could be used only once. Two main forms of Ancient Egyptian bread moulds have been found at Amarna. One was bowl-shaped, producing a round loaf; the other long and thin producing something akin to a short baguette. The experiments showed that the bread was often difficult to remove from the moulds; in the case of the long thin baguettes, impossible without breaking them. The association of bread moulds with the chapels led Samuel to suggest that moulds were mainly used for making bread for offerings to Aten. Depictions of offering tables often show objects looking remarkably similar to those of loaves in scenes from bakeries.

Several aspects of textile manufacture were also identified in the chapels. Today flax grown for fibre is often harvested before it sets seed; this may not have been necessarily the same in the past and plants may have been utilised for both oil and fibre. The stage at which flax is harvested affects the quality of the fibre. Harvest it just after flowering and the fibres are fine, harvest it fully ripe and the fibres are coarse, fit for sacks, mats and ropes. The textiles recovered from the village suggest that much flax for fibre was readily harvested before the plants set seed. It would seem reasonable to conclude then that the mature flax capsules found were brought to the village so that the seed could be processed into oil. Finds of flax capsule fragments, except from building X1 were commonest in some of the chapel samples and so may indicate that linseed was processed in these buildings. Flax seeds themselves were less common apart from in situations where they had been buried, probably because they were selectively eaten by insects.

92 *Desiccated botanical remains from the Workmen's Village, Amarna*

The chapels also produced evidence of flax being used for fibre. Some of the botanical samples from the chapels contained stems of flax, with some appearing to be in the process of having their fibres removed (**92d**). Some of the samples also contained terminal nodes; the top part of the stem that the capsule connects to. That fact these same samples contained lower numbers of capsules then suggested that the capsules had already been removed, perhaps in preparation for processing into fibre. A few immature capsules were also present, perhaps coming from crops that had been harvested for fibre before they had fully set seed.

A question remains concerning this material: was flax for fibres retted before it reached the village or within the village itself? Certainly no large tanks were found that might be expected for such purposes. However, flax does not always need to be retted. Investigations carried out by Barry Kemp suggest that it is possible to separate fibres through a method known as decortification. The flax is retted dry by crushing the stems, although they may be soaked for 24 hours prior to this. Bundles of stems found within the village do not appear to have been retted but rather beaten with hammers, while those found in the chapel do not even appear to have been pounded. Colour may have been added to the worked flax as seeds of safflower are common in the samples from the chapels, but not from elsewhere in the village. The dye is made from the flowers, giving a yellow or red colour. Several textiles were found that were red in colour. These were analysed using ultra-violet spectrometers and by a variety of other chemical tests, but as yet safflower dye has not been identified. Rather the dyes used were more probably of iron oxide and the root of madder, the former giving anything from yellow to red, the latter a strong brick-red colour. Blue dyes were also identified, being obtained either from woad or a species of the legume genera *Indogofera*. Some textiles were double-dyed to obtain purple. The presence of safflower seed, however, is not directly indicative of dyeing: if the plant was to be used for dye the flowers would be collected before they set seed. Perhaps a more logical use of safflower seeds was for their oil which could have been used for cooking.

Gardens

Garden plots of a number of different types were associated with several of the chapels. One chapel to the southeast contained garden plots that were enclosed within a small courtyard (**93**). Others, associated with a set of animal pens directly to the east of the village, were laid in a grid of 20 almost identical plots. The garden plots consisted generally of small squares, about a metre across, delineated by low ridges of mud-brick and/or stones. The soil in these plots was noticeably different from anything else excavated in the village consisting of thick, dark, alluvial silts, presumably brought in from the Nile floodplain. Garden plots lined with trees are a common feature of Egyptian tomb art where they appear to have been used for growing flowers and vegetables.

Sampling of the garden soils was undertaken with the anticipation that they would reveal evidence for some of the plants that had been grown. The soils were not rich in plant remains and many of the species identified had already been identified from other parts of the village. The botanical samples from the garden plots also contained other organic finds. Fragments of dung-covered seeds and chaff provide testimony to the fertilising of the plots using waste from the

93 *Reconstruction of the chapel gardens in the south-east of the Workmen's Village, Amarna*

pig pens. What would provide the link from gardens to elsewhere in the village was whether the seeds of those species recovered were relatively more common on the plots than elsewhere in the village. It is notable that only seeds were found int he garden samples. However, many of the plants that are likely to have been grown within the plots are likely to have been grown from seeds, but then removed either before they set seed or with the seeds intact. By far the most common seeds in the garden plots, especially compared with their frequency elsewhere in the village, were seeds of coriander (**92**A). Coriander is of course mainly cultivated for its seed, although the leaves may also be used in cooking. While the seeds are cultivated for human consumption it is ironically the taste that makes them desirable to humans that also protects them from consumption by insects. Celery seeds are also a common find within the garden plots. Today celery is grown mainly for its young petioles (leaf stalks) or storage organs. However, its earliest historically recorded cultivation is for its seeds, which can be used for their oil or as a spice. Although not well represented basil, dill and beetroot seeds were also recovered from plots whereas they were absent elsewhere in the village. Nigel Hepper reported wild basil from the Tomb of Tutankhamun. Dill was only represented by a single seed, but it may have been used for its seeds and leaves. Beetroot is best known for its purple root. It is, however, the same species that gives us sugar beet and leaf beet, and therefore it may have been first cultivated for its spinach-like leaves, rather than its roots. To this ancient spice rack we may also add mustard, as a few capsules of probable white mustard were recovered from the garden plots. Evidence of further spices were found elsewhere in the village, including black cumin seed, fenugreek and *Trachyspermum copticum*. The seeds of this latter species were of special interest. They were found in a tiny cloth bag, along with a sprig of tamarisk, in the village dump and identified by Delwen Samuel. Several of these small bags were found, although few contained any contents and it is possible that they were either used as charms or perhaps herbs were exchanged in such bags.

It is probable that the finds of watermelon and melon came through the manuring route rather than from having directly been planted or grown upon the plot. Some elements, such as eggs of water flea, no doubt came from the water used to irrigate them.

Building materials

Although houses, pens and chapels were made of mud brick, plants played an important role in the construction of the buildings. The bricks themselves contained a great quantity of added plant material used as a binding agent. The break down of these bricks revealed produced the oldest plant remains from the village. These were the plants available to the workers when the village was first constructed. The bricks are likely to have been made in the village itself rather than transported to the site. They contained cereal chaff, rachises of barley, straw fragments, whole spikelets and the weeds of cultivation and of the irrigation channels of the Nile. As much mud-brick has broken down in the thousands of years since the village was abandoned, mud-brick provides an important source of material in the archaeobotanical samples.

The structure of the roofs of the houses could be analysed not only by the remains themselves but also by looking at impressions in the mud-brick. They were formed from criss-crossed beams of tamarisk with matting made of palm and halfa grass laid across. On top of this reed poles and hand-sized bundles of halfa grass, tied with rope and twine probably made from halfa grass or date leaves were laid. The reeds were probably of arundo judging by the hollow stems left by impressions in the mud-brick. Around this structure mud brick was piled on to form the outer roof. The walls and roofs of the chapels were 'plastered' using mud brick containing barley rachises and straw that was then smoothed off and whitewashed.

Relationship of the Workmen's Village with the city

Finally we come to the economic relationships of the village inhabitants with the city itself. The village inhabitants received flax fibres, linseed, cereal crops, dates, dôm palm nuts and wood according to the plant remains recovered from the samples, while water must also have been brought in. It is possible that other produce was also brought to the village. Finds of grape pips (**92**B), castor oil seed and olive stones were all noticeably scarce in the village compared with the city. If wine, castor and olive oil were available in the village they certainly do not appear to have been manufactured there. This would accord with records from Deir el-Medina where wine was occasionally included in the workmen's rations but certainly not on a regular basis. The inhabitants of Deir el-Medina also received vegetables, firewood, oil and clothes, so we might suppose that the inhabitants of the Workmen's Village may have supplemented such rations rather than being entirely self-sufficient. Fuel is another resource that may have had to be brought in, although it is possible that animal dung was one such source. It is often speculated that dung may have been a common form of fuel in the past and it is also a source of charred material. Certainly the modern inhabitants of Amarna use dung for fuel. Chaff from threshing is added to the animal dung then the finished rows of dung cakes are laid out neatly to dry on the edge of the cultivation area. It should be noted, however, that the greatest source of identifiable material from such sources after it had been charred would not be what the animal had eaten, but rather the threshing waste added afterwards. Charred and desiccated dung certainly survives at very well at Amarna, and in both cases it contains seeds of species such as melon, fig and other remains that the animals had no doubt eaten. These finds are, though, very rare compared with the enormous amounts of wood charcoal found at the site. We can conclude then that like the inhabitants of Deir el-Medina those at Amarna also received firewood.

It may have been that a reciprocal arrangement existed whereby plant resources were brought into the Workmen's Village in return for pork. Whereas we know that pigs were kept at the village from the pens themselves ,no such structures were present in the city, suggesting that pigs were not reared. In the Workmen's Village Heckler noted that fore-limbs of pigs were much better represented than hind-limb bones. Clearly they had been removed, perhaps in the chapels and possibly transported to the city. If this was the case we might expect to find hind-limb bones of pigs in higher proportion to other material at the city. However, the studies of Luff have shown that was not the case, and hind- and fore-limb bones of pigs are equally well represented. This perhaps suggest that pigs were brought to the city and slaughtered there, or even as Kemp suggests perhaps wandered the city streets eating rubbish. While it was clear that pigs were raised at the village, the situation regarding goats is less certain. Heckler found that the hind-limbs of goats were also under-represented in the village. Only this time Luff found that hind-limbs of goats were better represented in the city. It appears that hind-limbs of goats were brought to the city perhaps as tax, tribute or in exchanged for more rations or other objects.

The bones of goats from the Workmen's Village showed the same methods of butchery by knife, as the pigs, which for reasons that we shall see shortly suggested to Luff that the animals were slaughtered and butchered within the village. The method of slaughter could be seen to be by the slitting of the throat from cut marks upon the cervical vertebrae or neck bones. It appears that like pork, goat meat was also exchanged, only this time the evidence suggests that the inhabitants of the city did receive goat meat from the Workmen's Village. What is less clear, however, is whether the villagers kept and bred goats at the village, or whether they were brought to the village as rations, slaughtered when needed and the hind-limb bones further exchanged. It is notable that most of the goats were slaughtered at less than a year old. Certainly the dung found in some areas such as site X1 was very small compared to modern goat dung, even taking into account considerable shrinkage caused by desiccation.

The villagers may also have kept cattle, but evidence suggests that the economic practices between village and city was very different for these animals to that seen for pigs and goats. On examining the cattle bones Rosemary Luff noted a very different pattern from that seen in the pig and goat remains. Many cattle were not butchered by a skilled knife, rather a cleaver had cut inaccurately through the bone, sometimes requiring several blows. Such patterns would seem more in tune with the mass-butchery of cattle carried out within temple abattoirs or animal-processing workshops based in the city.

In support of the import of beef into the Workmen's Village is the evidence from texts recovered from Deir el-Medina. These suggest that live cattle and joints of meat were brought from festivals to the inhabitants of the workmen's village at that location. Certainly the cult of Aten demanded that large numbers of cattle were sacrificed on a regular basis to fill the hundreds of offering tables around Amarna. It is not known whether these offerings were always of the prized *khepesh* cut (the cattle's right fore-limb), although certainly it is this cut that is shown adorning the offering tables of the temple of the Aten in depictions in the tomb of Panehsy. What is certain is that bones of the right fore-limb (the humerus, radius and ulna) were found in the village, so evidently such cuts did make it to the village. Some also show signs of knife marks, the preferred method of dismemberment for the *khepesh* cut. However, as knife marks were present on all cattle bones, it is difficult to ascribe any extra significance to the knife marks on the bones of the right fore-limb.

Other animals must have been caught in habitats outside the city and presumably reached the Workmen's Village via it. Waterfowl would have been caught in the marshes, while the plentiful fish remains found in the village must have been caught in the Nile. Bones of African catfish or schall (*Synodontis schall*) were recovered from both the Workmen's Village and the city. Fish frequently produce annual growth rings in some of their bones. By counting these in thin section the zooarchaeologist is able to calculate the age of the fish when it died. From depictions in tomb scenes, ancient Egyptians appear to have fished using nets. As fish get larger with age their chances of getting caught increase. Analysis of fish bones by Rosemary Luff and Geoff Bailey revealed that elder individuals were frequently present in the samples from Amarna indicating that net fishing was the most likely strategy.

Bioarchaeological interpretation of the Workmen's Village

The biological evidence from the Workmen's Village provides information that cannot be obtained from texts, as well as on occasion confirming such evidence. Without biological evidence our understanding of what was 'important' in everyday life, as contrasted to what was 'ideologically important' as approved by the state and seen within tomb art, would be slight.

Before we consider the relationship of the village with the main city further it is worth considering the relationships within the village as we have evidence for them. In terms of the resources needed to produce textiles, bread, vegetables, herbs, and meat it is clear that they were not distributed evenly between the houses and that some resources, such as certain meats, spices and herbs were probably channelled through the chapels, their associated workrooms and the garden plots. Further to this Kemp has also suggested that weaving and baking facilities may have been shared between individual households.

The village consisted of around 73 houses. The evidence from Deir el-Medina suggests may have consisted of anywhere between a single occupant and five individuals. The number of possibilities as to how the internal economy of the village may have been organised are many. It is possible that chapels, pens and gardens were state owned and controlled. However, this seems unlikely given the distinctive spatial separation of these elements. It is this that has led to the suggestion that they were rather privately owned and controlled by separate individuals. Given the role of these elements in the community, for instance the raising and distribution of pigs, it would give the individuals who built them and the priests who controlled them a great deal of power. Alternatively we can consider the example of Hekanakhte, a family head of the Middle Kingdom, where the extended family appears to be equal to, if not greater in, importance than the nuclear family. It is possible that extended families may have stretched across several houses at the Workmen's Village combining resources and labour. Chapels, garden plots and pens may then have actually belonged to individual families. This explanation would account for the uneven distribution of weaving equipment facilities for processing grain and cooking it.

Nevertheless, as is the case with other ancient societies, the economy of ancient Egypt was clearly part and parcel of religion and religious institutions. This is seen at Amarna in the importance of the city's temple slaughterhouses in providing cattle, and the village chapels and their gardens in supplying herbs and spices and probably in acting as the main distributor of animal protein in the village. Through these ultimately religious institutions the production and distribution of certain resources was controlled. Ultimately as the head of the state religion Akhenaten's position as protector of the faith and of the economic fecundity of Egypt was supreme.

Here and elsewhere we have stressed how economy and ideology are not separable in the way that many archaeologists have treated them previously. Before we leave this case study it is worth taking this point somewhat further and making some suggestions as to the location and nature of the village. The discovery of the pigsties and such positive evidence for the consumption of pigs in both the village and the main city is of some interest in the development and nature of pig taboos. As we saw in Section 4 such taboos were clearly in place some 800 years later according to Herodotus. He states that pig herders lived in separate communities. Other textual evidence from earlier periods also implies that cattle herders formed a distinctive, separated segment of the population, and that cattle were grazed in the marshes by a separate, mobile, community. On the basis of a single site it is difficult to say if the practice of pig herders living separately to agricultural workers was a feature of the 18th dynasty economy. What we can say is that the Workmen's Village was an isolated community which at least in part engaged in rearing pigs and that evidence for pigpens is absent from the main city. The Greeks recorded pigs, as being kept close to fields, trampling in the seed-corn and therefore presumably close to the Nile. The possibility is still an interesting one. Did the village represent a suitable place for such taboo activities? Or were pigpens a common feature of Egyptian agricultural village life? Peet and Wooley suggested that the village was so located because they believed that it may have been taboo for tomb workers to live amongst the everyday folk of Egypt. It is an interesting possibility that activities that rendered their practitioners unclean may have led to them being practised together.

Writing environmental archaeology

The product of each of the individual environmental archaeological studies conducted during a project is usually a text, or 'specialist report' to use the jargon – many such reports have been been produced for Amarna for example. These words are used advisedly, for not only is the text a product of work by a skilled and knowledgeable individual or team, but in many cases the reader will need to have a comparable knowledge in order to understand it. Hopefully the preceding pages of this book will have provided sufficient background for the reader to interpret reports of this nature.

Publication is the ultimate product of environmental archaeology, just as it is with any other aspect of archaeology, and indeed science as a whole. Only by communication in written form, be it book, journal article or web page, do data and new ideas become available to the wider community. No matter how good a piece of work may be, without publication it will amount to very little. As far as the environmental archaeologist is concerned, not only is it necessary to think hard about what bioarchaeological techniques will be employed on a site and how they will be combined with geoarchaeological investigations, but when the studies are completed, the results from each have to be combined in order to interpret changing environmental and economic situations. Not only this, but these interpretations have to be communicated in a clear and readily understandable manner suitable for those for whom the ideas and terminology of the natural sciences are alien, as well as in technical journal articles to fellow specialists. It is at this stage that many projects, which may have been carried out to the best standards of scientific rigour, fall down. It is all too easy when writing an account based on botanical, zoological and indeed geological data, to get carried away with the intricacies of the approach taken, the minutiae of identification criteria and potential of every possible comparative parallel, while forgetting exactly who the reader is. In most cases those accessing environmental archaeological publications – or at least those that are not specialist journal articles - are not environmental archaeologists, but archaeologists whose training has an arts or humanities base. Indeed the lack of accessibility of many environmental archaeological accounts is one of the main reasons that archaeologists such as Julian Thomas, complain about the sub-discipline. Certainly the scientific principles by which environmental archaeology is practised means that methodology, identification criteria and the relationship of the data to other studies, must be reported (all scientific studies must be repeatable and so reporting methodology is particularly important), but this need not be at the expense of, and certainly should not overshadow, interpretation. Indeed some of the best usage of environmental archaeological data has been by non-specialists precisely because it is easier for those without an expertise in any one area to take a step back and look at the wider picture.

A particularly good example of this, in which several sources of palaeoenvironmental and palaeoeconomic data have been combined to offer a detailed interpretation of an archaeological situation is the case of Romano-British utilisation of marginal intertidal land in North Somerset, south-west England (**94**). Prior to the investigations carried out by Stephen Rippon during the 1990s, the low lying, near-coastal areas of North Somerset was one of the least well understood areas of Roman Britain. One of the main reasons for this is that much of the area is blanketed by relatively thick layers of intertidal mud of post-Roman date. For this reason few upstanding features that pre-date this marine incursion exist above the present ground surface. However, what remains that were known, for example the villa of Wemberham (**94**) suggested that the area was intensively exploited in the Romano-British period. Nevertheless in order to use the area the Romano-British population would have had to somehow counter the sea, as now, during the Roman period the area would be subject to twice daily flooding from the adjacent Bristol Channel. Today sea defences between Middlehope and Clevedon prevent the sea invading the farmland that now characterises much of the area. But what approach had the Romano-British population taken? Had they simply **exploited** the intertidal marshland for the purposes of grazing and salt production on a seasonal basis, as is known, for example, to be the case in south-eastern England during the same period? Had they instead **modified** the upper saltmarshes with drainage ditches to allow agriculture, as in Roman East Anglia? Or, alternatively, had they adopted the same strategy as now, and by constructing a sea wall, **reclaimed** the saltmarshes? Whatever the answer to these questions, a further problem to address concerned the nature of the economy that supported both villas and lower status settlements. The use of environmental archaeological techniques to address these questions is a good example of the focused approach which we discussed earlier in this section, while the publications that have resulted from the research (see the references at the end of the book) are clear, readable and informative for all audiences.

The first suite of questions concerning landscape exploitation, modification or reclamation could only be addressed using the techniques of environmental archaeology, while similarly environmental archaeology would have to play a substantial role in the investigation of economy. However, before any samples could be recovered other archaeological techniques were needed to better characterise the area. Consequently aerial photography, fieldwalking, geophysics and small scale excavation were all used to locate, date and characterise archaeological features in three sample sites: Puxton, Banwell and Kenn Moor (**94**). At each of these sites rectangular field systems could be seen in the aerial photographs, extending from around palaeochannels representing former saltmarsh creeks, (**95**). Smaller clusters of paddock-like features and platforms seen on the aerial photographs were investigated using resistivity survey, the results indicating that they were the remains of small settlements or farmsteads. Fieldwalking of ploughed areas within the enclosure system suggested that artefact density decreased rapidly outside presumed settlement areas, while the majority of finds were of Romano-British date. However, gridded metal detector surveys provided a more precise chronology and the coins collected indicate that the landscape had been used between the late third to mid fourth centuries AD. The results of these various surveys, when coupled with evidence from previous (largely unpublished) archaeological investigations, allowed locations to be identified for targeted excavation (in such a wet environment, where soils and sediments are extremely heavy, large open area excavations are an extremely expensive – and destructive – option). A number of enclosure ditches in each of the three study areas were investigated in this way, together with a possible corn drier at Kenn Moor. Although there was some variation between sites and individual ditches, a broadly similar stratigraphy was revealed. The ditches forming the enclosures had been cut through blue-grey silty clays, which at Banwell appeared to have dried allowing a soil to develop. This immature palaeosol was associated with burnt clay, charcoal and stone fragments, while associated depressions contained Late Iron Age pottery (*c.* first century BC to first century AD), biquetage, ash and other by-products of salt production. The ditch sequences consisted of basal grey silts and clays, sealed by a dark blue-black layer representing a second (late Romano-British) palaeosol. Finds associated with this included stone, burnt clay, animal bone and later Roman pottery which cannot date from later than *c.* AD 335. The palaeosol was in turn capped by further silts and clays. In order to address the questions relating to exploitation/modification/reclamation, a series of hypotheses was set up for testing by the use of proxy faunal (mollusc shells, foraminiferal tests, insect exoskeletons, mammal bones) and floral (pollen, plant macro-remains, diatoms) data recovered from samples taken during the excavations:

1. Both exploited and modified environments imply an interaction with the intertidal zone to varying degrees. Therefore biological remains indicative of saline or brackish water environments would exist within any related sediments.
2. In reclaimed environments – such as those of the present day North Somerset Levels, as well as better-known areas such as the western Netherlands – sea walls act to exclude tidal waters. Therefore sediments accumulating in reclaimed zones should contain biological proxies entirely indicative of freshwater environments.
3. Exploited environments comprise saltmarsh, whereas modified and reclaimed environments are occupied by pasture and arable fields. Thus each will leave a characteristic floral signature in sediments from adjacent ditches.

The deposits through which the ditches were cut, together with the Iron Age palaeosols, contained flora and fauna characteristic of brackish water conditions. For example the Gastropod genus *Hydrobia*, which only inhabits intertidal water, was found in large numbers in direct association with the late Iron Age saltern at Banwell. The Foraminifera assemblage from Kenn Moor is dominated by *Elphidium williamsoni*, a species characteristic of the lower salt marshes or upper tidal mud flats. However, there is some indication that sea levels were rising immediately before the ditches were dug in the third century AD, as *Haynesina germanica*, a species more commonly found lower in the tidal frame increases in frequency upwards through the sequence. Given the inferred tidal environment of the Iron Age it is hardly surprising that a saltern was established at Banwell: Banwell is the topographically lowest of the sites investigated and therefore the most marginal with respect to fully terrestrial conditions. Indeed terrestrial mollusc species such as the obligate open country species, *Vertigo pygmaea* and *Pupilla muscorum* were found in samples from the saltern. The vegetation of the saltmarsh is indicated by pollen evidence as being dominated by sea plantain (*Plantago maritime*), buck's-horn plantain (*Plantago coronopus*) and members of the goosefoot (Chenopodiaceae) family. However, macro-remains of great fen-sedge (*Cladium mariscus*), found burnt with the saltern debris, suggest that freshwater peat was the source of fuel used in salt extraction.

In a background where Iron Age evidence is suggestive of increasingly brackish water, the fact that the biological remains from the third century AD ditch sequences are entirely indicative of freshwater conditions, strongly suggests that the saltmarshes were reclaimed by the Romano-British population. In a reclaimed environment, ditches would have been dug for drainage purposes. The evidence from ditches at Banwell is particularly impressive, and suggests that the fresh water in the ditches was filled with aquatic vegetation such as watercress (*Rorippa nasturtium-aquaticum*). The water would also appear to have been present for only part of the year as the algae stonewort (*Chara*), a genus that is often first to colonise new aquatic environments, and the slum species of gastropod (*Anisus leucostoma*), were both found in large numbers. Water tables are thus likely to have been higher than those of the present day, a factor which prevented excavation of some of the basal ditch deposits. Nevertheless, when water was present in the third century AD, it would have been relatively deep as diving beetles such as *Agabus* sp. and *Dysticus* sp., as well as the whirligig genus *Gyrinus* were all found in the ditch sediment. Away from the ditches the landscape would appear to have been more open than at present, with pollen evidence suggesting grassland predominated, while the mollusc species *Pupilla muscorum* and *Vallonia pulchella* suggest that the sward was both short and damp. Nevertheless willow (*Salix*) may have lined some of the ditches, as although no macrofossils were found, remains of the beetle *Phyllodeta vulgatissima*, which lives only in such trees, were. Similarly remains of small mammals such as woodmouse (*Apodemus* sp.), short-tailed vole (*Microtus agrestis*) and mole (*Talpa europaea*) also found in the ditch deposits suggest at least some scrub, the last also confirming a lower water table than now.

The palaeosol that is found at the top of the ditch sequences developed as a result of gradual drying, which suggests a further fall in the water table. Micromorphological examination of the palaeosol suggests that its dark colour is the result of microscopic charcoal, probably produced as the surrounding grassland was cleared, while charred plant macro-remains included the chaff of wheat and barley, suggesting that the cereals were being processed. Indeed having established that the North Somerset Levels were reclaimed during the third century AD, the most obvious follow-up question is why? Reclamation requires huge start-up costs, both financial and in terms of labour, to build coastal defences and drainage ditches, but also, and perhaps more significantly, maintenance costs. It is therefore highly likely that economic returns were high and that reclamation did not represent piecemeal efforts by individual farmers. Perhaps reclamation was undertaken under some regional authority and perhaps managed from the surrounding villas, with the farmsteads on the levels being the focus of day-to-day agricultural and maintenance activity. Palynological and plant macro-remain evidence suggest that the reclaimed lands were used for both pasture and arable purposes, as well as for production of hay for forage. The presence of grass pasture is indicated by associated weeds such as buttercup (*Ranunculus acris/repens/bulbosus*), daisy (*Bellis perennis*) and dandelion (*Taraxacum*), while finds of the dung beetle genus *Aphodius* indicate the presence of large herbivores. Bones recovered from pits on the edges of the settlements suggest that the pasture was for both cattle and sheep; while a horse mandible with tooth-wear that indicates that a bit from a harness had been the animal's frequent companion during life, suggest's horses were used for traction and/or transport. The best evidence for crops grown in the arable fields is the corn drier (used either to dry ears of cereals prior to storage or parch grain before milling), although the charred plant remains found are from an associated ditch deposit, formed largely of sweepings. Nevertheless these suggest that two and six-rowed hulled barley (*Hordeum* sp.) and spelt wheat (*Triticum spelta*) were the main crops processed in this feature. The associated weed species indicate that they were indeed grown in the surrounding reclaimed marsh. For example stinking chamomile (*Anthemis cotula*) is typical of the heavy soils that characterise former tidal areas. Other crop plants cultivated in the fields between the ditches would have included the celtic or horse bean (*Vicia faba*) and possibly oats (*Avena* sp.).

Some time in the mid- or early fourth century AD the North Somerset Levels returned to a saltmarsh environment after less than a century of exploitation. Biological indicators similar to those recovered from the Iron Age silts and clays were found in deposits sealing the Romano-British features. Clearly the coastal defences had been breached, although whether this is a result of non-maintenance or sea level rise (or both) is unclear. It is tempting to relate the changes to a coincident increase in oak pollen, which indicates that in the adjacent dryland woodland was expanding. If this interpretation is correct it would suggest that the landscape as a whole was being used less intensively and that an economic decline had set in. During the decline the reclaimed areas were once more of only marginal value.

The publications that report on the results of this environmental archaeological work operate at two levels. A summary of the results of the various bioarchaeological and geoarchaeological investigations and their implications for the Romano-British landscape and economy are presented in a text that is fully integrated with that relating to field survey, excavation and the artefacts that were recovered. In this Stephen Rippon has used the data obtained by his team of environmental archaeologists to develop his argument, just as he has done with pottery, earthwork and coin evidence. However, the detailed specialist reports suitable for a specialist audience are also included, but not, as with most archaeological reports, tucked away in appendices in an almost unreadable 8pt text size, but immediately following his discussion of their evidence. They must therefore be read, but in a context where their importance and relevance is also known.

Of course there are many examples of good practice in both integrating and publishing environmental archaeological data, of which that from Romano-British North Somerset is just one. A somewhat different approach, in a very different environment, and indeed with a very different theoretical perspective, has been adopted by Ian Hodder and his colleagues in their investigations at Çatalhöyük in the Konya Basin of central Anatolia, Turkey. The way the results allow a past landscape and the economy of its people to be reconstructed are impressive. Publication of the Çatalhöyük Project's results, including the palaeoenvironmental and palaeoeconomic aspects, has been unconventional when compared with the other projects that we have previously discussed. To date (October 2003) two edited volumes have been produced, together with several journal papers. However, with notable exceptions the chapters in the two books deal with methodology, and particularly with reflections by specialists on how their data relate to methodological constraints and field procedures. The actual results and interpretations are published, together with much more detailed specialist impressions on the Project's impressive website, while three interpretative volumes are due for publication late in 2003. Nevertheless, despite the piecing together of the largely web-based resources that is currently needed, it is manifest that the Çatalhöyük Project includes a well-resourced,

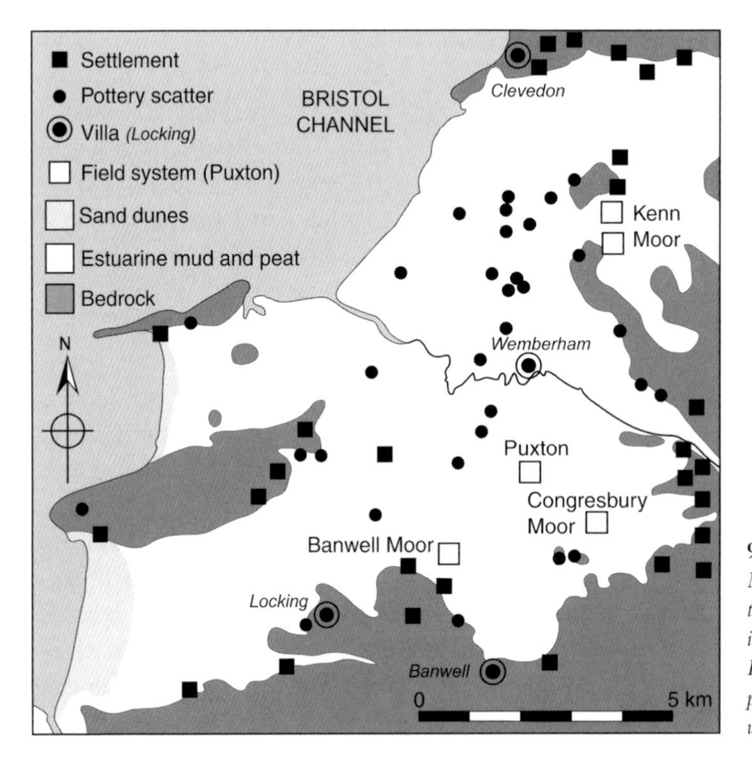

94 *The location of the North Somerset levels, the three sites investigated by Stephen Rippon and the position of Roman villas*

95 *Aerial photograph of Puxton showing Romano-British field systems and palaeochannels typical of those investigated.* Photograph courtesy of Stephen Rippon

but targeted palaeoenvironmental programme that is adding considerable new data and interpretations to a long-worked area of interest, the Neolithic of the Near East.

Çatalhöyük is a name that is well known to the majority of archaeologists, and with the possible exception of Jericho, is by far the best known Neolithic tell in the Near East. The reason for this fame is that the site's 1960s excavator, James Mellaart, suggested that Çatalhöyük was the earliest urban site anywhere in the world, while the myriad of houses that were found contained the earliest wall paintings and sculpture. These 'firsts' are not all now sustainable, but Mellaart's excavations of 1961-1965 also recovered an impressive array of utilitarian artefacts, unusual house structures (famous for having no doors! Entrance was by a ladder onto the roof) with the dead being buried beneath the house floors, and because of the high water tables and many hearths, well preserved plant and faunal remains. The recent investigations began in 1994 and are planned to last for 25 years. To this end the Project has constructed from scratch a series of laboratories, accommodation blocks, seminar rooms, a library and a visitor centre. As well as concentrating on methodology, by formalising interactions between specialists and excavation staff, the project aims to place the buildings and the art found in them within a full palaeoenvironmental, palaeoeconomic and social context. Thus the origin of Çatalhöyük is important, as is the social and economic organisation of the site and the reasons for the adoption of, and then subsequent intensification in, agriculture. In order to meet these aims specialist data must be well integrated with the stratigraphic record being compiled by the site workers. The approach taken – in contrasts to most other archaeological field projects – has been to accommodate the specialist staff in the laboratories that are located immediately adjacent to the site and for field seasons of two months, to ensure they work on samples straight from the site. The Project has a system of formalised seminars attended by all specialists and relevant excavation staff where each context is discussed, accompanied by photographs and video clips, and where each team member can forward their point of view concerning its interpretation.

One of the most important tasks of environmental archaeologists on the Çatalhöyük Project was to investigate the setting of the site and especially to suggest why the location was selected for habitation over others and what resources would have been available for human exploitation. This was tackled by a geomorphological survey of an area at a radius of approximately 60 km from the site, coupled with deep geological test pits around the site to fit its stratigraphy in with that of the Konya Basin as a whole. Bore holes were also drilled in nearby lakes, and diatoms, ostracods and molluscs were studied to reconstruct former lake depths and salinities, and pollen to provide a vegeta-

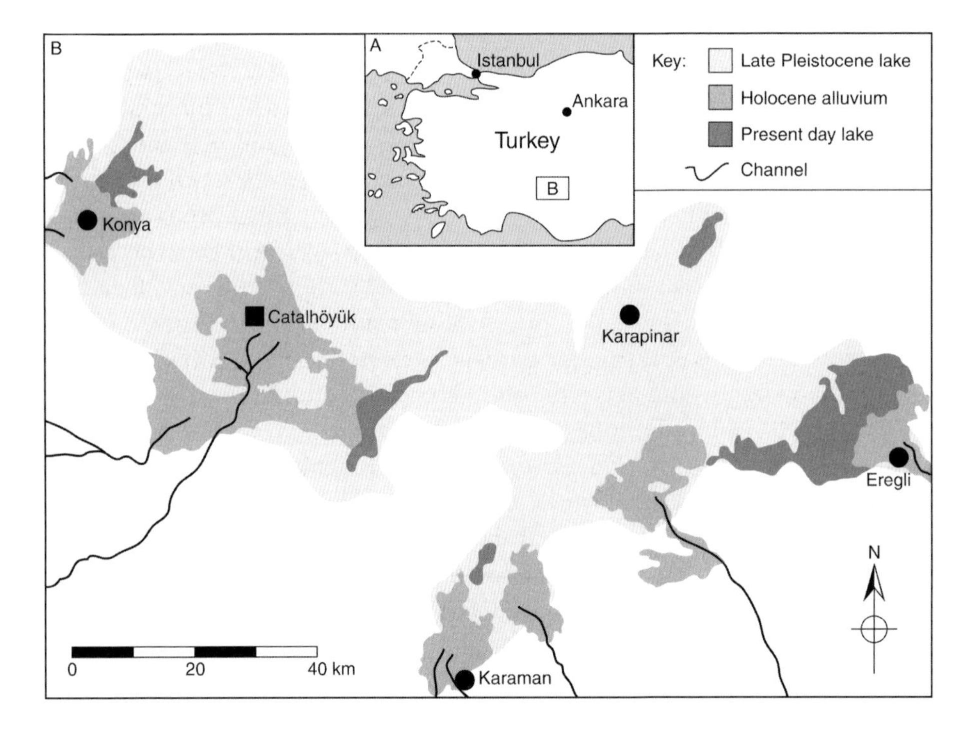

96 *The location of Çatalhöyük in relation to geomorphological phenomena*

tion history. Being focused on off-site locations, this part of the project (termed KOPAL – Konya Basin Palaeoenvironmental Research Programme) was an exception to the integrative procedures we have previously discussed, except in the excavation of the geological soundings around the Çatalhöyük tell. The Konya Basin seen at the present day is essentially the relic of a huge Pleistocene lake of which the last phase of high lake level has been dated to 23-17,000 BP (**96**). Except for minor expansions during the late Weichselian glacial the Konya Basin lake has contracted ever since, and by the early to middle Holocene a number of smaller and shallower lakes existed in depressions within the basins. Also during the early Holocene several rivers and seasonal wadis have expanded and have built up alluvial fans at the edge of the basin. It is on the largest of these, the Çarşamba Fan that Çatalhöyük was built, indeed the fan was active from around 7500-7000 BC until 2000-1500 BC, well after the occupation of Çatalhöyük had ceased. Eventually it deposited a thickness of about 5m of silts and clays around the tell, and on top of Pleistocene lake marls dated to the late glacial and on an overlying palaeosol dated to 7700-7300 BC. In other words the geoarchaeological studies of KOPAL had demonstrated that human activity on the Çatalhöyük tell took place in an environment where the Çarşamba fan was depositing silts and clays reworked from the surrounding uplands on a seasonal basis, and across a 20km wide floodplain. This is likely to have produced a mosaic of marshland, back swamp and other floodplain environments, supporting, according to charcoal recovered from the site, an open herbaceous vegetation of *Bolboschoenus maritimus* and *Phragmites australis*, together with stands of willow, poplar and elm woodland. Was this apparently fertile alluvium the key to the situation of Çatalhöyük? Pollen evidence from residual lakes in the Konya Basin suggests that the early Holocene before the occupation of Çatalhöyük saw the relatively rapid colonisation of terebinth (*Pistacia* sp.) and juniper (*Juniperus* sp.). Deciduous oak (*Quercus pubescens* and/or *Quercus robur*) was also in the process of colonising the region and together with terebinth and juniper, formed a parkland type of open woodland on the slopes of the Taurus mountains that surround the Konya Basin. It is of particular interest that there was a lag in time between increases in both warmth and rainfall in the early Holocene and the arrival and expansion of the oak. Neil Roberts (the director of the KOPAL programme) thinks this may in part be due to Neolithic people, such as those who inhabited Çatalhöyük, burning the parkland environment to promote the growth of grasses and/or to increase soil fertility thereby increasing agricultural productivity. Pollen analysis suggests that the vegetation of the Konya Basin at the time of occupation was comprised of grassland with occasional stands of hazel (*Corylus* sp.).

We must now return to the excavations carried out on the tell itself. The results from the first few seasons of excavation since 1994 indicate that the earliest phases of the site spanned 7300-7150 BC, in other words more or less the same period as when the Çarşamba fan began to deposit its fine-grained load. The site seems to have been occupied for a total of about 950–1150 years, represented on the site by 15 building levels, which means that each house in the site lasted for an average of 50–80 years. The houses all appear to have been constructed to a standard pattern (**85**). Hearths are located in the southern parts of buildings together with storage bins, while these parts of the houses also contained sweepings from the hearths. This area is collectively termed the 'dirty' part of the house. The remaining 'clean' part of the building was used for craft activities and was frequently separated from the dirty part by mud brick walls. Indeed different buildings seem to have had different craft specialisations such as basket manufacture, pot production and obsidian (a type of volcanic glass, used in the same way as flint for tool production) working (this activity seems to have taken place in the 'dirty' areas). Burials are found beneath 'clean' areas of the houses, some of which are associated with waterlogged wood, basketry and rope. Indeed some of the baby burials are in baskets made from wild grasses, including floral elements, indicating spring death. One burial was covered with owl droppings, while another had a mat of reeds and plank of hackberry on his chest, and had had his head removed – an act that is also seen in some of the artwork.

Although burials were associated with botanical remains that survived well in the archaeological record, these floral data could not be used in any reliable sense for the interpretation of Çatalhöyük's subsistence economy. Instead efforts in this respect had to be focused upon bulk samples taken for flotation processing (see section 3) from pits (numbering more than 4000 by 1999), hearths and middens, both within and outside the house structures, where less spectacular preservation was expected. The animal component of subsistence was tackled from the bones recovered by the excavators and those collected as heavy residue in flotation samples. Samples from the middens contained the most varied plant remains as they are the receptacle of the by-products of a variety of different activities. One of the most important seems to have been the burning of a dung-based fuel that is rich in the waste of cereal processing. Contrasting with this are samples taken from hearths and associated sweepings, and, storage areas, where diversity is low, and where cereal remains dominate. The cultivated plants are dominated in the samples by emmer wheat (*Triticum dicoccum*), which as it appeared to be of a rather large-seeded variety, suggested that good growing conditions were present. Einkorn wheat (*Triticum monococcum*) was also grown and was stored in separate areas from the emmer, suggesting that the two crops were grown separately within the Çatalhöyük landscape. All other cultivated crops (e.g. six row barley [*Hordeum vulgare*], lentil [*Lens* sp.] and pea [*Pisum sativum*]) were present in much lower quantities and were therefore probably grown for specialist purposes or to provide variety. The tacit assumption as we have already seen is that one major reason for locating Çatalhöyük on the Çarşamba fan was to make use of the floodplain soils for the cultivation of plant food resources. However, the weed seeds recovered from samples to date suggests that none of the cereal crop was grown in this location. Instead the samples contain weed seeds that are associated with dry conditions. Given the closed nature of the Konya Basin, and the consequent high water tables prior to recent drainage, it has been argued that the foothills of the Taurus range, some 10-12 km to the south of the site, are the most likely location of the fields on which Çatalhöyük's cereals were grown. The charcoal recovered from the site also suggests that this location was exploited for both oak and juniper which did not grow in the Konya Basin but were used for structural purposes. Despite the fact that the fields appear to have been so distant from the site, all stages of cereal processing appear to have been undertaken at Çatalhöyük, suggesting that the newly harvested crop was taken back to the site whole. Indeed large quantities of chaff were found in a geological trench dug by the KOPAL programme at the northern extreme of the site, suggesting that this area was used for winnowing the threshed cereal crop. The rather complex pattern of cultivation does not appear to have changed with time as the composition of plant remains in samples from early phases of the site dating to 7500-7000 BC are identical to those of 6500-6000 BC. Indeed the only change in plant remains is in the wood charcoal, which suggests that local wetland resources were exploited for firewood in the early phases of the site, while fuel from the Taurus foothills became more important in the later phases.

If the area immediately around Çatalhöyük was not being used for cultivating cereals, which all previous studies of the site have assumed was the basis of its subsistence, what was it being used for? One small-scale use may have been for the cultivation of legume crops such as pea and vetch. These could have been spring-sown and thereafter have little need for rain – a patchy commodity in a central Anatolian summer, and one which would have prevented the spring sowing of cereals in such a location. However, pasturing of animals seems to have been a far more important use of this landscape unit. Remains of animal dung were recovered in the vast majority of the flotation samples as they had been used for fuel in the settlement's hearths and were therefore analysed by the archaeobotanists. Plant remains in the dung contain by-products of cereal processing such as chaff and weed seeds, but dominant among the plant seed inclusions were *Bolboschoenus maritimus*, which as we have seen is found in wetland environments. Animal bones have also been recovered in profusion from the site. These suggest that sheep and/or goat was the mainstay of the pastoral economy, with cattle playing a secondary role. The evidence would therefore seem to suggest that these animals, together with some horses, were stabled within the settlement, but grazed on the alluvial lands surrounding the tell. Despite the dominance of domestic animals amongst the bone assemblages, remains of wild animals were found, the most significant of which was wild boar. Bones of this animal, which is likely to have occupied the upland, wooded areas at the margins of the Konya Basin, increase in later phases, suggesting, in the same way as the charcoal evidence, that these more distant territories became economically more important with time.

The results of palaeoeconomic study at Çatalhöyük have completely changed the view of the site from that resulting from the 1960s investigations. Mellaart's view was then that Çatalhöyük was essentially an agricultural community, sited to optimise cereal production, in an increasingly arid environment. The changes in interpretation as a result of the 1994 and later investigations have largely been as a result of the developments of methodology in the intervening decades and the particular approach and attention to detail of Hodder's team. If agricultural productivity were simply the explanation Çatalhöyük would have been located in the Taurus foothills. Instead, the recent data suggest that the settlement was located to exploit a range of different environments, and more controversially, that the site was not a permanently occupied at all (at least by all the population), but was part of a seasonal round of settlements. If correct this interpretation resulting from environmental archaeology would have major implications for the urban status of the settlement as advanced by Mellaart. The Çatalhöyük case study then, amply demonstrates the contributions that environmental archaeology can make to archaeological interpretation given a fully integrated investigative programme, where bioarchaeologists, zooarchaeologists and geoarchaeologists form part of the site team and where question-focused approach to archaeological investigation is implemented. We can only await further publication of the results with anticipation.

Conclusions

Books, as is the case with so much in life, evolve as they are written, often extending far beyond their original remit. This book is no exception. Our intention at the outset was to produce a short but simple textbook dealing with the methods of environmental archaeology, together with an account of what type of interpretation could (and could not) be expected from its data. We believe that we have done this in the preceding pages albeit in twice the space we intended. However, in tackling these aims we have been constantly pulled away from methodological considerations to look at other issues relating to interpretation. Many of these, we now realise as a result of researching the book, are of great importance to environmental archaeology. Indeed as we wrote we came to the conclusion that environmental archaeology is at a crossroads and it is by no means certain what turning it will take, if it does not simply split and take them all! With few exceptions, notably in the study of biomolecules, the techniques used by environmental archaeologists have not greatly developed in the last 20 years. We are effectively using 1960-1970s technology (albeit backed up by twenty-first century computing and communications) – which is useful from our point of view in easing our task of producing a text book that will not become rapidly out dated. It is difficult to see how these techniques can be revolutionised, although constant minor improvements will be made to speed data collection and make this process more accurate. We can therefore assume that in two decades' time environmental archaeologists will be carrying out their field and laboratory studies in a manner that is very familiar to their present-day counterparts. We have no problem with such a vision. However, environmental archaeologists, again with a few notable exceptions, are also interpreting their data according to the 1960-1970s paradigm as well. In the meantime most of the rest of archaeology has moved on and is asking different questions to those that it did in decades past. Additionally, we are not just talking about those of post-processual persuasion in this respect. The questions asked by twenty-first-century archaeologists have less to do with economics, and are instead concerned with past societies and their functioning. As a whole we hope we have demonstrated here that the methods of environmental archaeology can help address these new questions, but the interpretatory frameworks certainly can not. Therefore we consider the most important part of this book to be those parts which examine the interpretation of biological and geological material, *together* or *alongside* the artefacts and structural evidence of mainstream archaeology. If we have in any way prompted those using or otherwise considering environmental archaeological data to consider multiple viewpoints in their interpretations, even if they opt for the conventional functional approach, this book will have been a success. In order to do this, those who carry out and use environmental archaeology should have the archaeological knowledge that enables them to do this. So we end by agreeing - in part - with the Julian Thomas quote with which we started: 'there should simply be archaeologists who specialise in the analysis of snails, pollen and seeds', or, rather, archaeologists who do environmental archaeology.

FURTHER READING

Note: Where we consider one paper of relevance in an edited book, we have listed it below by reference to the author of that paper. However, where two or more papers in an edited book are of interest, we have included a single reference to the book and the papers are not listed individually. Listings are in these cases by reference to the book editors. In such cases we list the names of the authors making important contributions in parentheses at the end of each bibliographic entry.

SECTION 1

Albarella, U. (ed.) (2000) *Environmental archaeology: meaning and purpose.* Kluwer, Dordrecht. (papers by Albarella, Driver and Thomas).

Bell, M.G., Fowler, P. J. and Hillson, S.W. (eds), (1996) *The experimental earthwork project 1960-1991.* Council for British Archaeology Research Report 100, London.

Bell, M.G. and Walker, M.J.C. (1992) *Late Quaternary environmental change: physical and human perspectives.* Longman, London.

Butzer, K.W. (1982) *Archaeology as human ecology.* Cambridge University Press, Cambridge.

Dincauze, D.F. (2000) *Environmental archaeology: principles and practice.* Cambridge University Press, Cambridge.

Delcourt, H.R. and Delcourt, P.A. (1991) *Quaternary ecology: a palaeoecological perspective.* Chapman and Hall, London.

Evans, J.G. (1978) *An introduction to environmental archaeology.* Paul Elek, London.

Evans, J.G. and O'Connor, T.P. (1999) *Environmental archaeology: principles and methods.* Alan Sutton, Stroud.

Gould, S.J. (1965) Is uniformitarianism necessary? *American Journal of Science* **263**, 223-228.

Lowe, J.J. and Walker, M.J.C. (1997) *Reconstructing Quaternary environments.* Second edition. Longman, London.

Lyell, C. (1830-1833) *Principles of geology.* Murray, London (reprinted by Penguin).

O'Connor, T.P. (1998) Environmental archaeology: a matter of definition. *Environmental Archaeology* **2**, 1-6.

Rapp, G. Jr. and Hill, C.L. (1998) *Geoarchaeology: the earth-science approach to archaeological interpretation.* Yale University Press, Newhaven.

Schiffer, M.B. (1987) *Formation processes of the archaeological record.* University of New Mexico Press, Albuquerque (NM).

Thomas, J. (1990) Silent running: the ills of environmental archaeology. *Scottish Archaeological Review* **7**, 2-7.

Waters, M.R. (1992) *Principles of geoarchaeology: a North American perspective.* University of Arizona Press, Tucson.

SECTION 2

Allen, T.G. and Robinson, M.A. (1993) *The prehistoric landscape and Iron Age enclosed settlement at Mingies Ditch, Hardwick-with-Yelford, Oxon.* Oxford University Committee for Archaeology, Oxford.

Anderson, S.T. (1973) The differential pollen productivity of trees and its significance for the interpretation of a pollen diagram from a forested region. In Birks, H.J.B. and West, R.G. (eds) *Quaternary plant ecology.* Blackwell, Oxford, 109-115.

Ashbee, P., Bell, M.G. and Proudfoot, E. (1989) *Wilsford Shaft: excavations 1960-62.* English Heritage Archaeological Report 11, London (chapter 6).

Balaam, N.D., Bell, M.G., David, A.E.U., Levitan, B., Macphail, R.I., Robinson, M. and Scaife, R.G. (1987) Prehistoric and Romano-British sites at Westward Ho!, Devon. Archaeological and palaeo-environmental surveys 1983 and 1984. In Balaam, N.D., Levitan, B. and Straker, V. (eds) *Studies in palaeoeconomy and environment in south west England*. British Archaeological Reports British Series 181, Oxford, 163-264.

Barber, K.E., Chambers, F.M., Maddy, D., Stoneman, R. and Brew, J.S. (1994) A sensitive high-resolution record of late Holocene climatic change from a raised bog in northern England. *The Holocene* **4**, 198-205.

Batterbee, R.W. (1988) The use of diatom analysis in archaeology: a review. *Journal of Archaeological Science* **15**, 621-644.

Bell, M. G. (1983) Valley sediments as evidence of Prehistoric land-use on the South Downs. *Proceedings of the Prehistoric Society* **49**, 119-150.

Bell, M.G. and Boardman, J. (eds.) *Past and present soil erosion: archaeological and geographical perspectives*. Oxbow Monograph 22, Oxford. (papers by Allen, Bell, Bintliff, Boardman, Preece and Zangger).

Berendsen, H.J.A. and Stouthamer, E. (2001) *Palaeogeographic development of the Rhine-Meuse delta, the Netherlands*. Van Gorcum, Assen.

Berglund, B.E. (ed.) *Handbook of Holocene palaeoecology and palaeohydrology*. New York, Wiley. (papers by Coope and Lôzek).

Briggs, D., Smithson, P., Addison, K. and Atkinson, K. (1997) *Fundamentals of the physical environment*. Second Edition. Routledge, London.

Brown, A.G. (1997) *Alluvial geoarchaeology: floodplain archaeology and environmental change*. Cambridge University Press, Cambridge.

Buckland, P.C. and Coope, G.R. (1991) *A bibliography and literature review of Quaternary entomology*. Sheffield Archaeological Press, Sheffield.

Butzer, K.W. (1983) Urban geoarchaeology in Medieval Alzira (Prov. Valencia, Spain). *Journal of Archaeological Science* **10**, 333-349.

Coles, B. (1998) Doggerland: a speculative survey. *Proceedings of the Prehistoric Society* 64, 45-81.

Courty, M.-A., Goldberg, P. and Macphail, R.I. (1989) *Soils and micromorphology in archaeology*. Cambridge University Press, Cambridge.

Davidson, D.A., Grieve, I.C., Tyler, A.N., Barclay, G.J. and Maxwell, G.S. (1998) Archaeological sites: assessment of erosion risk. *Journal of Archaeological Science* **25**, 857-860.

Delcourt, H.R. and Delcourt, P.A. (1991) *Quaternary ecology: a palaeoecological perspective*. Chapman and Hall, London.

English Heritage (2002) *Environmental archaeology: a guide to the theory and practice of methods, from sampling and recovery to post-excavation*. Centre for Archaeology, English Heritage, London.

Elias, S.A. (1994) *Quaternary insects and their environments*. Smithsonian Institution Press, Washington.

Evans, J.G. (1972) *Land snails in archaeology*. Seminar Press, London.

Girling, M.A. and Greig, J. (1985) A first fossil record for *Scolytus scolytus* (F.) (Elm bark beetle): its occurrence in elm decline deposits from London and the implications for Neolithic elm decline. *Journal of Archaeological Science* **12**, 347-351.

Harris, D.R. and Thomas, K.D. (eds.) *Modelling ecological change*. Institute of Archaeology, London. (papers by Bradshaw, Edwards and Evans)

Hillson, S. (1986) *Teeth*. Cambridge University Press, Cambridge.

Howard, A.J. (1999) A generic geomorphological approach to archaeological interpretation and prospection in British river valleys: a guide for archaeologists investigating Holocene landscapes. *Antiquity* **73**, 527-541.

Jacobsen, G.L. and Bradshaw, R.H.W. (1981) The selection of sites for paleovegetational studies. *Quaternary Research* **16**, 80-96.

Kraft, J.C., Rapp, G.J., Szemler, G.J., Tzianos, C. and Kase, E.W. (1987) The pass at Thermopylae, Greece. *Journal of Field Archaeology* **14**, 181-198.

Lister, A.M. (1992) Mammalian fossils and Quaternary biostratigraphy. *Quaternary Science Reviews* **11**, 329-344.

Lowe, J.J. and Walker, M.J.C. (1997) *Reconstructing Quaternary environments*. Second Edition. Longman, London. (chapter 4).

Moore, P. D., Webb, J. A. and Collinson, M.D. (1991) *Pollen analysis*. Second Edition. Blackwell, Oxford.

Munsell Color (2000) *Munsell soil color charts*. Gretaf Macbeth, New Windsor (NY).

Reineck, H.-E. and Singh, I.B. (1980) *Depositional sedimentary environments*. Second Edition. Springer Verlag, Berlin.

Roberts, M.B. and Parfitt, S.A. (1999) *Boxgrove: a Middle Pleistocene hominid site at Eartham Quarry, Boxgrove, West Sussex*. English Heritage, London (chapters 2, 3 and 4).

Roberts, N. (1999) *The Holocene: an environmental history*. Second Edition. Blackwell, Oxford.

Tooley, M.J. and Shennan, I. (eds) (1987) *Sea level changes*. Blackwell, Oxford.

Vita-Finzi, C. (1969) *The Mediterranean valleys: geological changes in historical times*. Cambridge: Cambridge University Press.

Waters, M.R. (1992) *Principles of geoarchaeology: a North American perspective*. University of Arizona Press, Tucson.

Wilkinson, K.N. (2003) 'Colluvial deposits in dry valleys of southern England as proxy indicators of paleoenvironmental and land-use change' *Geoarchaeology 18*, 725-755

SECTION 3

Champion, T.C. and Collis, J.R. (eds) (1996) *The Iron Age in Britain and Ireland: recent trends*. J.R. Collis Publications, Department of Archaeology and Prehistory, University of Sheffield, Sheffield. (papers by Jones and Maltby).

Copley, M.S., Berstan, R., Dudd, S.N., Docherty, G., Mukherjee, A.J., Straker, V., Payne, S. and Evershed, R.P. (2003) Direct chemical evidence for widespread dairying in prehistoric Britain. *Proceedings of the National Academy of Sciences*, **100**, 1524-1529.

Davis, S.J.M. (1987) *The archaeology of animals*. Batsford, London.

DeRoche, C.D. (1997) Studying Iron Age production. In Gwilt, A. and Haselgrove, C. (eds) *Reconstructing Iron Age societies*. Oxbow Monograph 71, Oxford, 19-25.

Dickson, C. and Dickson, J. (2000) *Plants and people in ancient Scotland*. Tempus, Stroud.

Fowler, P.J. (1983) *The farming of prehistoric Britain*. Cambridge University Press, Cambridge.

Hastorf, C.A. and Popper, V.S. (eds.) (1988) *Current palaeoethnobotany: analytical methods and cultural interpretations of archaeological plant remains*. University of Chicago Press, Chicago.

Hather, J.G. (1993) *An archaeobotanical guide to root and tuber identification. Volume 1: Europe and South-west Asia*. Oxbow Monograph 28, Oxford.

Hillman, G.C. (1984) Interpretation of archaeological plant remains: the application of ethnographic models from Turkey. In van Zeist, W. and Casparie, W.A. (eds) *Plants and ancient man: studies in the palaeoethnobotany*. Balkema, Rotterdam, 1-42.

Hubbard, R.N.L.B and Clapham, A.R. (1992) Quantifying macroscopic plant remains. *Review of Palaeobotany and Palynology* **73**, 73-132.

Jones, G.E.M. (1987) A statistical approach to the archaeological identification of crop processing. *Journal of Archaeological Science* **14**, 311-323.

Jones, M.K. (ed.) (1988) *Archaeology and the flora of the British Isles*. Oxford University Committee for Archaeology.

Jones, M.K. (2002) *The molecule hunt. Archaeology and the search for ancient DNA*. Arcade Publishing, London.

Legge, A., Payne, S. and Rowley-Conwy, P. (1998) The study of food evidence in British prehistory. In Bayley, J. (ed.) *Science in archaeology*. English Heritage, London, 89-94.

Mercer, R. (ed.) (1981) *Farming practice in British prehistory*. Edinburgh University Press, Edinburgh.

Murphy, P.J. (1989) Carbonised Neolithic plant remains from the Stumble, an intertidal site in the Blackwater Estuary, Essex, England. *Circaea* **6**, 21-38.

O'Connor, T. (2000) *The archaeology of animal bones*. Sutton Publishing, Stroud.

Pearsall, D. (1989) *Palaeoethnobotany: handbook of procedures*. San Diego Academic Press, San Diego.

Powers-Jones, A. (1994) The use of phytolith analysis in the interpretation of archaeological deposits: an Outer Hebridean example. In Luff, R. and Rowley-Conwy, P. (eds.) *Whither environmental archaeology?* Oxbow Monograph 38, Oxford, 41-50.

Pryor, F. (1996) Sheep, stockyards and field systems: Bronze Age livestock populations in the Fenlands of East Anglia. *Antiquity* **70**, 313-324.

Pryor, F. (1999) *Farmers in prehistoric Britain*. Tempus, Stroud.

Rackham, D.J. (1994) *Interpreting the past: animal bones*. British Museum Press, London.

Renfrew, J.M. (ed.) (1991) *New light on early farming*. Edinburgh University Press, Edinburgh (papers by Murphy, and van der Veen).

Richards, M.P. and Hedges, R.E.M. (1999) A Neolithic revolution? New evidence of diet in the British Neolithic. *Antiquity* **73**, 891-897.

Salisbury, Sir E. (1961) *Weeds and aliens*. Collins, London.

Samuel, D. (1996) Investigation of ancient Egyptian brewing and baking methods by correlative microscopy. *Science* **273**, 488.

Smith, B.D. (1995) *The emergence of agriculture*. Scientific American Library, New York.

Stevens, C.J. (2003) An investigation of agricultural consumption and production models for prehistoric and Roman Britain. *Environmental Archaeology* **8**, 61-76.

Troy, C.S. MacHugh, D.E., Bailey, J.F., Magee D.A., Loftus, R.T., Cunningham, P., Chamberlain, A.T., Sykes, B.C. and Bradley, D.G. (2001) Genetic evidence for Near-Eastern origins of European cattle. *Nature* **410**, 1080-1091.

van der Veen, M. (1992) *Crop husbandry regimes: an archaeobotanical study of farming in northern England 1000 BC-AD 500*. Sheffield Archaeological Monographs 3, J.R. Collis Publications, Department of Archaeology and Prehistory, University of Sheffield, Sheffield.

SECTION 4

Audouze, F. and Büchsenschütz, O. (1989) *Villes, villages et campagnes de L'Europe Celtique*. Hachette, France (chapters 8-10) [translated into English by Cleere, H. (1991) as *Towns, villages and countryside of Celtic Europe*. Batsford, London].

Bell, C. (1998) *Ritual: perspectives and dimensions*. Oxford University Press, Oxford.

Bradley, R. (1998) *The significance of monuments*. Routledge, London.

Brennand, M. and Taylor, M. (2003) The survey and excavation of a Bronze Age timber circle at Holme-next-the-Sea, Norfolk, 1998-9. *Proceedings of the Prehistoric Society* **69** (forthcoming at the time of publication).

Burkert, W. (1987) *Homo necans: the anthropology of ancient Greek sacrificial ritual and myth*. Harvard University Press, Harvard.

Frazer, J.G. (1922) *The golden bough: a study of magic and religion*. Macmillan Press, London.

Isaakidou, V., Halstead, P., Davis, J. and Stocker, S. (2002) Burnt animal sacrifice at the Mycenaean 'Palace of Nestor', Pylos. *Antiquity* **76**, 86–92.

Green, M. (1992) *Animals in Celtic life and myth*. Routledge, London.

Lee, R.B. and Daly, R. (ed.) (1999) *The Cambridge encyclopedia of hunters and gatherers*. Cambridge University Press, Cambridge.

Lévi-Strauss, C. (1962) *Totemism*. Pelican, London.

Lévi-Strauss, C. (1966) *The savage mind*. Weidenfeld and Nicholson, London.

Renfrew, C. and Bahn, P. (2000) *Archaeology: theory, methods and practice*. Third Edition. Thames and Hudson, London. (chapter 10).

Scarre, C. (ed.) (2002) *Monuments and landscape in Atlantic Europe*. Routledge, London (papers by Scarre, Bradley and Whittle).

SECTION 5

Albarella, U. (ed.) (2000) *Environmental archaeology: meaning and purpose*. Kluwer, Dordrecht. (papers by Albarella and Driver)

Allen, M.J. (2002) The chalkland landscape of Cranborne Chase: a prehistoric human ecology. *Landscapes* **2**, 55-69.

Cuff, E.C., Sharrock, W.W. and Francis, D.W. (1998) *Perspectives in sociology*. Fourth Edition. Routledge, London.

Hodder, I. (1986) *Reading the past: current approaches to interpretation in archaeology*. Cambridge University Press, Cambridge.

Johnson, M. (1999) *Archaeological theory: an introduction*. Blackwell, Oxford.

Layton, R. (1997) *An introduction to theory in anthropology*. Cambridge University Press, Cambridge.

Luff, R. and Rowley-Conwy, P. (1994) The (dis)integration of environmental archaeology. In Luff, R. and Rowley-Conwy, P. (eds) *Whither environmental archaeology?* Oxbow Monograph 38, Oxford, 1-6.

Renfrew, C. and Bahn, P. (2000) *Archaeology: theory, methods and practice*. Third Edition. Thames and Hudson, London (chapter 12)

Shanks, M. and Tilley, C. (1987) *Social theory and archaeology*. Polity Press, Oxford.

Tilley, C. (1994) *A phenomenology of landscape: places, paths and monuments*. Berg, Oxford.

Tilley, C. (1999) *Metaphor and material culture*. Blackwell, Oxford

Trigger, B.G. (1989) *A history of archaeological thought*. Cambridge University Press, Cambridge.

Trigger, B.G. (1998) *Sociocultural evolution*. Blackwell, Oxford.

SECTION 6

Çatalhöyük: Excavations of a Neolithic Anatolian höyük (2002) http://catal.arch.cam.ac.uk/catal/catal.html.

Fairbairn, A., Asouti, E., Near, J. and Martinoli, D. (2002) Macro botanical evidence for plant use at Neolithic Çatalhöyük, south-central Anatolia, Turkey. *Vegetation History and Archaeobotany* **11**, 41-54.

Hodder, I. (ed.) (1996) *On the surface: Çatalhöyük 1993-95*. McDonald Institute for Archaeological Research and the British Institute of Archaeology at Ankara, Cambridge.

Hodder, I. (ed.) (2000) *Towards reflexive method in archaeology: the example of Çatalhöyük*. McDonald Institute for Archaeological Research and the British Institute of Archaeology at Ankara, Cambridge.

Kemp, B.J. (1989) *Ancient Egypt: anatomy of a civilisation*. Routledge, London (chapter 7).

Kemp, B.J., Samuel, D. and Luff, R. (1994) Food for an Egyptian city: Tell el-Amarna. In Luff, R. and Rowley-Conwy, P. (eds.) *Whither environmental archaeology?* Oxbow Monograph 38, Oxford, 133-170.

Kemp, B.J. and Vogelsang-Eastwood, G. (2001) *The ancient textile industry at Amarna*. Egypt Exploration Society 68th Excavation Memoir, London.

Luff, R. M. and Bailey, G.N. (2000) Analysis of size changes and incremental growth structures in African catfish *Synodontis schall* (schall) from Tell el-Amarna, Middle Egypt, *Journal of Archaeological Science* **27**, 821–835

Panagiotakopulu, E. (1999) An examination of biological materials from coprolites from XVIII Dynasty Amarna, Egypt, *Journal of Archaeological Science* **26**, 547–551

Panagiotakopulu, E. (2001) New Records for Ancient Pests: Archaeoentomology in Egypt, *Journal of Archaeological Science* **28**, 1235–1246

Rippon, S. (2000) The Romano-British exploitation of coastal wetlands: survey and excavations on the North Somerset Levels 1993-7. *Britannia* **31**, 69-200.

Further information

Organisations
The following organisations are concerned to a greater or lesser extent with environmental archaeology. Website addresses have been given to get around the problem of changes in addresses of membership secretaries.

Association for Environmental Archaeology – (http://www.envarch.net/)
This UK-based organisation also has members from continental Europe and North America. It organises conferences, produces quarterly newsletters and publishes the journal *Environmental Archaeology: The Journal of Human Palaeoecology.*

Quaternary Research Association – (http://www.qra.org.uk/)
Another UK-based organisation that primarily concerns itself with Pleistocene geology and palaeoecology, although many members also have Holocene interests. The QRA organises an annual residential field visit, shorter field excursions, conferences and day meetings. It also has a quarterly newsletter.

Society of American Archaeologists, Geoarchaeological Interest Section –
(http://www.saa.org/membership/I-GEO/)
As the name suggests this sub-branch of the SAA is concerned with geoarchaeology. The vast majority of members are from North America, but also include a few from the UK. The Geoarchaeology Interest Section runs a symposium held at the SAA annual conference and also has an e-mail discussion forum on which the latest jobs, conferences, workshops and courses are listed.

Information and guidance
Guidance from the Association for Environmental Archaeology on teaching environmental archaeology and using environmental archaeology during archaeological field evaluations
(http://www.envarch.net/publications/papers/papers.html)

Bibliography of environmental archaeology in England compiled by Dominique de Moulins.
(http://www.english-heritage.org.uk/default.asp?wci=Link&wce=136)

Journals
The following journals frequently publish papers on environmental archaeology.

Environmental Archaeology (http://www.envarch.net/publications/envarch/index.html)
Journal of Archaeological Science (http://www.elsevier.com/locate/issn/0305-4403)
Geoarchaeology (http://www.interscience.wiley.com/jpages/0883-6353/)
Vegetation History and Archaeobotany
(http://www.springerlink.com/app/home/journal.asp?wasp=524cyvtyyk1rhd6vkyvm&referrer=parent&backto=linkingpublicationresults,1,1)
The Holocene (http://www.holocenejournal.com/)

INDEX

Figures in italics denote page numbers of illustrations